Mathematics and Visualization

More information about this series at http://www.springer.com/series/4562

Michael Breuß • Alfred Bruckstein •
Petros Maragos • Stefanie Wuhrer
Editors

Perspectives in Shape Analysis

Springer

Editors
Michael Breuß
Institute for Applied Mathematics
and Scientific Computing
Brandenburg University of Technology
Cottbus, Germany

Alfred Bruckstein
Israel Institute of Technology
Technion
Haifa, Israel

Petros Maragos
School of Electrical and Computer
Engineering
National Technical University of Athens
Athens, Greece

Stefanie Wuhrer
INRIA, Grenoble Rhône-Alpes
Saint Ismier, France

ISSN 1612-3786 ISSN 2197-666X (electronic)
Mathematics and Visualization
ISBN 978-3-319-24724-3 ISBN 978-3-319-24726-7 (eBook)
DOI 10.1007/978-3-319-24726-7

Library of Congress Control Number: 2016949751

Mathematics Subject Classification (2010): 65D18, 65D19, 68.06

Cover illustration: Cover illustration from "Shape Compaction" by Honghua Li and Hao Zhang with kind permission by the authors

Printed on acid-free paper

This Springer imprint is published by Springer Nature
The registered company is Springer International Publishing AG Switzerland

To our families and students

Preface

In everyday life, geometric shapes surround us, and thus the field of shape analysis has a growing variety of applications, including ergonomic design, virtual shopping, scientific and medical visualization, realistic simulation, photo-realistic rendering, the design of natural user interfaces, and semantic scene understanding. The efficient processing of shapes and the discovery and investigation of informative representations for shapes are core tasks in the context of shape analysis research.

Traditionally, the notion of shape has been studied either by analyzing a sparse set of marker positions on three-dimensional (3-D) shapes, primarily for medical imaging applications, or by analyzing projections of shapes in 2-D images, chiefly for image processing and computer vision applications. New challenges in the analysis and processing of such data arise with the increasing amount of data captured by sensors used to acquire shapes, and with modern applications such as natural user interfaces that require real-time processing of the input shapes. Recently, it has also become increasingly affordable to digitize 3-D shapes using multiple modalities, such as laser-range scanners, image-based reconstruction systems, or depth cameras like the Kinect sensor. Using these dense 3-D shapes in the above mentioned applications calls for processing and describing the shapes in an efficient and informative way.

The purpose of this book is to highlight recent advances that address these challenges from different perspectives with the help of the latest tools for geometric, algorithmic, and numerical concepts. As the analysis of 3-D shapes and *deformable shape models* has received considerable attention recently, classic shape analysis tools from differential geometry now have a fresh influence on the field. As they address the issue of how to represent shapes efficiently, the research areas of *sparse data representation* and *machine learning* have begun to influence shape analysis modeling and the numerics. Especially in the context of three-dimensional data (or even higher-dimensional data sets), efficient optimization methods will certainly become increasingly important, since many shape analysis tasks can be formulated as optimization problems. As the efficiency of shape analysis methods is of general importance in all of these fields as well as for the evolution of classic approaches, we also examine *numerical computing* as an important research theme.

As is typical for many fields within image processing, it is often impossible to distinguish a single aspect of the topics mentioned above as a given article's main contribution. For instance, a model for deformable shapes can be at the heart of a novel optimization approach derived from a sparsity concept, which in turn gives rise to an efficient computational model. Especially, the development of new model formulations is closely related to computational approaches. Naturally, this is also the case with regard to the contributions in this volume. Nevertheless, we have sought to emphasize the importance and originality of certain core developments by grouping the content into the following main areas:

Part I: Numerical computing for shape analysis
Part II: Sparse data representation and machine learning for shape analysis
Part III: Deformable shape modeling

Let us now give a brief account of the themes represented in the respective chapters.

A typical task in shape analysis is the segmentation of objects in images. With regard to the numerical computation of proper image segmentations, various approaches have been proposed in the literature. In the first chapter, a classical method for this purpose is reexamined, using the screened Poisson equation as a computational basis. The model is given via a *partial differential equation (PDE)* and, in this chapter, is employed for the purpose of ornament analysis.

The book then goes on to explore another PDE-based approach, this time to the classical shape-from-shading (SFS) problem. In contrast to the first chapter, where the elliptic Poisson equation was at the heart of the developments, here hyperbolic PDEs are addressed. The semi-Lagrangian method employed for the computations in the second chapter offers an efficient method for this purpose.

In turn, the third chapter shows how to make use of robust variational approaches to deal with the SFS problem. The computational problem that arises here in the corresponding *energy minimization* problem is based on solving a parabolic PDE for its elliptic steady state, and in the discrete scheme, also typical components from hyperbolic numerics are also employed. Therefore, the first three chapters nicely show that in today's shape analysis, models of all three fundamental types of PDEs and corresponding computational approaches are important.

Turning to the fourth chapter, which concerns morphological amoebas, again a method useful for segmentation is investigated. Let us emphasize that here the mathematical basis is provided by differential geometry. From a computational point of view, the adaptive amoeba construction shows a strong relation to nonstandard discretizations of the arising PDEs. Here, the task of the numerical description of image content such as texture, and of shapes themselves, is intimately related to segmentation ideas. These aspects are addressed in both Chaps. 4 and 5, while methodically we now turn from PDE-based methods to other approaches.

Chapter 5 especially focuses on numerical shape characteristics that have an intuitive meaning and are useful for building shape discriminators and classifying shapes. The latter issue is also part of the objective of Chap. 6, where the effect of shape distances in an energy minimization method is studied. At a technical level, in this chapter we arrive very obviously at *optimization methods*. Here a specific trust region scheme is at the heart of the investigation, and the different shape distances are shown to define distinct trust regions. In the spectral method used for segmentation of point clouds in RGB-D data in Chap. 7, a large optimization problem is tackled via the numerical solution of the generalized eigenvalue problem for a specific graph Laplacian.

Summarizing Part I, we conclude that computational methods for PDEs and optimization problems are essential tools in the field of shape analysis. It is extremely difficult to say which techniques are the most prominent ones, as a large variety of different problems appear. We feel this is an intriguing aspect of shape analysis, as it leaves the field open for introducing advances from many branches of computational mathematics. Let us also note that, on a technical level, approaches for reducing the size of computational problems are also often employed (as especially apparent in Chap. 7). All these modeling and computational tools naturally resurface in several parts of this book.

Turning to Part II of this volume, a prominent aspect in the corresponding chapters when compared to those of Part I is that *fully discrete concepts* that lead to efficient shape representations are investigated, e.g., with the aim of reducing storage or for constructing shape abstractions (Chap. 8). In comparison to the works of Part I, this is especially important for analyzing 3-D shapes, which appears natural as the additional third dimension leads to large data sets, making it all the more important to reduce the data load. The topology of 3-D shapes is investigated in Chap. 9 via Morse theory, while the correspondence between sparse 3-D shapes, exploring shape similarity and also referring to *deformable shape models*, is the subject of Chap. 10. Concepts from machine learning and related optimization tools in the context of deformable shapes are also explored in Chaps. 11 and 12. Lastly, Chap. 12 discusses the extensive use of machine learning techniques for 2-D images.

Summarizing the key points from Part II, one may note that the sparsity and machine learning techniques explored here are naturally intimately related to the arising optimization methods and numerical computing, while some of the models are concerned with deformable shapes, as they deal with shape correspondence methods. However, we chose to group these works here in a separate section, given the high significance of introducing machine learning and sparsity concepts into these fields.

Coming finally to Part III, we show here that the field of deformable shapes is already rich in terms of the different aspects that can be explored, complementing and completing the previous works in the other parts of this volume. Beginning with correspondences between deformable shapes in the spectral domain (Chap. 13), which also connects to Chaps. 10 and 11, we turn to the use of morphable shape models in computer vision in Chap. 14. The use of multimodal data for shape recognition, as well as a related use of machine learning methods, is explored in Chap. 15. Moreover, one may find that the concept of a morphable model is related to the template fitting approach explored for shape analysis in MRI data in Chap. 16.

Summarizing important concepts from Part III, one may note that deformable shape modeling includes the important problems of correspondence computation and statistical analysis and directly relates to a diverse range of applications. One recent trend is to extend classic applications of shape analysis techniques to cover new applications in other fields of science. An example of this is the application of shape analysis methods in computational linguistics and speech science in Chaps. 15 and 16.

The content of this book represents the contributions of respected experts in the field of shape analysis that highlight different new perspectives on the mentioned tasks. A key aspect of this book that sets it apart from other volumes is that it includes both discrete and continuous settings in shape analysis, as both are relevant for the modeling and processing of shape representations.

This volume originated in the inspiring research discussions that took place at a Dagstuhl seminar in February 2014. Both new scientific results and tutorial-style chapters that survey recent aspects in the field are included. As the demands in the individual fields are high, the research groups in which the most interesting techniques are proposed are highly specialized. This not only holds true for discrete and continuous-scale modeling and numerical computing but also for the areas of sparsity and machine learning highlighted here. Thus, in spite of the strong interconnections between the works as they are represented in this volume, at the moment there is no regular conference that could produce such a dedicated book.

It was a great pleasure to exchange scientific ideas with all of our colleagues who participated in the Dagstuhl seminar on New Perspectives in Shape Analysis and contributed to this volume. We hope that this collection will inspire new research ideas and promote further collaboration.

Cottbus, Germany — Michael Breuß
Haifa, Israel — Alfred Bruckstein
Athens, Greece — Petros Maragos
Grenoble, France — Stefanie Wuhrer

Acknowledgments

A number of people supported the publication of this book. First, we thank the staff of Schloss Dagstuhl for their help in all aspects of the seminar. The collaborative atmosphere at Dagstuhl made our seminar such an enjoyable and successful scientific meeting.

This book has benefited greatly from a rigorous peer-review process that would not have been possible without the help of numerous researchers who volunteered their time to improve the scientific quality of this collection. We thank Mirela Ben-Chen, Jonathan Boisvert, Ali Sharifi Boroujerdi, Michael Bronstein, Anastasia Dubrovina, Yong Chul Ju, Oliver van Kaick, Margret Keuper, Maryam Khanian, Ron Kimmel, Artiom Kovnatsky, Jochen Lang, Leonid Pishchulin, Ronald Römer, Silvia Tozza, Chang Shu, William A. P. Smith, Matthias Vestner, Martin Welk, Ying Yang, and Ashkan Mansouri Yarahmadi for their thorough and constructive help in this process.

Furthermore, we would like to thank the editors of the Springer series *Mathematics and Visualization* for the opportunity to publish this volume in their series. We sincerely thank Ruth Allewelt from Springer for her practical support and patience throughout the preparation of this book.

Contents

Contributors

Jonathan Aflalo Technion, Israel Institute of Technology, Haifa, Israel

Mehmet Ali Aktaş Computer Science, Toros University, Mersin, Turkey

Ismail Ben Ayed University of Western Ontario, London, Canada

École de Technologie Supérieure, University of Quebec, Montreal, QC, Canada

Yuri Boykov Computer Science Department, University of Western Ontario, London, ON, Canada

Michael Breuß Institute for Applied Mathematics and Scientific Computing, Brandenburg University of Technology, Cottbus, Germany

Alexander M. Bronstein School of Electrical Engineering, Tel Aviv University, Tel Aviv, Israel

Michael M. Bronstein Faculty of Informatics, Institute of Computational Science, University of Lugano, Lugano, Switzerland

Andrés Bruhn Institute for Visualization and Interactive Systems, University of Stuttgart, Stuttgart, Germany

Thomas Brox Computer Science Department and BIOSS Centre for Biological, University of Freiburg, Freiburg, Germany

Daniel Cremers Technical University Munich, Munich, Germany

Anastasia Dubrovina Technion, Israel Institute of Technology, Haifa, Israel

Maurizio Falcone Dipartimento di Matematica "G. Castelnuovo", Università "Sapienza" di Roma, Rome, Italy

Leila de Floriani Department of Computer Science, Bioengineering, Robotics, and Systems Engineering, Università degli Studi di Genova, Genova, Italy

Ulderico Fugacci Department of Computer Science, Bioengineering, Robotics, and Systems Engineering, University of Genova, Genova, Italy

Lena Gorelick Computer Science Department, University of Western Ontario, London, ON, Canada

Alexander Hewer Deutsches Forschungsinstitut für künstliche Intelligenz, Saarbrücken, Germany

Federico Iuricich Department of Computer Science and UMIACS, University of Maryland, College Park, MD, USA

Yong Chul Ju Institute for Visualization and Interactive Systems, University of Stuttgart, Stuttgart, Germany

Athanasios Katsamanis School of Electrical and Computer Engineering, National Technical University of Athens, Athens, Greece

Margret Keuper University of Freiburg, Freiburg im Breisgau, Germany

Ron Kimmel Technion, Israel Institute of Technology, Haifa, Israel

Iasonas Kokkinos Ecole Centrale Paris, Paris, France

Center for Visual Computing, Centrale-Supélec and INRIA-Saclay, Grande Voie des Vignes, Chatenay-Malabry, France

Honghua Li Simon Fraser University, Burnaby, BC, Canada

Petros Maragos School of Electrical and Computer Engineering, National Technical University of Athens, Athens, Greece

Daniel Maurer Institute for Visualization and Interactive Systems, University of Stuttgart, Stuttgart, Germany

George Pavlakos Computer and Information Science, University of Pennsylvania, Philadelphia, PA, USA

Vassilis Pitsikalis School of Electrical and Computer Engineering, National Technical University of Athens, Athens, Greece

Jonathan Pokrass School of Electrical Engineering, Tel Aviv University, Tel Aviv, Israel

Korin Richmond Centre for Speech Technology Research, University of Edinburgh, Edinburgh, UK

Emanuele Rodolà Technische Universität München, Munich, Germany

Paul L. Rosin School of Computer Science & Informatics, Cardiff University, Cardiff, UK

Samuel Rota Bulò Fondazione Bruno Kessler, Trento, Italy

Guillermo Sapiro School of Electrical and Computer Engineering, Duke University, Durham, NC, USA

Frank R. Schmidt Computer Science Department and BIOSS Centre for Biological Signalling Studies,University of Freiburg, Freiburg, Germany

William A.P. Smith Department of Computer Science, University of York, York, UK

Pablo Sprechmann School of Electrical and Computer Engineering, Duke University, Durham, NC, USA

Ingmar Steiner Deutsches Forschungsinstitut für künstliche Intelligenz, Saarbrücken, Germany

DFKI Language Technology Lab, Saarbrücken, Germany

Cluster of Excellence Multimodal Computing and Interaction, Saarland University, Saarbrücken, Germany

Sibel Tari Middle East Technical University, Ankara, Turkey

Stavros Theodorakis School of Electrical and Computer Engineering, National Technical University of Athens, Athens, Greece

Silvia Tozza Dipartimento di Matematica "G. Castelnuovo", Sapienza – Università di Roma, Roma, Italy

York University, York, UK

Matthias Vestner Technical University Munich, Munich, Germany

Martin Welk Biomedical Image Analysis Division, Department of Biomedical Computer Science and Mechatronics, University for Health Sciences, University for Medical Informatics and Technology, Hall/Tyrol, Austria

Thomas Windheuser Technical University Munich, Munich, Germany

Stefanie Wuhrer Inria Grenoble Rhône-Alpes, Saint Ismier, France

Hao Zhang Simon Fraser University, Burnaby, BC, Canada

Joviša Žunić Mathematical Institute, Serbian Academy of Sciences and Arts, Belgrade, Serbia

University of Exeter, Exeter, UK

Part I
Numerical Computing for Shape Analysis

Chapter 1
Ornament Analysis with the Help of Screened Poisson Shape Fields

Sibel Tari

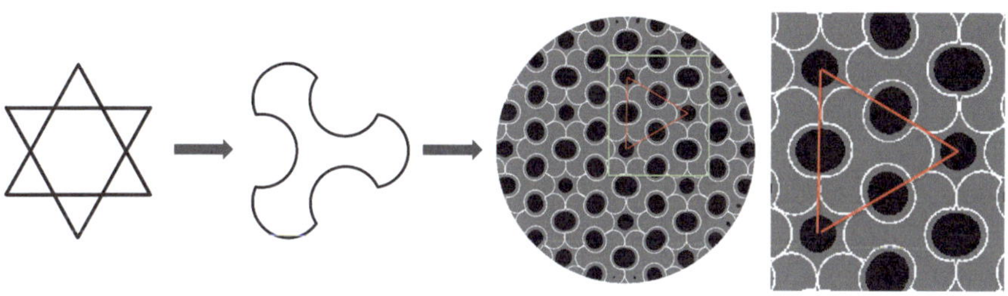

Abstract In this chapter, some thought-provoking application problems in Ornament Analysis are examined. Fields constructed via Screened Poisson Equation are used as intermediate level representations towards developing solutions. In the considered problems, the fields serve to a variety of purposes – i.e., to embed critical point detection process into a suitable morphological scale space, to regularise an ill-posed search problem, and finally to integrate features in a context – extending the visual functions of the Screened Poisson Equation based shape fields.

1.1 Introduction

Ornaments are a part of human culture, independent of time and location. In modern times, as important as ever, an ornament is a beautiful link between art and science. In this chapter, I consider some problems in Ornament Analysis.

Towards developing solutions to selected problems, shape *fields* governed by the Screened Poisson Equation serve as intermediate level representations. Hence,

S. Tari (✉)
Middle East Technical University, 06800 Ankara, Turkey
e-mail: stari@metu.edu.tr

M. Breuß et al. (eds.), *Perspectives in Shape Analysis*, Mathematics and Visualization, DOI 10.1007/978-3-319-24726-7_1

before discussing the problems, let us define Screened Poisson Shape Fields. Imagine a drawing on a frame. Let the frame be $\Omega \subset R^2$, and the drawing be $g : \Omega \to \{0, 1\}$. Consider $v : \Omega \to R$ governed by

$$\Delta v - \frac{v}{\rho^2} = 0 \tag{1.1}$$

$$v|_{\{(x,y)\ s.t.\ g(x,y)=1\}} = 1 \text{ and}$$

$$\frac{\partial v}{\partial n}|_{\partial\Omega} = 0$$

where $\frac{1}{\rho^2}$ is the screening parameter. The value of v is approximately equal to $e^{-\frac{|d|}{\rho^2}}$, d is the distance from (x, y) to the nearest point where $g(x, y) = 1$. Thus, $-\rho\sqrt{\ln v}$ defines a smooth distance field where the parameter ρ controls both the smoothing and the decay. It has been further shown that v is an implicit coder of the level curve curvature [12]:

$$v(x, y) \approx \rho\left(1 + \frac{\rho}{2}\, curv(x, y)\right)\frac{\partial v}{\partial n} + O\left(\rho^3\right) \tag{1.2}$$

Implicitly coding curvature is the main difference of the v-field (Screened Poisson Shape Field) from the simple distance transform.

The v-field has been employed for solving shape related tasks since mid 1990s [10, 11]. It has been proposed as an intermediate level representation that bridges filtering and segmentation with shape abstraction [12]. In modified forms, it has been used to address local-to-global integration issues and multi perspective partitioning of shapes [8, 9].

In the ornament analysis problems considered in this chapter, the v-fields are used to embed critical point detection process into a suitable morphological scale space [5], to regularise an ill-posed search problem where the goal is to locate a subpart in crowded drawings [2, 6], and finally to integrate features in a context. The considered problems significantly extend the utility of Screened Poisson based shape fields.

The rest of the material is organised as two sections. In Sect. 1.2, parts in an ornament – be it a single shape drawn with a creational brush or a tile containing parts hidden in a context – are searched. In Sect. 1.3, structure discovery in tiles are addressed by integrating features from local to global.

1.2 Creative Design with Shapes: Seeing a Part

Consider the task of creating new designs starting from an existing one. In [7], Stiny argues that creative design is merely a cut-and-paste process. The creative element is in the process of seeing the right part to be cut (Fig. 1.1). That is towards the grand goal of modelling creative design, a basic question is which part of the whole is to be selected.

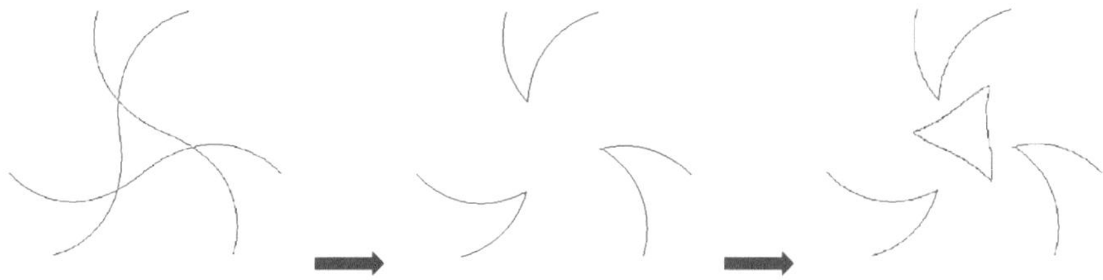

Fig. 1.1 From one design to another. Design as cut-and-paste (Adopted from [7])

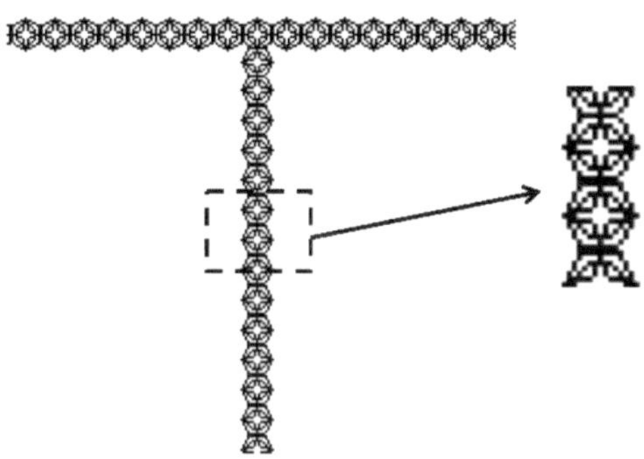

Fig. 1.2 A letter T drawn using an ornamental font

1.2.1 Natural Break Up Locations

A highly popular view since Attneave [1] is that the wholes break up to natural parts at the critical points such as corners, curvature extrema, intersection points and end points. Following this line of thinking, several computational methods for detecting curvature related criticalities in line drawings are developed. These computational methods can be used for cutting out the curved triangle in Fig. 1.1. But they fail if the drawing is produced using creational brushes or ornamental fonts (Fig. 1.2).

In [5], we proposed an alternative route for detecting natural break points. Rather than measuring bending, we measured the deviation at each point on the drawing from a reference drawing that is provided externally. The reference drawing is a straight cut out from the original drawing as in Fig. 1.2. It is also possible to select the reference drawing externally as a regular straight line. In such case, due to the selection of the reference drawing, the deviation correlates with the curvature.

Our method is as follows: we compute v fields for both the original drawing and the reference drawing, i.e. straight line segment drawn in ornamental pattern. Offline, by placing ellipse and disk shaped windows of varying sizes centered at the middle of the reference drawing, we collect statistics on v. The search for critical break up points is performed as a two step iterative refinement process. In the first step, starting with a large size ellipse-shaped window, each location on the drawing for which the field statistics deviates from the stored reference statistics is marked as critic. Then the critic points are grouped into chunks based on connectivity (Fig. 1.3 the top row). Then at the centre of each chunk a disk-shaped window of which size is proportional to the area of the chunk is placed (Fig. 1.3 the bottom row). In the second step, contents of the field inside the disk-shaped windows are compared to the contents of the respective windows on the reference field. If the deviation is significant, the process returns to the first step with a reduced size for the ellipse-shaped window; the size reduction is proportional to the deviation. If, however, the deviation in the second step is not significant, the iterations stop. The size of the disk-shaped window of the last iteration is taken as the critical window size meaning the smallest window size below which the pattern as seen from an aperture no longer deviates from the so called straight line. This procedure returns quite consistent results if the goal is to break up the whole via the so called critical points (Figs. 1.4 and 1.5).

Even if the method is applied in the usual setting where the drawing is drawn using regular pen, our approach has significant advantages from a computational point of view. First of all, through the introduction of straight-line reference as an external drawing, we eliminate the need to develop discrete counterparts of the continuous concepts of curvature, bending, and points being on a line. Second of all, because both the analyzed drawing and the external reference are objects of the same type, i.e., images, it becomes easier to compare them. They can also be compared in a scale space. In the suggested framework there are three intertwined scales. One being related to the local scope (aperture), the other two to the diffusion of the pattern. Inside each window, the level curves of the v-field are successively

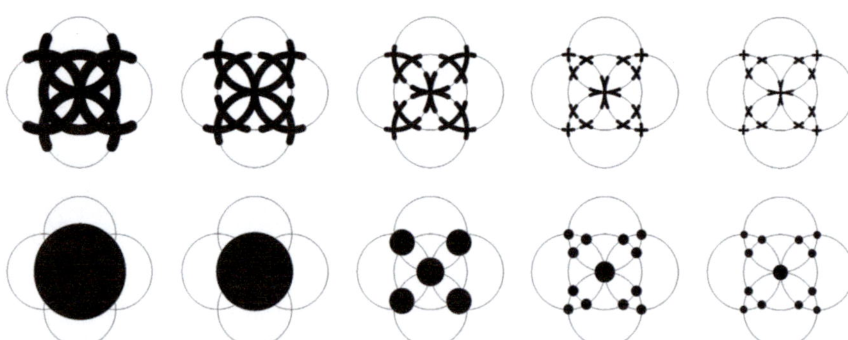

Fig. 1.3 A demonstration of the iterative process [5]

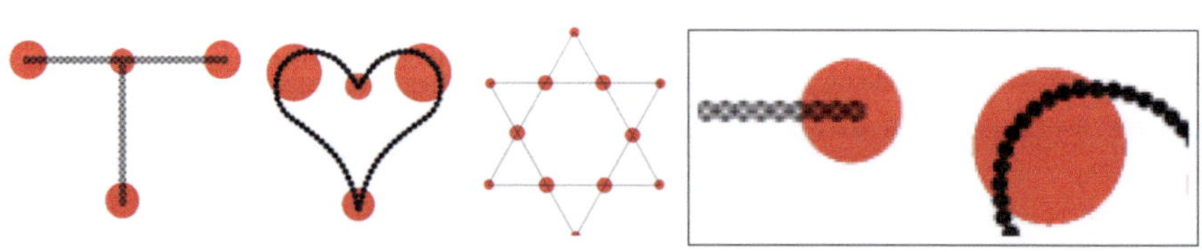

Fig. 1.4 Sample results [5]

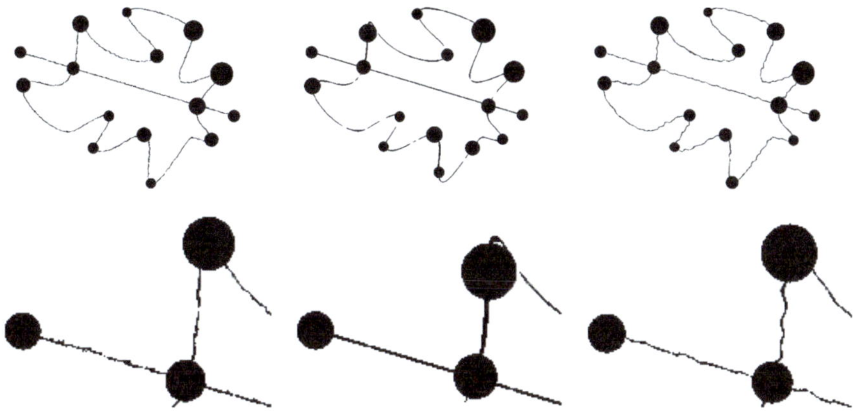

Fig. 1.5 Sample results with Photoshop brush effects. The *top row* depicts critical locations detected from a *line* drawing drawn using a variety of effects via Photoshop. The *bottom row* depicts zoomed cuts from the *top left* corner of each figure in the *top row*. The effects are Ripple (*left*), Glass (*middle*), and Stroke (*right*). More examples are in [5]

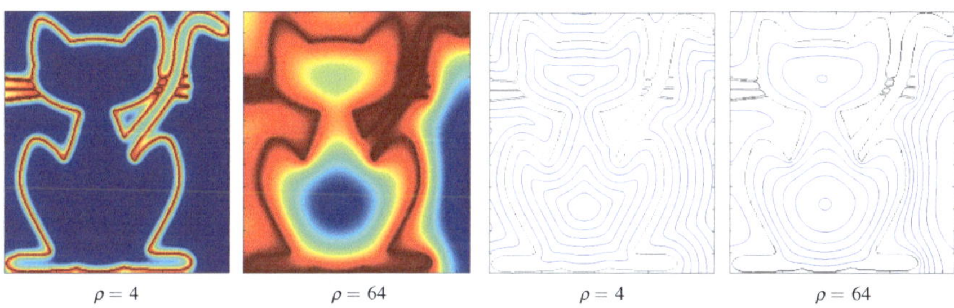

Fig. 1.6 For a cat drawn on a rectangular frame, two v fields with $\rho = 4$ and 64. The one parameter family defines a 2D scale space representation of the cat boundary – coarsening in the direction of increasing ρ and increasing v. The plots on the *right* are the *level curves* of v. The cat boundary is depicted by the *black curve*

evolved (mimicking diffusive erosion) versions of the drawing with the speed of diffusion related to the screening parameter ρ (Fig. 1.6). The diffusive effect is to a large extent responsible for the robustness of the method.

On convergence, each detected location is equipped with a representative "scale", the aperture size below which the pattern as seen through the aperture no longer deviates from the reference. This final scale is quite consistent within critical points of the same type. For example, in Fig. 1.4, observe that the final windows of equivalent criticalities are consistent. For sharp points, the lower the curvature the larger is the window. Furthermore, the three end points of letter T, the two round corners of the heart shape, the six corners the six-pointed star in the hexagram, or the six corners of the inscribed hexagon are all consistent. These consistently detected points may facilitate a passage from one motif to another as the sample demonstration in the front page depicts. The $D3$ shape (dihedral symmetry group of order 3) in the second column is constructed using very simple rules. Each pair of straight line segments linking the six corners the six-pointed star – one type of criticality in the initial motif – is replaced with a curved line and the line segments connecting the six points of the hexagon – the second type of criticality in the initial motif – are deleted. In the third column, the $D3$ motif is repeated to fill the plane and the resulting ornament is coloured with the help of the level curves of the v-field.

1.2.2 Embedded Parts

In some situations, the so-called natural break points lose relevance (Fig. 1.7). There are many examples both in art and Gestalt psychology where simple shapes are embedded in more complex organisations [4], a variety of visual puzzles for children and a whole genre of artistic expression around the basic idea of intentionally hidden shapes. Psychologists study embedded shapes for possible correlation between a person's creativity and his/her ability to eliminate the distracting influence of context.

In Fig. 1.7, two sample designs containing embedded shapes are depicted. The figure on the left has been designed by a colleague M. Ozkar based on an actual carving from Seljuk times in Anatolia [6]. The figure on the right contains a hidden clover [2].

In the case of embedded parts, because the natural break points lose relevance, the most viable option appears to be to know what one is looking for. Technically, this reduces the problem to template matching. The intended fragment needs to be matched to the entire drawing. The search for the pose, scale and location can be formulated as an optimisation where the right pose, scale and location parameters are the ones that maximize the match between the fitted target and the respective part

Fig. 1.7 Embedded shapes. Natural break points become irrelevant

of the original drawing. But this is an ill-posed problem because local information is insufficient. Consider searching a pentagon in the design shown in the left in Fig. 1.7. Let us consider four different hypothesis as shown in Fig. 1.8. Brown hypotheses yield lower matching cost compared to blue hypotheses, even though the later ones are better fits. This is because most of the pixels on the design plane are background pixels and contain no information. A point is either on the drawing or not.

Our solution to this problem is to assign values coding features of the design to the information*less* background pixels. We achieved this via v-fields since a v-field value at a point depends on the distance of the point to the nearest point on the design as well as the curvature at the nearest point [11]. A sample result where we searched a winding line in the previous design is shown in Fig. 1.9. The key point, here, is to use empty background pixels to code contextual clues. Value of a v-field at any background location is an aggregation of information from a certain context

Fig. 1.8 Comparing four hypothesis when searching a pentagon

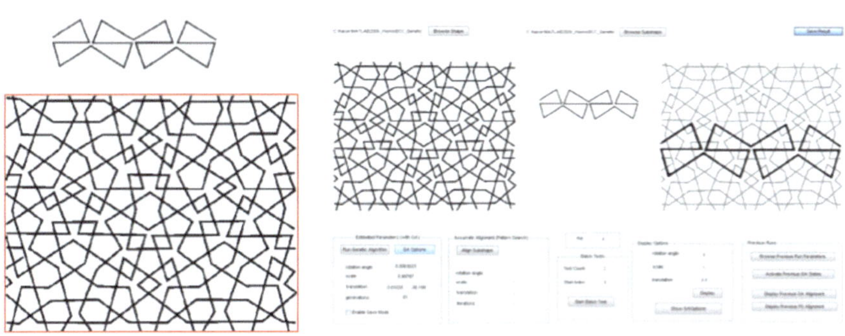

Fig. 1.9 Finding an embedded part. The *winding line* on the *top left* is being searched on the pattern shown in *red box*. The best outcome of the search process is shown on the *right* [6]

capturing interaction among a pair of locations and their surrounding; it codes the distance to the nearest point on the drawing as well as the curvature feature at that point. We have also experimented with alternative approximate fields in figure hunt [2, 3].

In the hands of a skilled design person, our method and the accompanying software [6] is a powerful tool (Fig. 1.9). But it can also be useful for the not-so-skilled, as demonstrated in the next section for the task of structure discovery in tiles.

1.3 Structure Discovery in Tiles

Tiles are stunning ornaments constructed by repeating one or more shapes. A tile has two complementary sides, one being mechanic, the other one being artistic. The shapes drawn (motifs) form the artistic side, and the way they are repeated form the mechanic side. The scientific study of tiles is the study of the repetition structure, the mechanic side.

1.3.1 Repeated Search

In Fig. 1.10, the outcomes of two experiments where two target figures are repeatedly searched in the previous ornament. Repeated embeddings weighted by the number of hits are displayed. Neither of the target figures are special. The first target is a small cut-out from the original design. The second target is a pentagram, which is not even a part of the original design. (Indeed, use of the pentagram as a target is suggested by M. Ozkar, the colleague who had drawn the design.) Yet, repeated embedding reveals structures and nodes that are not easily perceivable to the naked eye, hence, sheds light into the repetition process.

Fig. 1.10 Repeated embedding reveals structure. The probe in the first experiment (*middle*) is a cut-out from the original design (*left*) whereas the one in the second experiment (*right*) is a generic shape with five-fold symmetry which is not even an embedded part of the original

1.3.2 Integrating Mid-level Features in a Context

Each v-field is a member of a one parameter family parameterized by the screening parameter (Fig. 1.11). Converting a shape (or a design) to v-field is a kind of feature integration. Individual pixel level information is gathered to form higher-level information. What is represented by v-value is a complex interaction among a group of locations and their surrounding. As ρ increases, the extent of the surrounding increases. This multi-scale information can be gathered to form higher level integration where the v-values are integrated. Technically, this can be done in a number of ways.

That being said, a preliminary experiment is depicted in Fig. 1.12. In this preliminary experiment, we have collected first and second order statistics and gradient magnitude information from each of the four Screened Poisson fields with varying ρ (Fig. 1.11) to form a 12-dimensional feature vector. We then re-organize this data by applying dimensionality reduction. For simplicity, we use multidimensional scaling. As the parameter ρ can be seen as a spectral variable, what this organization achieves is spatial-spectral integration. In the figure, the second component and its thresholded negative values are depicted to convey some

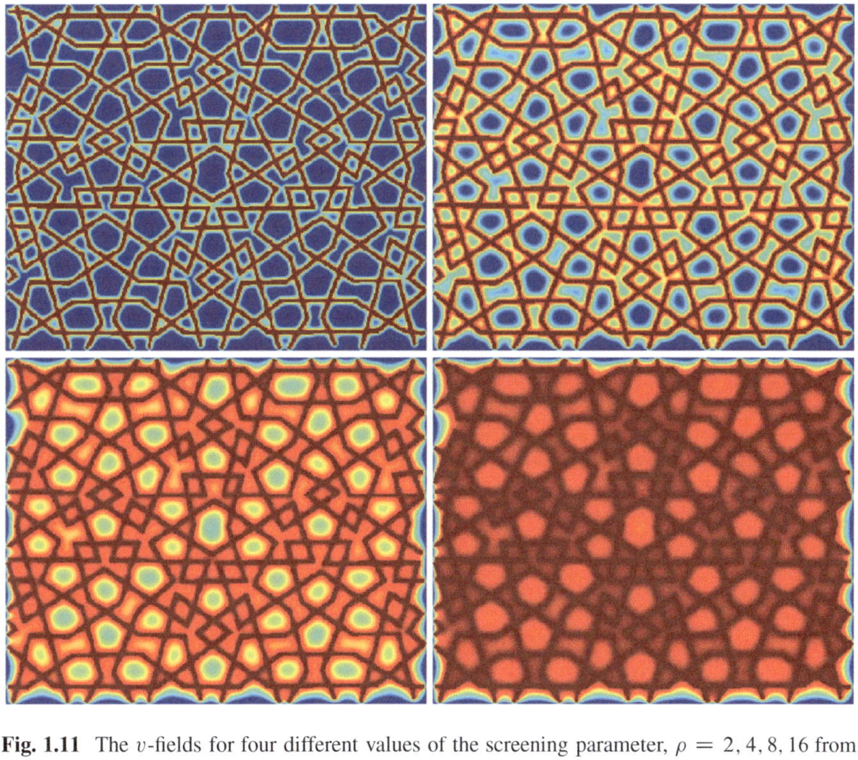

Fig. 1.11 The v-fields for four different values of the screening parameter, $\rho = 2, 4, 8, 16$ from *left* to *right*, *top* to *bottom*. As ρ increases global features prevail

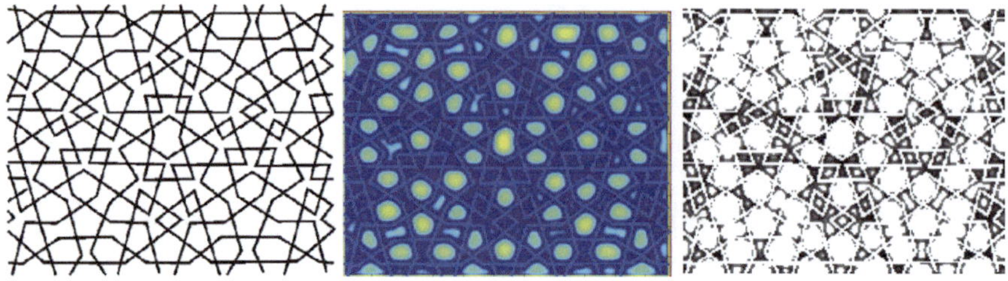

Fig. 1.12 The second component obtained by multi-dimensional scaling of 4 scale 3 measurement feature vector. The image on the middle depicts the second component in false color. The image on the *right* depicts the thresholded negative values of the second component. Best viewed from a distance

impression. Especially when viewed from a distance, both the second component and its thresholded form highlight nodes of construction not easily noticeable in the input design. This is just a preliminary illustration shown as a proof of concept. In general, the choice of scales as well as the measurements that are employed could be task-dependent.

1.4 Summary

Two major problems in Ornament Analysis, defining and detecting parts and discovering structure, are addressed. For each problem, two separate strategies all using Screened Poisson fields are suggested. For the first problem, the first strategy is to seek for natural parts whereas the second one is to search a given target. Both strategies involve the designer in the loop, respectively via the *reference line* and the *embedded part*. The second strategy can be used for structure discovery in tiles via repetitive search, thus, links the two problems. The key idea in structure discovery is the integration of local and global information, whether be it using a single field or a collection of fields with varying screening parameter.

Acknowledgements I thank the reviewers for their meticulous reading and supportive feedback. The work reported in the second half of this chapter is partially funded by TUBITAK grants 112E208 and 114E204.

References

1. Attneave, F.: Some informational aspects of visual perception. Psychol. Rev. **61**(3), 183–193 (1954)
2. Bergbauer, J., Tari, S.: Wimmelbild analysis with approximate curvature coding distance images. In: Scale Space and Variational Methods in Computer Vision (SSVM), pp. 489–500. Springer, Berlin/New York (2013)

3. Diebold, J., Tari, S., Cremers, D.: The role of diffusion in figure hunt. J. Math. Imaging Vis. (2015). doi:10.1007/s10851-014-0548-6
4. Gottschald, K.: Ueber den Einfluss der Erfahrung auf die Wahrnehmung von Figuren. Psychologische Forschung **8**, 261–317 (1926)
5. Keles, H.Y., Tari, S.: A robust method for scale independent detection of curvature-based criticalities and intersections in line drawings. Pattern Recognit. **48**(1), 140–155 (2015)
6. Keles, H.Y., Ozkar, M., Tari, S.: Weighted shapes for embedding perceived wholes. Environ. Plan. B: Plan. Des. **39**, 360–375 (2012)
7. Stiny, G.: Shape: Talking About Seeing and Doing. MIT, Cambridge (2006)
8. Tari, S.: Fluctuating distance fields, parts, three-partite skeletons. In: Breuss, M., Bruckstein, A., Maragos, P. (eds.) Innovations for Shape Analysis. Mathematics and Visualization, pp. 439–466. Springer, Berlin/New York (2013)
9. Tari, S., Genctav, M.: From a non-local Ambrosio-Tortorelli phase field to a randomized part hierarchy tree. J. Math. Imaging Vis. **49**(1), 69–86 (2014)
10. Tari, S., Shah, J.: Local symmetries of shapes in arbitrary dimension. In: ICCV, pp. 1123–1128 (1998)
11. Tari, Z.S.G., Shah, J., Pien, H.: A computationally efficient shape analysis via level sets. In: Mathematical Methods in Biomedical Image Analysis, San Francisco, pp 234–243 (1996)
12. Tari, S., Shah, J., Pien, H.: Extraction of shape skeletons from grayscale images. Comput. Vis. Image Underst. **66**(2), 133–146 (1997)

Chapter 2
A Comparison of Non-Lambertian Models for the Shape-from-Shading Problem

Silvia Tozza and Maurizio Falcone

Abstract In this paper we present in a unified approach Shape-from-Shading models under orthographic projection for non-Lambertian surfaces and compare them with the classical Lambertian model. Those non-Lambertian models have been proposed in the literature by various authors in order to take into account more realistic surfaces such as rough and specular surfaces. The advantage of our unified mathematical model is the possibility to easily modify a single differential model to various situations just changing some control parameters. Moreover, the numerical approximation we propose is valid for that general model and can be easily adapted to the real situation. Finally, we compare the models on some benchmarks including real and synthetic images.

2.1 Introduction

The three dimensional reconstruction of an object is a topic of great interest in many different fields of application: from the digitization of curved documents [12] to the reconstruction of archaeological finds [18]. Other examples come from astronomy for the characterization of properties of planets or other astronomical entities [20, 31, 47]. Facial recognition of individuals [45] is useful for application to security.

This problem has always attracted a great attention because there is still no global method for its resolution under realistic assumptions despite the fact that its formulation is rather simple. The pioneering work of Horn [22] and his activity with the collaborators at MIT [23, 24] produced first formulation of the Shape from Shading (SfS) problem in mathematical terms, via a partial differential equation (PDE) and variational problem. These inspiring works gave rise to many other contributions (see e.g. the two surveys [15, 61] for an extensive list of references).

S. Tozza (✉)
Dipartimento di Matematica "G. Castelnuovo", Sapienza – Università di Roma, Roma, Italy
e-mail: tozza@mat.uniroma1.it

M. Falcone
Dipartimento di Matematica "G. Castelnuovo", Sapienza – Università di Roma, Roma, Italy
e-mail: falcone@mat.uniroma1.it

M. Breuß et al. (eds.), *Perspectives in Shape Analysis*, Mathematics and Visualization, DOI 10.1007/978-3-319-24726-7_2

Several approaches to the SfS problem for classical Lambertian surfaces have been proposed in order to compute a solution. These models mainly belong to two classes: methods based on partial differential equations (PDEs) and optimization methods based on the variational approach. In the first class we can find rather old works based on the method of characteristics and recent works based on the approximation of viscosity solutions for first order Hamilton-Jacobi equations (for a comprehensive presentation of the theory of viscosity solutions we refer the interested reader to the book [4]).

In this work we use the differential approach based on Hamilton-Jacobi equations trying to solve some non-Lambertian models which have been proposed in the literature to overcome some of the limitations of the Lambertian model. It is well known that the classical approach leads to a nonlinear partial differential equation of the first order (of Hamilton-Jacobi type) and it has been shown that this problem is ill-posed even in the framework of viscosity solutions (see the seminal papers by Lions, Rouy and Tourin [30, 43] and also [6, 39]). In fact, there can be many viscosity solutions (no matter which regularity is required for the solutions) unless additional conditions/informations are added to the problem or an a-priori choice is made to compute the maximal solution of the Hamilton-Jacobi equation (see [8, 9, 15]). This explains the growing importance of a generalization of this classical problem in order to obtain uniqueness of the solution while reducing the assumptions on the physical reflectance properties of the objects.
A continuous effort has been made by the scientific community to take into account more realistic reflectance models [2, 3, 42, 56], different scenarios including perspective deformations [1, 11, 34, 38, 49, 57] and/or multiple images of the same object [59, 60]. The images can be taken from the same point of view but with different light sources as in the photometric stereo method [29, 32, 48, 58] or from different points of view but with the same light source as in stereo vision [10]. Recent works have considered more complicated scenarios, e.g. when the light source is not at the optical center under perspective camera projection [26]. It is possible to consider in addition other supplementary issues, as the estimation of the albedo [5, 45, 46, 62] or of the direction of the light source that are usually considered known quantities for the model but in practice are hardly available for real images. Depending on what we know the model has to be adapted leading to a calibrated or uncalibrated problem (see [19, 41, 59, 60] for more details). In this work we will assume that the albedo and the light direction are given.

Our Contribution We want to take into account more realistic models for generic surfaces with nonuniform reflection properties, which means that the light intensity of the image does not depend only on the angle between the outgoing normal to the surface and the light source as in the *Lambertian model*. In particular, we will focus our attention on two non-Lambertian models under orthographic projection originally proposed by Oren-Nayar [35, 36] and by Phong [37]. These models have been introduced to deal respectively with rough or shiny surfaces and are not well suited for other surfaces such as objects with multiple materials, human skin or glass. Typical examples of rough materials are clay and plaster works whereas bronze and plastic are shiny materials.

We should mention that other authors have contributed to the SfS problem for non-Lambertian surfaces. We mention in particular the contributions in [2], who derived the PDEs associated to several models solving them via a Lax-Friedrichs Sweeping (LFS) method and in [26] where the Hamilton-Jacobi equations based on the Oren-Nayar reflectance model appear in spherical coordinate under perspective camera projection. As we said, here we work in Cartesian coordinate under orthographic projection to derive the Hamilton-Jacobi equations for the above mentioned models under general light directions. Some preliminary results just for the Oren-Nayar problem have appeared in [51] and the Lambertian SfS problem with oblique light direction has been studied in [17]. Extending these results to another non-Lambertian model (the Phong model), we will show that the three models share the same fixed point form so that we can have a unified approach to their analysis and approximation. Moreover, we propose a semi-Lagrangian approximation scheme for that general first order PDE, we give evidence that this scheme converges to the weak solution (in the viscosity sense) of that equation and we compare the performances of this approximation scheme with other finite difference solvers. The scheme is also used to test the models on a number of real and synthetic images in order to understand if the introduction of non-Lambertian models can be really effective.

Organization of the paper In Sect. 2.2 we present an overview of the most relevant non-Lambertian models and derive their Hamilton-Jacobi formulation. In Sect. 2.3 we present the semi-Lagrangian schemes for these equations and shortly the Fast Marching and the Fast Sweeping schemes based on finite difference solver. In Sect. 2.4 we compare these methods and algorithms on a series of benchmarks on synthetic and real images. Finally, we conclude with some comments and future perspectives.

2.2 Some Non-Lambertian Models for the Orthographic SfS

Let us consider a surface given as a graph $z = u(\mathbf{x}), \mathbf{x} \in \mathbb{R}^2$. We will denote by Ω the region inside the silhouette and we will assume (just for technical reasons) that Ω is an open and bounded subset of $\mathbb{R}^2$. We assume that $u(\mathbf{x}) \geq 0$ and the surface is standing on a flat background (hence $u(\mathbf{x}) = 0$ on $\partial\Omega$). Note that non homogeneous Dirichlet boundary condition like $u(\mathbf{x}) = g(\mathbf{x})$ can be easily handled in our approach. The function $g(\mathbf{x})$ will represent the height of the surface at the boundary of the silhouette. Clearly, this is an additional information which in general is not available but can be derived, for example, for rotational surfaces or by symmetry arguments.

It is well known that the Shape-from-Shading problem is described by the image irradiance equation introduced by Bruss [7]

$$I(\mathbf{x}) = R(\mathbf{N}(\mathbf{x})), \tag{2.1}$$

where $I(\mathbf{x})$ is the normalized brightness of the given grey-value image, $\mathbf{N}(\mathbf{x})$ is the unit normal to the surface at the point $(\mathbf{x}, u(\mathbf{x}))$ and $R(\mathbf{N}(\mathbf{x}))$ is the reflection map giving the value of the light reflection on the surface as a function of its orientation (i.e., of the normal) at each point. Note that a more general formulation of the reflectance function R present in the irradiance equation (2.1) consists of adding a dependence on $\mathbf{x}$ too, in order to include several features like e.g. non uniform ambient light depending on some diffuse lights in the ambient (that can be generated by other light sources at finite distance). We will not consider this generalization in this paper.

For the analysis of the different models, it would be useful to introduce a representation of the brightness function $I(\mathbf{x})$ in which we can distinguish different terms representing the contribution of ambient, diffused reflected and specular reflected light. We will write then

$$I(\mathbf{x}) = k_A I_A(\mathbf{x}) + k_D I_D(\mathbf{x}) + k_S I_S(\mathbf{x}), \tag{2.2}$$

where $I_A(\mathbf{x})$, $I_D(\mathbf{x})$ and $I_S(\mathbf{x})$ are respectively the above mentioned components and k_A, k_D and k_S indicate the percentages of these components such that their sum is equal to 1 (we do not consider absorption phenomena). Note that the diffuse or specular albedo is inside the definition of $I_D(\mathbf{x})$ or $I_S(\mathbf{x})$, respectively. This will allow to switch on and off the different contributions depending on the model. Let us note that the ambient light term represents light present everywhere in a given scene. As we will see in the following sections, the intensity of diffusely reflected light in each direction is proportional to the cosine of the angle θ_i between surface normal and light source direction, without taking into account the point of view of the observer, but another diffuse model (the Oren–Nayar model) will consider it in addition. The amount of specular light reflected towards the viewer is proportional to $(\cos\theta_s)^\alpha$, where θ_s is the angle between the ideal (mirror) reflection direction of the incoming light and the viewer direction, α being a constant modelling the specularity of the material. In this way we have a more general model and, dropping the ambient and specular component, we retrieve the Lambertian reflection as a special case. In order to underline the differences, let us briefly sketch the classical Lambertian model (L–model) and two non-Lambertian models: the Oren-Nayar model (ON–model) and the Phong model (PH–model). Our goal in this section is to derive the nonlinear PDEs corresponding to each model.

2.2.1 Lambertian Model

Let us consider a single light source located at infinity in the direction of the unit vector $\boldsymbol{\omega}$. For a *Lambertian surface*, which generates a purely diffuse model, the specular component does not exist. So, the general Eq. (2.2) becomes

$$I(\mathbf{x}) = k_A I_A(\mathbf{x}) + k_D I_D(\mathbf{x}), \tag{2.3}$$

whose diffuse component $I_D(\mathbf{x})$ is

$$I_D(\mathbf{x}) = \gamma_D \, \mathbf{N}(\mathbf{x}) \cdot \boldsymbol{\omega}, \tag{2.4}$$

where γ_D is the diffuse albedo. Neglecting the ambient component that can be considered as a constant (i.e. setting $k_A = 0$), recalling that the sum $k_A + k_D + k_S$ must be equal to 1, we obtain that necessarily $k_D = 1$ and we can omit it in the following. Then, for a Lambertian surface the image irradiance equation (2.1) becomes

$$I(\mathbf{x}) = \gamma_D \, \mathbf{N}(\mathbf{x}) \cdot \boldsymbol{\omega}, \tag{2.5}$$

where we assume to know γ_D (in the sequel we suppose uniform albedo and we put $\gamma_D = 1$, that is all the points of the surface reflect completely the light that hits them). For Lambertian surfaces [23, 24], just considering an orthographic projection of the scene, we can write the model for SfS via a first order nonlinear PDE which describes the relation between the surface $u(\mathbf{x})$ (our unknown) and the brightness function $I(\mathbf{x})$. The data are the grey-value image $I(\mathbf{x})$, the direction of the light source $\boldsymbol{\omega}$ and the albedo γ_D.

Recalling that the normal to a graph is given by

$$\mathbf{N}(\mathbf{x}) = (-\nabla u(\mathbf{x}), 1) / \sqrt{1 + |\nabla u(\mathbf{x})|^2}, \tag{2.6}$$

we can write (2.1) as

$$I(\mathbf{x}) \sqrt{1 + |\nabla u(\mathbf{x})|^2} + \tilde{\boldsymbol{\omega}} \cdot \nabla u(\mathbf{x}) - \omega_3 = 0 \text{ in } \Omega, \tag{2.7}$$

where $\tilde{\boldsymbol{\omega}} := (\omega_1, \omega_2)$. This is a Hamilton-Jacobi type equation which does not admit in general a regular solution. It is known that the mathematical framework to describe its weak solutions is the theory of viscosity solutions as in [30].

It is important to note that if the light is oblique we have shadows in the image since the object projects its shadow on the flat background. Then, we can divide the image into subdomains

$$\Omega_l \equiv \{\mathbf{x} : I(\mathbf{x}) > 0\}, \qquad \Omega_s \equiv \{\mathbf{x} : I(\mathbf{x}) = 0\}, \tag{2.8}$$

which represent respectively the "light" and the "black shadow" regions. Naturally, $\Omega = \Omega_l \cup \Omega_s$ and we assume for simplicity that the projection of the shadows on the background also falls in Ω.

In Ω_l the equation is always the same, whereas in the "shadow" region the surface can have any shape since the model is naturally not able to describe the real surface there. One approach is to deal only with the "light" region setting the equation only on Ω_l, however this will require to use oblique boundary conditions (e.g., Neumann boundary conditions) on $\partial\Omega_l$ to treat the problem in Ω_l because the height there is not known on $\partial\Omega_l$. This can in turn create difficulties in the construction of the numerical algorithm since the curved boundary of Ω_l can be nonsmooth and can be efficiently approximated only via a triangulation (which collides with the use of a simple structured grid).

Our approach (see [17] for details) includes the region Ω_s in the computation by defining there a virtual surface which replaces the unknown surface corresponding to the "black shadow" region. Conventionally, we will substitute it to the surface generated by the "separation plane" (or "shadow plane"), i.e. the plane separating light from shadow. That plane has the same direction of $\boldsymbol{\omega}$. This means that in Ω_s we have to solve the equation

$$(\omega_1, \omega_2) \cdot \nabla u(\mathbf{x}) - \omega_3 = 0, \quad \mathbf{x} \in \Omega_s. \tag{2.9}$$

Note that the irradiance equation coincides with (2.9) since $I = 0$ in Ω_s. Then, we can use the same equation everywhere in Ω avoiding in this way the use of boundary conditions on Ω_l, i.e. we can write the global problem as

$$\begin{cases} I(\mathbf{x})\sqrt{1+ \mid \nabla u(\mathbf{x}) \mid^2} + (\omega_1, \omega_2) \cdot \nabla u(\mathbf{x}) - \omega_3 = 0, & \mathbf{x} \in \Omega, \\ u(\mathbf{x}) = 0 & \mathbf{x} \in \partial\Omega. \end{cases} \tag{2.10}$$

Writing the surface as $S(\mathbf{x}, z) = z - u(\mathbf{x}) = 0$ for $\mathbf{x} \in \Omega$, $z \in \mathbb{R}$, we can obtain a more compact form for (2.10). In fact, $\nabla S(\mathbf{x}, z) = (-\nabla u(\mathbf{x}), 1)$ and (2.10) becomes

$$\begin{cases} I(\mathbf{x}) \mid \nabla S(\mathbf{x}, z) \mid - \nabla S(\mathbf{x}, z) \cdot \omega = 0, & \mathbf{x} \in \Omega, \\ u(\mathbf{x}) = 0 & \mathbf{x} \in \partial\Omega. \end{cases} \tag{2.11}$$

Using the equivalence $\mid \nabla S(\mathbf{x}, z) \mid \equiv \max_{a \in \partial B_3(0,1)} \{a \cdot \nabla S(\mathbf{x}, z)\}$, we get

$$\max_{a \in \partial B_3(0,1)} \{ (I(\mathbf{x})a_1 - \omega_1, I(\mathbf{x})a_2 - \omega_2, I(\mathbf{x})a_3) \cdot \nabla S(\mathbf{x}, z)\} = \omega_3. \tag{2.12}$$

For analytical and numerical reasons it is useful to introduce the exponential Kružkov transform $\mu v(\mathbf{x}) = 1 - e^{-\mu u(\mathbf{x})}$. By this change the variable $v(\mathbf{x})$ will assume values only in $[0, 1/\mu]$ whereas u is in principle unbounded. So the change of variable avoids the risk of an overflow in the approximation. Note that here μ is a free positive parameter without a specific physical meaning, but it is important because varying its value it is possible to modify the slope (the slope increases for increasing values of μ). Clearly, once v is obtained we can always get back to the original surface u simply setting $u(\mathbf{x}) = -\ln(1-\mu v(\mathbf{x}))/\mu$. By the above approach we can write (2.10) in a fixed point form in the new variable v as

$$\begin{cases} \mu v(\mathbf{x}) = \min\limits_{a\in\partial B_3} \{\mathbf{b}^L(\mathbf{x}, a)\cdot\nabla v(\mathbf{x}) + f^L(\mathbf{x}, a, v(\mathbf{x}))\}, & \text{for } \mathbf{x}\in\Omega, \\ v(\mathbf{x}) = 0, & \text{for } \mathbf{x}\in\partial\Omega, \end{cases} \tag{2.13}$$

where $\mathbf{b}^L : \Omega\times\partial B_3(0,1)\to\mathbb{R}^2$ and $f^L : \Omega\times\partial B_3(0,1)\times[0,1/\mu]\to\mathbb{R}$ are defined as

$$\mathbf{b}^L(\mathbf{x}, a) := \frac{1}{\omega_3}\left(I(\mathbf{x})a_1 - \omega_1, I(\mathbf{x})a_2 - \omega_2\right), \tag{2.14}$$

$$f^L(\mathbf{x}, a, v(\mathbf{x})) := -\frac{I(\mathbf{x})a_3}{\omega_3}(1-\mu v(\mathbf{x}))\} + 1, \tag{2.15}$$

where B_3 denotes the unit ball in $\mathbb{R}^3$ and $\partial B_3(0,1)$ its boundary.

2.2.2 *Oren-Nayar Model*

The diffuse reflectance ON–model [35, 36] is an extension of the previous L-model which explicitly allows to handle *rough* surfaces. The idea of this model is to represent a rough surface as an aggregation of V-shaped cavities, each with Lambertian reflectance properties (see Fig. 2.1a).

The I_D *brightness equation* for the ON–model [36] is given by

$$I_D(\mathbf{x}) = \gamma_D\cos(\theta_i)(A + B\sin(\alpha)\tan(\beta)\max[0,\cos(\varphi_r-\varphi_i)]) \tag{2.16}$$

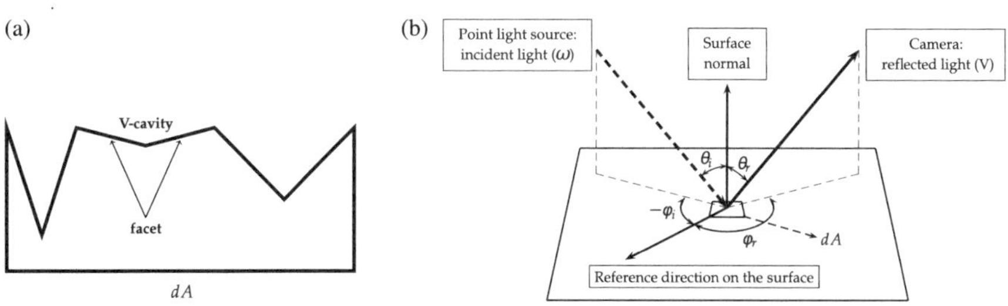

Fig. 2.1 Description of the ON–model (Figure adapted from [25]). (**a**) Facet model for surface patch *dA* consisting of many V-shaped Lambertian cavities. (**b**) Diffuse reflectance for the ON–model

where

$$A = 1 - 0.5\,\sigma^2(\sigma^2 + 0.33)^{-1} \tag{2.17}$$

$$B = 0.45\sigma^2(\sigma^2 + 0.09)^{-1}. \tag{2.18}$$

Note that A and B are two nonnegative constants depending on the statistics of the cavities via the roughness parameter σ that we can imagine to take values between 0 and $\pi/2$, representing the slope of the roughness for the surface considered. In this model (see Fig. 2.1b), θ_i represents the angle between the unit normal to the surface $\mathbf{N}(\mathbf{x})$ and the light source direction $\boldsymbol{\omega}$, θ_r stands for the angle between $\mathbf{N}(\mathbf{x})$ and the observer direction $\mathbf{V}$, φ_i is the angle between the projection of the light source direction $\boldsymbol{\omega}$ and the x_1 axis onto the (x_1, x_2)-plane, φ_r denotes the angle between the projection of the observer direction $\mathbf{V}$ and the x_1 axis onto the (x_1, x_2)-plane and the two variables α and β are given by

$$\alpha = \max\,[\theta_i, \theta_r] \text{ and } \beta = \min\,[\theta_i, \theta_r]\,. \tag{2.19}$$

For smooth surfaces, we have $\sigma = 0$ and the ON–model becomes identical to the L–model. In the particular case $\boldsymbol{\omega} = \mathbf{V} = (0, 0, 1)$, or, more precisely, when $\cos(\varphi_r - \varphi_i) \leq 0$, the equation simplifies and reduces to a L–model scaled by the coefficient A. This happens for example when the unit vectors $\boldsymbol{\omega}$ and $\mathbf{V}$ are perpendicular so that $\cos(\varphi_r - \varphi_i) = -1$ or, more in general, when the scalar product between $\tilde{\boldsymbol{\omega}} = (\omega_1, \omega_2)$ and $\tilde{\mathbf{V}} = (V_1, V_2)$ is equal to zero. Therefore the ON–model is more general and flexible than the L–model.

Also for this diffuse model we neglect the ambient component. Then, we get $k_D = 1$ and, as a consequence, in the general Eq. (2.2) the total light intensity $I(\mathbf{x})$ is equal to the only diffuse component $I_D(\mathbf{x})$, in this case described by the Eq. (2.16). Hence, for what follows, we will write $I(\mathbf{x})$ instead of $I_D(\mathbf{x})$.

To deal with this equation one has to resolve the *min* and *max* operators which appear in (2.16) and (2.19). In general, several cases must be considered but here we just take one to illustrate the technique. Namely, we consider the particular case where the position of the light source $\boldsymbol{\omega}$ and of the observer $\mathbf{V}$ coincide in a general oblique direction (see [50, 52] for the other cases and compare with [26] in order to note that we obtain the same cases). This choice implies $\max[0, \cos(\varphi_i - \varphi_r)] = 1$, then defining $\theta := \theta_i = \theta_r = \alpha = \beta$ and putting for simplicity the albedo $\gamma_D = 1$, the Eq. (2.16) simplifies to

$$I(\mathbf{x}) = \cos(\theta)\ \left(A + B\sin(\theta)^2\cos(\theta)^{-1}\right) \tag{2.20}$$

and we arrive to a first order nonlinear Hamilton-Jacobi equation

$$(I(\mathbf{x}) - B)(\sqrt{1 + |\nabla u(\mathbf{x})|^2}) + A(\tilde{\boldsymbol{\omega}} \cdot \nabla u(\mathbf{x}) - \omega_3) + B\frac{(-\tilde{\boldsymbol{\omega}} \cdot \nabla u(\mathbf{x}) + \omega_3)^2}{\sqrt{1 + |\nabla u(\mathbf{x})|^2}} = 0, \tag{2.21}$$

where $\tilde{\boldsymbol{\omega}} = (\omega_1, \omega_2)$. Following [51], we write the surface as $S(\mathbf{x}, z) = z - u(\mathbf{x}) = 0$, for $\mathbf{x} \in \Omega$, $z \in \mathbb{R}$, and $\nabla S(\mathbf{x}, z) = (-\nabla u(\mathbf{x}), 1)$, so (2.21) becomes

$$(I(\mathbf{x}) - B)|\nabla S(\mathbf{x}, z)| + A(-\nabla S(\mathbf{x}, z) \cdot \boldsymbol{\omega}) + B\left(\frac{\nabla S(\mathbf{x}, z)}{|\nabla S(\mathbf{x}, z)|} \cdot \boldsymbol{\omega}\right)^2 |\nabla S(\mathbf{x}, z)| = 0. \tag{2.22}$$

Defining $d(\mathbf{x}, z) := \nabla S(\mathbf{x}, z)/|\nabla S(\mathbf{x}, z)|$ and $c(\mathbf{x}, z) := I(\mathbf{x}) - B + B(d(\mathbf{x}, z) \cdot \boldsymbol{\omega})^2$, using the equivalence $|\nabla S(\mathbf{x}, z)| \equiv \max_{a \subset \partial B_3}\{a \cdot \nabla S(\mathbf{x}, z)\}$ we get

$$\max_{a \in \partial B_3}\{c(\mathbf{x}, z)\, a \cdot \nabla S(\mathbf{x}, z) - A\boldsymbol{\omega} \cdot \nabla S(\mathbf{x}, z)\} = 0. \tag{2.23}$$

Defining the vector field for the ON-model

$$\mathbf{b}^{ON}(\mathbf{x}, a) := \frac{1}{A\omega_3}\left(c(\mathbf{x}, z)a_1 - A\omega_1, c(\mathbf{x}, z)a_2 - A\omega_2\right), \tag{2.24}$$

introducing the exponential Kružkov transform $\mu v(\mathbf{x}) = 1 - e^{-\mu u(\mathbf{x})}$ as already done for the L–model, we can finally write the Dirichlet problem in the new variable v

$$\begin{cases} \mu v(\mathbf{x}) + \max_{a \in \partial B_3}\{-\mathbf{b}^{ON}(\mathbf{x}, a) \cdot \nabla v(\mathbf{x}) + \dfrac{c(\mathbf{x}, z)a_3}{A\omega_3}(1 - \mu v(\mathbf{x}))\} = 1, & \mathbf{x} \in \Omega, \\ v(\mathbf{x}) = 0, & \mathbf{x} \in \partial\Omega. \end{cases} \tag{2.25}$$

Note that the simple homogeneous Dirichlet boundary condition is due to the flat background behind the object but a condition like $u(\mathbf{x}) = g(\mathbf{x})$ can also be considered if necessary.

In the particular case when $\cos(\varphi_r - \varphi_i) = 0$, the Eq. (2.16) simply reduces to

$$I(\mathbf{x}) = A\cos(\theta) \tag{2.26}$$

and, as a consequence, the Dirichlet problem in the variable v is equal to (2.25) with $c(\mathbf{x}, z) = I(\mathbf{x})$.

2.2.3 *Phong Model for Specular Surfaces*

The PH–model introduces a specular component to the brightness function $I(\mathbf{x})$. As we said at the beginning of this section, this can be described in general as the sum $I(\mathbf{x}) = k_A I_A(\mathbf{x}) + k_D I_D(\mathbf{x}) + k_S I_S(\mathbf{x})$, where $I_A(\mathbf{x})$, $I_D(\mathbf{x})$ and $I_S(\mathbf{x})$ are the ambient, diffuse and specular light component, respectively. We will set for simplicity $k_A = 0$ and represent the diffuse component $I_D(\mathbf{x})$ as the Lambertian reflectance model.

The most simple specular model is obtained putting the incidence angle equal to the reflection one and $\boldsymbol{\omega}$, $\mathbf{N}(\mathbf{x})$ and $\mathbf{R}(\mathbf{x})$ belong to the same plane. The PH–model is an empirical model that was developed by Phong [37] in 1975. This model describes the specular light component $I_S(\mathbf{x})$ as a power of the cosine of the angle between the unit vectors $\mathbf{V}$ and $\mathbf{R}(\mathbf{x})$ (it is the vector representing the reflection of the light $\boldsymbol{\omega}$ on the surface), then for the Phong model

$$I_S^{PH}(\mathbf{x}) = \gamma_S(\mathbf{R}(\mathbf{x}) \cdot \mathbf{V})^\alpha \tag{2.27}$$

where α expresses the specular reflection characteristics of a material.

Hence, the brightness equation for the PH–model is

$$I(\mathbf{x}) = k_D \gamma_D(\mathbf{N}(\mathbf{x}) \cdot \boldsymbol{\omega}) + k_S \gamma_S(\mathbf{R}(\mathbf{x}) \cdot \mathbf{V})^\alpha, \tag{2.28}$$

where γ_D and γ_S represent the diffuse and specular albedo, respectively.

We will illustrate in details the PH–model and the numerical scheme to which we arrive in the case of a general oblique light source $\boldsymbol{\omega}$ and observer $\mathbf{V} = (0, 0, 1)$.

Assuming that $\mathbf{N}(\mathbf{x})$ is the bisector of the angle between $\boldsymbol{\omega}$ and $\mathbf{R}(\mathbf{x})$, we obtain

$$\mathbf{N}(\mathbf{x}) = \frac{\boldsymbol{\omega} + \mathbf{R}(\mathbf{x})}{||\boldsymbol{\omega} + \mathbf{R}(\mathbf{x})||} \text{ which implies } \mathbf{R}(\mathbf{x}) = ||\boldsymbol{\omega} + \mathbf{R}(\mathbf{x})||\mathbf{N}(\mathbf{x}) - \boldsymbol{\omega}. \tag{2.29}$$

From the parallelogram law, taking into account that $\boldsymbol{\omega}$, $\mathbf{R}(\mathbf{x})$ and $\mathbf{N}(\mathbf{x})$ are unit vectors, we can write $||\boldsymbol{\omega} + \mathbf{R}(\mathbf{x})|| = 2(\mathbf{N}(\mathbf{x}) \cdot \boldsymbol{\omega})$, then we can derive the unit vector $\mathbf{R}(\mathbf{x})$ as follow:

$$\begin{aligned}\mathbf{R}(\mathbf{x}) &= 2(\mathbf{N}(\mathbf{x}) \cdot \boldsymbol{\omega})\mathbf{N}(\mathbf{x}) - \boldsymbol{\omega} = 2\left(\frac{-\tilde{\omega} \cdot \nabla u(\mathbf{x}) + \omega_3}{\sqrt{1 + |\nabla u(\mathbf{x})|^2}}\right)\mathbf{N}(\mathbf{x}) - (\omega_1, \omega_2, \omega_3) \\ &= \left(\frac{-2\tilde{\omega} \cdot \nabla u(\mathbf{x}) + 2\omega_3}{1 + |\nabla u(\mathbf{x})|^2}\right)(-\nabla u(\mathbf{x}), 1) - (\omega_1, \omega_2, \omega_3).\end{aligned} \tag{2.30}$$

For $\mathbf{V} = (0, 0, 1)$ we have

$$\mathbf{R}(\mathbf{x}) \cdot \mathbf{V} = \frac{-2\tilde{\omega} \cdot \nabla u(\mathbf{x}) + 2\omega_3}{1 + |\nabla u(\mathbf{x})|^2} - \omega_3 = \frac{-2\tilde{\omega} \cdot \nabla u(\mathbf{x}) + \omega_3(1 - |\nabla u(\mathbf{x})|^2)}{1 + |\nabla u(\mathbf{x})|^2}. \tag{2.31}$$

Then, putting $\alpha = 1$, Eq. (2.28) becomes

$$\begin{aligned}I(\mathbf{x})(1 + |\nabla u(\mathbf{x})|^2) &- k_D \gamma_D(-\nabla u(\mathbf{x}) \cdot \boldsymbol{\omega} + \omega_3)(\sqrt{1 + |\nabla u(\mathbf{x})|^2}) \\ &- k_S \gamma_S\left(-2\tilde{\omega} \cdot \nabla u(\mathbf{x}) + \omega_3(1 - |\nabla u(\mathbf{x})|^2)\right) = 0,\end{aligned} \tag{2.32}$$

to which we add a Dirichlet boundary condition equal to zero assuming that the surface is standing on a flat background. As we have done for the previous models, we write the surface as $S(\mathbf{x},z) = z - u(\mathbf{x}) = 0$, for $\mathbf{x} \in \Omega$, $z \in \mathbb{R}$, and $\nabla S(\mathbf{x},z) = (-\nabla u(\mathbf{x}), 1)$, so (2.32) will be written as

$$I(\mathbf{x})|\nabla S(\mathbf{x},z)|^2 - k_D\gamma_D(\nabla S(\mathbf{x},z)\cdot\boldsymbol{\omega})(|\nabla S(\mathbf{x},z)|) - k_S\gamma_S(2\nabla S(\mathbf{x},z)\cdot\boldsymbol{\omega} - |\nabla S(\mathbf{x},z)|^2\omega_3) = 0. \tag{2.33}$$

Dividing both the terms by $|\nabla S(\mathbf{x},z)|$, defining $d(\mathbf{x},z) := \nabla S(\mathbf{x},z)/|\nabla S(\mathbf{x},z)|$ as in the ON–model and $c(\mathbf{x}) := I(\mathbf{x}) + \omega_3 k_S\gamma_S$, we get

$$c(\mathbf{x})|\nabla S(\mathbf{x},z)| - k_D\gamma_D(\nabla S(\mathbf{x},z)\cdot\boldsymbol{\omega}) - 2k_S\gamma_S(d(\mathbf{x},z)\cdot\boldsymbol{\omega}) = 0. \tag{2.34}$$

By the equivalence $|\nabla S(\mathbf{x},z)| \equiv \max_{a\in\partial B_3}\{a\cdot\nabla S(\mathbf{x},z)\}$ we obtain

$$\max_{a\in\partial B_3}\{c(\mathbf{x})\,a\cdot\nabla S(\mathbf{x},z) - k_D\gamma_D(\boldsymbol{\omega}\cdot\nabla S(\mathbf{x},z)) - 2k_S\gamma_S(d(\mathbf{x},z)\cdot\boldsymbol{\omega})\} = 0. \tag{2.35}$$

Defining the vector field

$$\mathbf{b}^{PH}(\mathbf{x},a) := \frac{1}{Q^{PH}(\mathbf{x},z)}\,(c(\mathbf{x})a_1 - k_D\gamma_D\omega_1, c(\mathbf{x})a_2 - k_D\gamma_D\omega_2) \tag{2.36}$$

where

$$Q^{PH}(\mathbf{x},z) := 2k_S\gamma_S(d(\mathbf{x},z)\cdot\boldsymbol{\omega}) + k_D\gamma_D\omega_3, \tag{2.37}$$

and using the exponential Kružkov transform $\mu v(\mathbf{x}) = 1 - e^{-\mu u(\mathbf{x})}$ as done for the previous models, we can finally write the nonlinear problem corresponding to the PH–model

$$\begin{cases} \mu v(\mathbf{x}) + \max_{a\in\partial B_3}\{-\mathbf{b}^{PH}(\mathbf{x},a)\cdot\nabla v(\mathbf{x}) + \dfrac{c(\mathbf{x})a_3}{Q^{PH}(\mathbf{x},z)}(1-\mu v(\mathbf{x}))\} = 1, & \mathbf{x}\in\Omega, \\ v(\mathbf{x}) = 0, & \mathbf{x}\in\partial\Omega. \end{cases} \tag{2.38}$$

Again, note that the simple homogeneous Dirichlet boundary condition considered is due to the flat background behind the object but a different boundary condition can also be considered.

2.3 Numerical Approximation

Let us describe some numerical schemes for the solution of the problems described in the previous section. Here we will focus our attention on semi-Lagrangian (SL) schemes which have shown to be very effective for first order problems since they try to mimic at the discrete level the method of characteristics (see [16] for more details). Other approaches based on finite differences or finite volumes are feasible. As we have seen there are basically two main problems related to the vertical light case and the oblique light case. In the vertical case, we have to solve an eikonal-type equation for each model. In the oblique case, we get the more general first-order Hamilton-Jacobi (HJ) equations (2.7), (2.21), and (2.32) where the nonlinear term is also coupled with linear terms. The general framework for these type of problems is the theory of viscosity solutions which guarantees (under appropriate assumptions) existence and uniqueness results for the vertical light case. A similar approach can also be applied to the case of an oblique light source when the surface is not smooth and black shadows are present in the image [17]. It should be noted that to have uniqueness when the eikonal equation is degenerate (i.e. when the right-hand side vanishes at some points) one has to add additional assumptions or more informations (like the height at maximum brightness points or the fact that we select to approximate the maximal solution, as introduced in [8]). General convergence results for the approximation scheme to the maximal solution of the degenerate eikonal equation can be found in [9, 15].

There are two types of algorithms based on the semi-Lagrangian approach. The first type of algorithm is global and gives an approximation of the fixed point problem on the whole grid at every iteration till the stopping rule is satisfied. Some acceleration methods, like the Fast Sweeping method [27, 28], can be introduced to speed up convergence. The second type of method is local and tries to concentrate the numerical effort only in a neighborhood of a region which is considered to be already exact (the so called Accepted region). The Fast Marching method (extensively described in [13, 44]) is a typical example of this class of methods.

The algorithms corresponding to the models presented in the previous section compute the maximal solution in the domain without additional information of the surface and with a single boundary condition which can be either homogeneous $u = 0$ or not (but to set $u = g$ on the boundary of the mask one has to know or guess the right solution there). This is due to the monotonicity properties of the discrete operator corresponding to the schemes. The interested reader can find in [16] a detailed presentation of the properties of semi-Lagrangian schemes and in [17] an application to the Shape-from-Shading problem with black shadows.

As already stated in Sect. 2.2, we suppose a surface given as a graph. In the case of vertical light, for such a surface we do not have shadows covering an open domain (i.e. the points where $I(\mathbf{x}) = 0$ are either isolated or curves in the plane). If the light is oblique, we usually have shadows so that we can divide the support of the surface (the domain of u) into two regions, $\Omega_l \equiv \{\mathbf{x} : I(\mathbf{x}) > 0\}$ and $\Omega_s \equiv \{\mathbf{x} : I(\mathbf{x}) = 0\}$, which represent respectively the "light" and the "shadow" regions. Typically they have both nonempty interior and, naturally, $\Omega = \Omega_l \cup \Omega_s$. Note that Ω now represents the new mask which also includes black regions. Moreover, we assume that $\Omega \subset \overline{Q}$, where Q is the rectangular domain corresponding to the image.
As we already explained, we can use the same equation everywhere in our computational domain Q and we do not need to introduce any boundary condition on $\partial\Omega_l$ (see [17] for more details).

Now, look at the discrete schemes for the described models.

Let $W_i = w(x_i)$ so that W will be the vector solution giving the approximation of the height of u at every node x_i of the grid. Following [16], the semi-Lagrangian scheme for the above models can be written in a fixed point form. In general, we will write it as

$$W_i = T_i^M(W), \tag{2.39}$$

where M is the acronym identifying the model, then $M = L, ON$ or PH. Denoting by G the global number of nodes in the grid, the operator for the L–model $T^L : \mathbb{R}^G \to \mathbb{R}^G$ is defined componentwise by

$$T_i^L(W) := \min_{a \in \partial B_3} \{e^{-\mu h} w(x_i + hb^L(x_i, a)) - \tau \frac{I(x_i)a_3}{\omega_3}(1 - \mu w(x_i))\} + \tau, \tag{2.40}$$

where $\tau := (1 - e^{-\mu h})/\mu$ and $w(x_i + hb^L(x_i, a))$ is obtained interpolating on W.

It has been shown in [17] that the corresponding operator T^L has three important properties: it is monotone, is a contraction mapping in $[0, 1/\mu)^G$ and $0 \leq W \leq \dfrac{1}{\mu}$ implies $0 \leq T(W) \leq \dfrac{1}{\mu}$.

Similarly, the *SL fully discrete scheme for the ON–model* at a node x_i will be given by the discrete operator

$$T_i^{ON}(W) := \min_{a \in \partial B_3} \{e^{-\mu h} w(x_i + hb^{ON}(x_i, a)) - \tau \frac{c(x_i, z)a_3}{A\omega_3}(1 - \mu w(x_i))\} + \tau. \tag{2.41}$$

The *SL fully discrete scheme for the PH–model* at a node x_i is given by the discrete operator T^{PH} defined as

$$T_i^{PH}(W) := \min_{a \in \partial B_3} \{e^{-\mu h} w(x_i + hb^{PH}(x_i, a)) - \tau \frac{c(x_i)a_3}{Q^{PH}(x_i, z)}(1 - \mu w(x_i))\} + \tau, \tag{2.42}$$

with $Q^{PH}(x_i, z) := 2k_S \gamma_S (d(x_i, z) \cdot \boldsymbol{\omega}) + k_D \gamma_D \omega_3$.
Although the operators T^{ON} and T^{PH} present some differences and additional terms, they converge and have similar properties of the operator T^L (see [50, 52] for the analytical proof of these properties).

In the numerical tests we will also compare results obtained with Fast Sweeping (FS) and Fast Marching (FM) methods so we briefly sketch here their properties.

2.3.1 *Fast Marching [21, 44, 55]*

For the implementation of Fast-Marching algorithm, the grid defined on the image is divided into three sets at every iteration n:

the set of Accepted nodes ACCEPTED(n), whose value has been already computed and accepted;
the set of Considered nodes CONSIDERED(n), or Narrow Band, for which the value has to be computed at the present iteration;
the set of Far nodes FAR(n), that are the nodes which will be computed in future iterations.

The engine of the method is the local fixed point operator. ACCEPTED(0) at the first iteration is the set of nodes where we have to apply boundary conditions (which are known). Then, at iteration n, the set CONSIDERED(n) contains the neighboring nodes of ACCEPTED(n) and FAR(n) are the remaining nodes where we do nothing at that iteration. The algorithm computes the value in CONSIDERED(n). The node x_j where the minimum is achieved is marked ACCEPTED (i.e. ACCEPTED$(n+1)$=ACCEPTED$(n) \cup \{x_j\}$), the set CONSIDERED(n) is updated adding the neighboring nodes to x_j and we compute the solution in CONSIDERED$(n+1)$. The algorithms accepts only one node for each iteration and ends only when the FAR region is empty. The method converges in a finite number of iterations and has a complexity of $O(G \ln(G))$ where G is the cardinality of the grid nodes. Unfortunately, its application is limited to eikonal type equations.

2.3.2 *Fast Sweeping [14, 40, 54]*

FS is another popular method for solving HJ equations. The main advantage of this method is its implementation, which is extremely easy (easier than that of Fast Marching). FS method is basically the classical iterative (fixed-point) method, since each node is visited in a predefined order, until convergence is reached. Here, the visiting directions (sweeps) are alternated in order to follow all the possible characteristic directions, trying to exploit causality. In two-dimensional problems, the grid is visited sweeping in four directions: $S \to N$ & $W \to E$, $S \to N$ & $E \to W$, $N \to S$ & $E \to W$ and $N \to S$ & $W \to E$.

The key point is the Gauss-Seidel-like update of grid nodes, which allows one to compute a relevant part of the grid nodes in only one sweep. Indeed, it is well known that in the case of eikonal equations FS converges in only four sweeps.

2.4 Numerical Experiments

We call G the discrete grid of points x_{ij}, with size $card(G) = n \times m$. We define $G_{in} := \{x_{ij} : x_{ij} \in \Omega\}$ as the set of grid points inside Ω; $G_{out} := G \setminus G_{in}$. The boundary $\partial\Omega$ is defined as the set $G_b \subset G_{out}$ such that at least one of the neighboring points belongs to G_{in}. For each image we define a map, called *mask or silhouette*, where the pixels $x_{ij} \in G_{in}$ are white and the pixel $x_{ij} \in G_{out}$ are black. In this way it is easy to distinguish the nodes that we have to use for the reconstruction (the nodes inside Ω) and the nodes on the boundary $\partial\Omega$ (see e.g. Fig. 2.4b, d).

In all our numerical experiments, we neglect the ambient component that we consider as a constant (i.e. setting for simplicity $k_A = 0$). Our work is mainly focusing on the semi-Lagrangian approach and also our intent is to analyze the behavior of the parameters involved in the two non-Lambertian models. For these reasons, the simulations focus the attention on SL performances and on the behavior of the parameters.

2.4.1 Synthetic Images

The synthetic tests are useful for a quantitative analysis on the behavior of the parameters and also because it is possible to compute the error on the surface. The synthetic image that we are going to present here is defined on the domain G, that is a rectangle containing the support of the image Ω, $G \equiv [-1, 1] \times [-1, 1]$. We can easily modify the number of the pixels choosing different values for the steps in space Δx and Δy. In this case we will use 256×256 pixels. X and Y represent the real size (e.g. for $G \equiv [-1, 1] \times [-1, 1]$, $X = 2$, $Y = 2$). As already said in Sect. 2.2, we can use homogeneous Dirichlet boundary condition but it is possible to define, if useful, the function $g(\mathbf{x})$, that is the height of the surface at the boundary of the silhouette. In this test we will use this general boundary condition that we can easily derive being the object a solid of rotation. In fact, if we denote by $c := (c_x, c_y)$ the center of the circle with ray R at the bottom of the vase we can write

$$(x - c_x)^2 + (y - c_y)^2 = R^2 \tag{2.43}$$

and then

$$y = \sqrt{R^2 - (x - c_x)^2} + c_y \tag{2.44}$$

that gives us the values of $g(\mathbf{x})$.

In iterative methods, the method stops when we have reached the required tolerance η or when we have exceeded the maximum number of iterations allowed. In an iterative method of fixed point, the point is reached when $||W^{n+1} - W^n|| < \eta$.

2.4.1.1 Synthetic Vase

We use this test in order to analyze and compare the performances of the SL scheme with respect to the three different models by varying the values of the parameters involved.

The synthetic vase is defined as

$$\begin{cases} u(x,y) = \sqrt{P(\bar{y})^2 - x^2} & (x,y) \in G_{in}, \\ u(x,y) = g(x,y) & (x,y) \in G_{out}, \end{cases} \tag{2.45}$$

where $\bar{y} = y/Y$,

$$P(\bar{y}) = (-10.8\,\bar{y}^6 + 7.2\,\bar{y}^5 + 6.6\,\bar{y}^4 - 3.8\,\bar{y}^3 - 1.375\,\bar{y}^2 + 0.5\,\bar{y} + 0.25)\,X$$

and

$$G_{in} = \{(x,y) | P(\bar{y})^2 > x^2\}.$$

The input images generated by L–model, ON–model and PH–model with a vertical light source ($\boldsymbol{\omega} = (0,0,1)$) are visible in Fig. 2.2. We show in Table 2.1 the values of the parameters related to some numerical tests performed. It is possible to compute the error in L^1, L^2, L^∞ norm on the image ($L^p(I)$) and on the surface ($L^p(S)$) because for synthetic images we know the real surface (for the vase this is given by (2.45)). Given a vector $\mathbf{T}$ representing the exact solution (or the original image) on the grid and a vector $\tilde{\mathbf{T}}$ representing its approximation, we define the error vector as

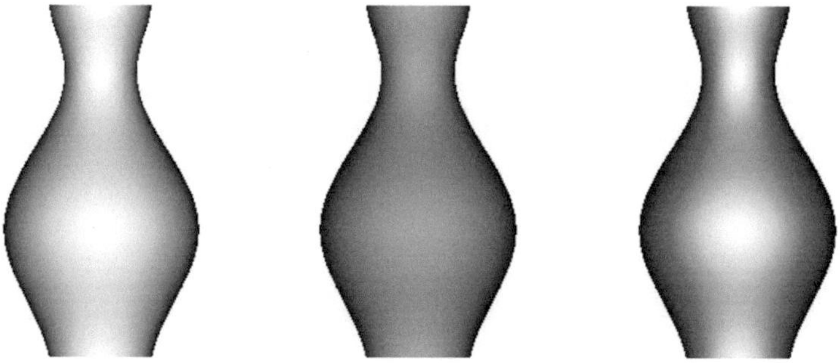

(a) L–model (b) ON–model (c) PH–model

Fig. 2.2 Input vase images by L–model, ON–model ($\sigma = 0.6$), PH–model ($k_S = 0.3$) with $\boldsymbol{\omega} = (0,0,1)$

Table 2.1 Synthetic vase: parameter values used in the models. When a parameter doesn't exist for a model we put a dash

Model	k_A	k_D	k_S	α	σ
LAM	0	1	–	–	–
ON-00	0	1	–	–	0
ON-04	0	1	–	–	0.4
ON-06	0	1	–	–	0.6
ON-10	0	1	–	–	1
PH-00	0	1	0	1	–
PH-03	0	0.7	0.3	1	–
PH-07	0	0.3	0.7	1	–
PH-10	0	0	1	1	–

$\mathbf{e} = \mathbf{T} - \tilde{\mathbf{T}}$ and

$$\begin{aligned} err_1 &= ||\mathbf{e}||_{L^1} &&= \frac{1}{N}\sum_i |e_i| \\ err_2 &= ||\mathbf{e}||_{L^2} &&= \left\{\frac{1}{N}\sum_i |e_i|^2\right\}^{1/2} \\ err_\infty &= ||\mathbf{e}||_{L^\infty} &&= \max_i \{|e_i|\} \end{aligned}$$

where N is the total number of grid points used for the computation, i.e. the grid points belonging to G_{in}.
The reconstructions of the surfaces and the output images obtained with the three models, starting from the input images in Fig. 2.2, are visible in Fig. 2.3.

In Table 2.2 we can observe the performances of the SL–scheme. In details, we reported the number of iterations, the CPU time (in seconds) and the error estimates in three different norms between the input image and the image reconstructed from the u approximation. Note that to obtain the reconstructed image we need an approximation of the gradient of u which is obtained via a centered finite difference which guarantees a second order accuracy. Looking at these errors, we note that the ON–model performs better increasing the parameter σ both on the image I and on the surface, with the same error order than the L–model but always lower. Instead, for the PH–model we can see that the errors on the surface decrease increasing the parameter k_S, except for the case completely specular ($k_S = 1$), but it is not true with respect to the errors on the image that increase for increasing values of k_S.

The errors on the images in different norms show how well the reprojection fits the input data. The L^∞ errors on the image I, that indicate the maximum over all the pixels of the difference in absolute value between the input image and the image computed as said before, are so big because if in only one point the reconstruction is not good, e.g. in a point on the boundary of the domain, then the error will be so big.

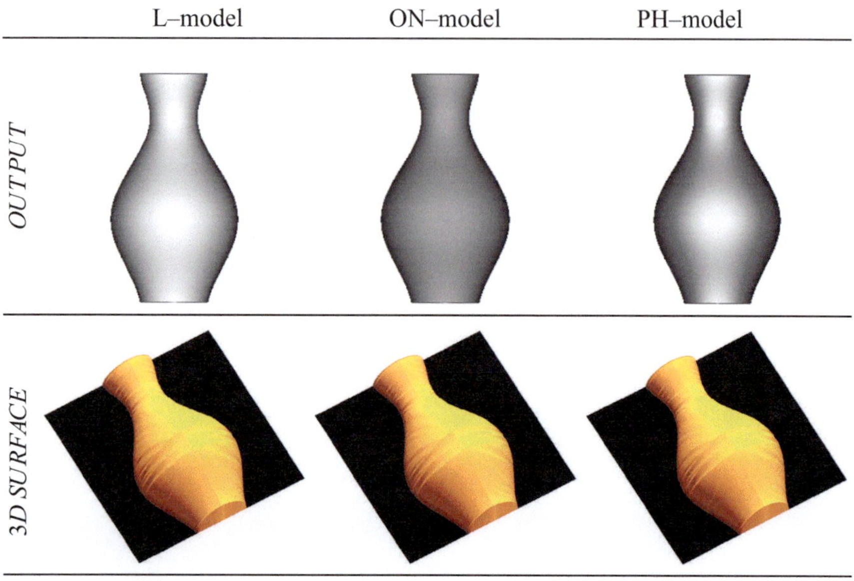

Fig. 2.3 Synthetic vase: output images and 3D reconstructions for the three models

Table 2.2 Synthetic vase: numerical results for $\omega = (0, 0, 1)$ with the errors on the image and on the surface

SL–schemes	Iter.	[*sec.*]	$L^1(I)$	$L^2(I)$	$L^\infty(I)$	$L^1(S)$	$L^2(S)$	$L^\infty(S)$
LAM	1337	0.73	0.0063	0.0380	0.7333	0.0267	0.0286	0.0569
ON-00	1337	0.72	0.0063	0.0380	0.7333	0.0267	0.0286	0.0569
ON-04	1341	0.73	0.0054	0.0316	0.6118	0.0263	0.0282	0.0562
ON-06	1344	0.75	0.0049	0.0277	0.5373	0.0259	0.0277	0.0553
ON-10	1334	0.74	0.0044	0.0229	0.4510	0.0254	0.0274	0.0547
PH-00	1337	0.76	0.0063	0.0380	0.7333	0.0267	0.0286	0.0569
PH-03	1331	0.73	0.0068	0.0396	0.8078	0.0264	0.0283	0.0561
PH-07	1356	3.81	0.0075	0.0419	0.9098	0.0235	0.0252	0.0496
PH-10	737	0.40	0.0081	0.0472	0.9961	0.1496	0.1590	0.2309

Table 2.3 Synthetic vase: errors on the surface via ON–model changing the size of the input image with vertical light source $\omega = (0, 0, 1)$

SL–schemes	Size	$L^1(S)$	$L^2(S)$	$L^\infty(S)$
ON-04	64×64	0.0459	0.0496	0.0898
ON-04	128×128	0.0347	0.0384	0.0819
ON-04	256×256	0.0263	0.0282	0.0562
ON-04	512×512	0.0177	0.0187	0.0360
ON-04	1024×1024	0.0121	0.0129	0.0280

In order to confirm that the non-Lambertian models converge to the surface depth, we reported in Tables 2.3 and 2.4 the errors on the surface with respect to

Table 2.4 Synthetic vase: errors on the surface via PH–model changing the size of the input image with vertical light source $\omega = (0, 0, 1)$

SL–schemes	Size	$L^1(S)$	$L^2(S)$	$L^\infty(S)$
PH-03	64×64	0.0462	0.0499	0.0904
PH-03	128×128	0.0349	0.0386	0.0828
PH-03	256×256	0.0264	0.0283	0.0561
PH-03	512×512	0.0177	0.0187	0.0356
PH-03	1024×1024	0.0120	0.0127	0.0267

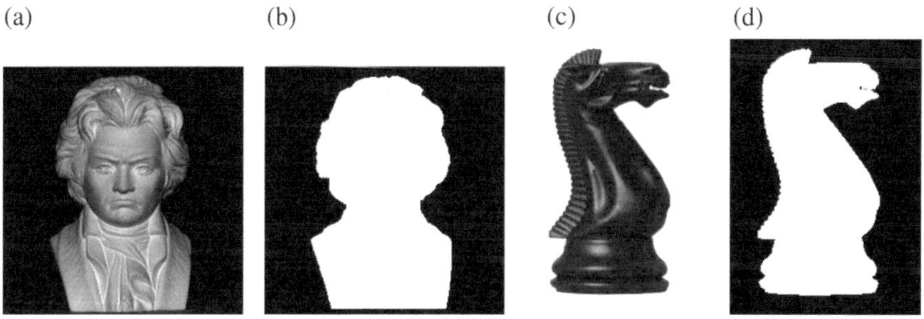

Fig. 2.4 Real input images and masks. (**a**) Beethoven input. (**b**) Beethoven mask. (**c**) Horse input. (**d**) Horse mask

different size of the vase image, from 64×64 to 1024×1024 obtained doubling the size. What we can note is that increasing the number of the pixels, hence considering a smaller and smaller space step, the errors decrease for both the models.

2.4.2 *Real Images*

In this section we will consider two real input images: the bust of Beethoven (size (256×256)) and the black horse (size (184×256)), both visible in Fig. 2.4a, c.

Unless otherwise specified, the value of η for the stopping criterion of the iterative method is fixed to 10^{-8} and the maximum number of allowed iterations is 9000. If a scheme arrives to the maximum of 9000 iterations, we put a $*$ before it in the table.

Obviously, for the real tests we do not know the real depth hence we cannot compute the error on the surface. The only quantity that is available is the image I, our data, and we added Tables in order to give a quantitative support to the qualitative analysis visible from the Figures reported below in the paper.

2.4.2.1 Beethoven

In this case, we have compared the performances of the SL–scheme applied to the three models using the parameters reported in Table 2.5 with two different cases for

Table 2.5 Beethoven: parameter values used in the models. When a parameter doesn't exist for a model we put a dash

Model	k_A	k_D	k_S	α	σ
LAM	0	1	–	–	–
ON-00	0	1	–	–	0
ON-01	0	1	–	–	0.1
ON-02	0	1	–	–	0.2
ON-03	0	1	–	–	0.3
ON-04	0	1	–	–	0.4
ON-06	0	1	–	–	0.6
PH-00	0	1.0	0	1	–
PH-01	0	0.9	0.1	1	–
PH-02	0	0.8	0.2	1	–
PH-03	0	0.7	0.3	1	–
PH-04	0	0.6	0.4	1	–

Table 2.6 Beethoven: numerical results for $\boldsymbol{\omega}_{vert} = (0, 0, 1)$ with the errors on the image

SL–schemes	Iter.	[*sec.*]	$L^1(I)$	$L^2(I)$	$L^\infty(I)$
LAM vertical	2920	1.68	0.0325	0.0605	0.4118
ON-00 vertical	2920	2.24	0.0325	0.0605	0.4118
ON-01 vertical	2885	2.89	0.0325	0.0605	0.4118
ON-02 vertical	2790	2.23	0.0326	0.0605	0.4118
ON-06 vertical	2264	1.94	0.0355	0.0628	0.4157
PH-00 vertical	2920	2.29	0.0325	0.0605	0.4118
PH-01 vertical	2676	2.12	0.0329	0.0609	0.4118
PH-02 vertical	2423	1.92	0.0333	0.0613	0.4118
PH-03 vertical	2160	1.92	0.0337	0.0617	0.4118
PH-04 vertical	1887	1.72	0.0337	0.0619	0.4118

the light source: the vertical case ($\boldsymbol{\omega}_{vert} = (0, 0, 1)$) and the oblique case ($\boldsymbol{\omega}_{obl} = (0.0168, 0.198, 0.9801)$). This test will show better the crucial role of the parameters involved for the convergence. As we can see in Table 2.6, in the vertical case all the models converge in less than 3 s with the same order of iteration. Looking at the errors on the images, they are of the same order for all the cases, $L^\infty(I)$ is a little bit higher for the ON–model with $\sigma = 0.6$. We can note that in the case of $\sigma = 0$ for the ON–model and $k_S = 0$ for the PH–model we obtain the same errors and number of iterations too because the three models coincide as expected. With respect to the ON–model, by increasing the value of σ the errors grow slightly or remain unchanged. The same behavior has the PH–model with respect to the parameter k_S. In fact, by increasing the value of k_S the errors tend to increase, remaining of the same order.

Looking at Table 2.7, we can note that the oblique cases require higher CPU time with respect to the corresponding vertical cases due to the fact that the equations are more complex because of additional terms involved. Analyzing the errors on the images, as noted just before, the cases of $\sigma = 0$ for the ON–model and $k_S = 0$ for the PH–model coincide with the L–model in terms of number of iterations used and

Table 2.7 Beethoven: numerical results for $\boldsymbol{\omega}_{obl} = (0.0168, 0.198, 0.9801)$ with the errors on the image

SL–schemes	Iter.	[*sec.*]	$L^1(I)$	$L^2(I)$	$L^\infty(I)$	η
LAM oblique	3129	234.60	0.0397	0.0659	0.4039	10^{-8}
LAM oblique	236	40.85	0.0464	0.0696	0.4039	10^{-3}
ON-00 oblique	236	46.85	0.0464	0.0696	0.4039	10^{-3}
ON-01 oblique	242	50.90	0.0439	0.0656	0.4118	10^{-3}
ON-02 oblique	262	53.43	0.0484	0.0699	0.4196	10^{-3}
ON-03 oblique	270	53.76	0.0550	0.0763	0.4039	10^{-3}
ON-04 oblique	314	65.63	0.0604	0.0830	0.4314	10^{-3}
ON-04 oblique	3598	709.80	0.0672	0.0890	0.4314	10^{-4}
ON-06 oblique	362	75.91	0.0722	0.0989	0.5647	10^{-3}
PH-00 oblique	236	47.42	0.0464	0.0696	0.4039	10^{-3}
PH-01 oblique	237	44.59	0.0712	0.0917	0.4510	10^{-3}
PH-02 oblique	303	58.04	0.1095	0.1291	0.4784	10^{-3}
PH-03 oblique	513	97.09	0.1506	0.1743	0.5333	10^{-3}
PH-04 oblique	9000*	1149.00	0.1701	0.2041	0.5765	10^{-3}

*Indicates the maximum number of iterations

error estimations. With respect to the ON–model, the errors increase by increasing the parameter σ. The same holds for the PH–model with respect to k_S. Because of additional terms involved in the oblique case, in Table 2.7 we have reported the results obtained using a value of the tolerance η for the stopping rule of the iterative method equal to 10^{-3}. This is the maximum accuracy achieved by the non-Lambertian models since roundoff errors coming from several terms occur and limit the accuracy since the schemes are first order accurate. Only for the ON–model with $\sigma = 0.4$ we have reported the result also for $\eta = 10^{-4}$ and for the L–model with $\eta = 10^{-8}$. Lastly, we can note that choosing $k_S = 0.4$ the PH–model not converges in the maximum number of allowed iterations, i.e. in 9000 iterations.

The 3D reconstructions and the output images for the three models are visible in Fig. 2.5. The first two rows refer to the vertical case, the others to the oblique case. The reconstructions in the vertical case are more accurate than the corresponding in the oblique case also because obtained with a tolerance $\eta = 10^{-3}$ instead of $\eta = 10^{-8}$ as in the vertical case. Moreover, it is important to note that there is a concave/convex inversion in the reconstructed surface due to the classical ambiguity of the SfS model. This typically depends also on the correctness of the Dirichlet boundary conditions (as one can see in the synthetic vase test where we have applied a correct boundary condition $u = g$ imposing a circular shape on that part of the boundary).

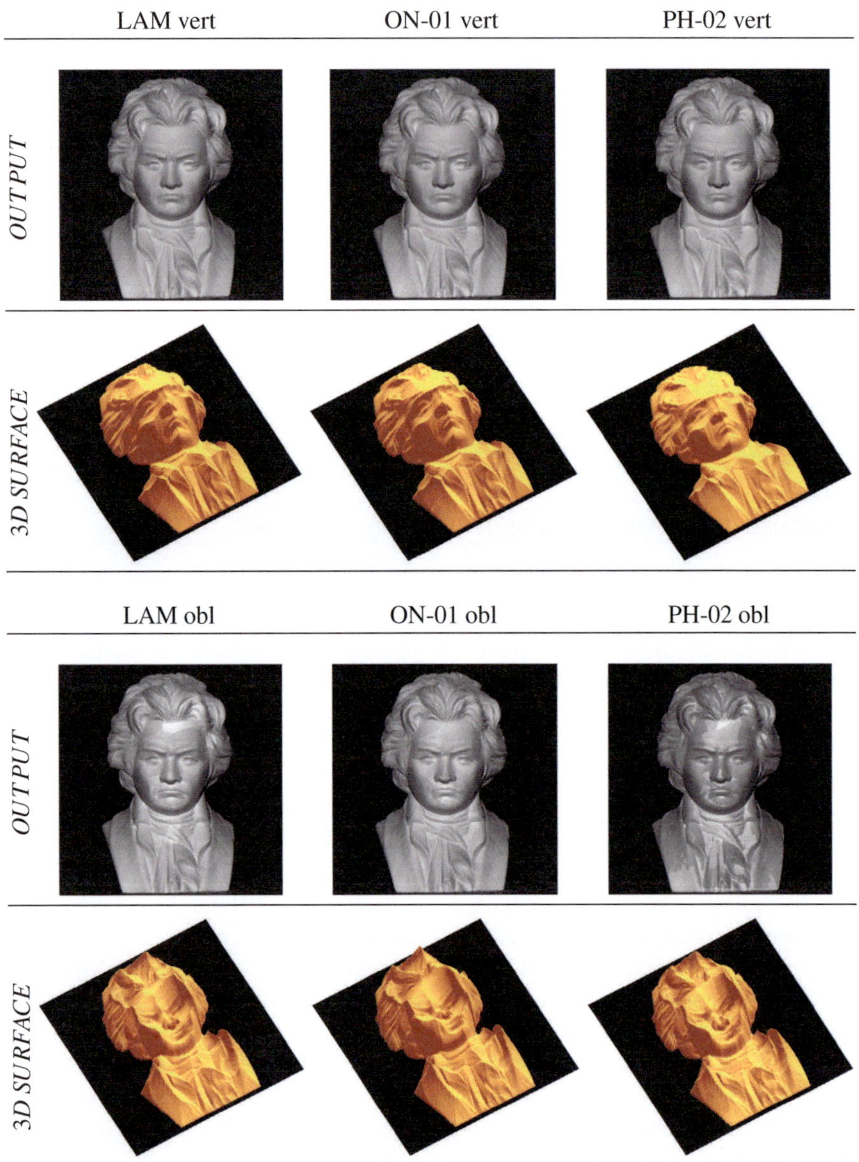

Fig. 2.5 Beethoven: output images and 3D reconstructions for the three models

2.4.2.2 Black Horse

We use this test to compare the performances of the global SL–scheme with respect to the acceleration schemes (FM and FS) based on a finite difference (FD) solver (FM-FD, FS-FD). The comparison will be made for all the models (L–model, ON–model, PH–model) with the parameter values reported in the second and third column of Table 2.8. Note that the SL–scheme, that is slower than FM-FD and FS-FD methods as expected, however it is more accurate with respect to the schemes based on FD. This confirms that the SL approach is competitive with other numerical techniques. We can also note that the parameters play an important role in these models. For example, in the PH–model passing from $k_S = 0.4$ to $k_S = 0.8$ the errors change significantly in L^1 and L^2 norm for the FM-FD and the FS-FD methods. In Fig. 2.6 one can see the output images and the 3D reconstructions of the surface obtained by the SL–schemes applied to the three models. Note that the reconstruction obtained by the PH–model recognizes better the object in the picture and this is coherent with the fact that the surface is shiny, so the PH–model seems to be the more realistic in this case.

Table 2.8 Black horse: parameters, CPU time and errors on the image with vertical light source. When a parameter doesn't exist for a model we put a dash

Model	k_S	σ	[*sec.*]	$L^1(I)$	$L^2(I)$	$L^\infty(I)$
LAM-FM	–	–	0.08	0.0363	0.0610	0.6902
LAM-FS	–	–	0.08	0.0362	0.0607	0.6902
LAM-SL	–	–	2.62	0.0346	0.0590	0.6863
ON-02-FM	–	0.2	0.07	0.0363	0.0611	0.6902
ON-02-FS	–	0.2	0.02	0.0362	0.0608	0.6902
ON-02-SL	–	0.2	2.49	0.0347	0.0591	0.6902
ON-03-FM	–	0.3	0.14	0.0364	0.0611	0.6941
ON-03-FS	–	0.3	0.14	0.0363	0.0609	0.6941
ON-03-SL	–	0.3	2.39	0.0348	0.0592	0.6902
PH-04-FM	0.4	–	0.28	0.0441	0.0677	0.6902
PH-04-FS	0.4	–	0.77	0.0439	0.0674	0.6902
PH-04-SL	0.4	–	1.06	0.0358	0.0606	0.6863
PH-08-FM	0.8	–	0.16	0.0788	0.1132	0.7098
PH-08-FS	0.8	–	0.63	0.0788	0.1132	0.7098
PH-08-SL	0.8	–	0.53	0.0463	0.0736	0.7059

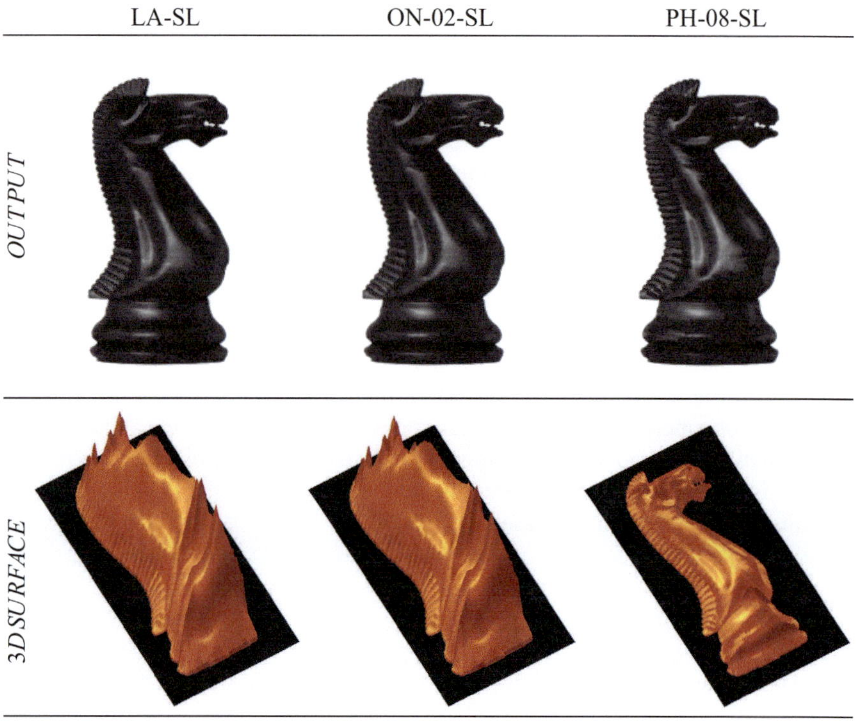

Fig. 2.6 Black horse: output images and 3D reconstructions for the three models

2.5 Conclusions and Perspectives

In this paper we derived the Hamilton-Jacobi equations related to three reflectance models and we presented some numerical methods to solve them. In our formulation the models share the same mathematical structure and this allows to switch on and off the different terms related to ambient, diffuse and specular reflection in a very simple way. This general model is very flexible to treat different light conditions with vertical and oblique light sources. As we noted in some of our tests, via non-Lambertian models it is not possible to solve the classical concave/convex ambiguity of the Lambertian SfS problem based on a single image despite the fact that these models can deal with more general surfaces (see [50, 52] for more details on the analysis performed on non-Lambertian models). This ambiguity can be eliminated only adding additional information on the image or dealing with more than one image as already done via the Photometric Stereo technique in the case of the Lambertian model [32, 33]. The application of the Photometric Stereo technique to models with a specular component goes beyond the scopes of this paper. Some results of this work can be found in [53].

From the numerical point of view, comparing the numerical methods we noted that the SL–schemes approximate in a rather effective way the equations related to non-Lambertian models and they are more accurate with respect to FD schemes. Looking at the numerical experiments of the three models, the PH–model seems to be the model more sensible to the values of the parameters involved, as visible in Table 2.8. This model recognizes better the object with respect to the L–model and the ON–model, although the errors computed on the image are higher. However, our numerical tests showed that all the schemes are consistent and we obtain good results for synthetic and real input images. Looking at the test performed with an oblique light source, we have some comments that are common to the PH–model and the ON–model. The equations corresponding to this case have additional terms and the corresponding discrete operators become more complex and require more iterations to converge. This produces an accumulation of floating point errors which reduces the accuracy of the approximation. Moreover, for real images, we do not know the exact direction of the light source and this introduces another perturbation in the model which affects the results. A possible improvement, at least when we know the light source direction, could be the use of second order schemes.

Another interesting direction would be to mix the models, e.g. coupling the ON–model with the PH–model. To this end, we need to verify if the new mixed model still can be written in the same fixed point form in order to apply the same approach.

Acknowledgements The first author wishes to acknowledge the support obtained by Gruppo Nazionale per il Calcolo Scientifico (GNCS-INdAM).

References

1. Ahmed, A.H., Farag, A.A.: A new formulation for Shape from Shading for non-Lambertian surfaces. In: Proceedings of the IEEE Conference on Computer Vision and Pattern Recognition (CVPR), New York, pp. 1817–1824 (2006)
2. Ahmed, A.H., Farag, A.A.: Shape from Shading under various imaging conditions. In: IEEE International Conference on Computer Vision and Pattern Recognition CVPR'07, Minneapolis, pp. X1–X8, June 2007
3. Bakshi, S., Yang, Y.-H.: Shape-from-Shading for non-Lambertian surfaces. In: Proceedings of the IEEE International Conference on Image Processing (ICIP), pp. 130–134 (1994)
4. Barles, G.: Solutions de viscosité des equations de Hamilton-Jacobi. Springer (1994)
5. Biswas, S., Aggarwal, G., Chellappa, R.: Robust estimation of albedo for illumination-invariant matching and shape recovery. IEEE Trans. Pattern Anal. Mach. Intell. **31**(5), 884–899 (2009)
6. Breuß, M., Cristiani, E., Durou, J.-D., Falcone, M., Vogel, O.: Perspective Shape from Shading: ambiguity analysis and numerical approximations. SIAM J. Imaging Sci. **5**(1), 311–342 (2012)
7. Bruss, A.: The image irradiance equation: its solution and application. Technical report ai-tr-623, Massachusetts Institute of Technology, June 1981
8. Camilli, F., Falcone, M.: An approximation scheme for the maximal solution of the Shape-from-Shading model. In: Proceedings of the International Conference on Image Processing, vol. 1, pp. 49–52, Sept 1996
9. Camilli, F., Grüne, L.: Numerical approximation of the maximal solutions for a class of degenerate Hamilton-Jacobi equations. SIAM J. Numer. Anal. **38**(5), 1540–1560 (2000)

10. Chambolle, A.: A uniqueness result in the theory of stereo vision: coupling Shape from Shading and Binocular information allows Unambiguous depth reconstruction. Annales de l'Istitute Henri Poincar **11**(1), 1–16 (1994)
11. Courteille, F., Crouzil, A., Durou, J.-D., Gurdjos, P.: Towards Shape from Shading under realistic photographic conditions. In: Proceedings of the of the 17th International Conference on Pattern Recognition, Cambridge, vol. 2, pp. 277–280 (2004)
12. Courteille, F., Crouzil, A., Durou, J.-D., Gurdjos, P.: Shape from shading for the digitization of curved documents. Mach. Vis. Appl. **18**(5), 301–316 (2007)
13. Cristiani, E., Falcone, M.: Fast semi-Lagrangian schemes for the eikonal equation and applications. SIAM J. Numer. Anal. **45**(5), 1979–2011 (2007)
14. Danielsson, P.: Euclidean distance mapping. Comput. Graph. Image Process. **14**, 227–248 (1980)
15. Durou, J.-D, Falcone, M., Sagona, M.: Numerical methods for Shape from Shading: a new survey with benchmarks. Comput. Vis. Image Underst. **109**(1), 22–43 (2008)
16. Falcone, M., Ferretti, R.: Semi-Lagrangian approximation schemes for linear and Hamilton-Jacobi equations. SIAM, Philadelphia (2013)
17. Falcone, M., Sagona, M., Seghini, A.: A scheme for the Shape-from-Shading model with "black shadows". In: Brezzi, F., Buffa, A., Corsaro, S., Murli, A. (eds.) Numerical Mathematics and Advanced Applications – ENUMATH2001, pp. 503–512. Springer, Milano/New York (2003)
18. Fassold, H., Danzl, R., Schindler, K., Bischof, H.: Reconstruction of archaeological finds using Shape from Stereo and Shape from Shading. In: Proceedings of the 9th Computer Vision Winter Workshop, Piran, pp. 21–30 (2004)
19. Favaro, P., Papadhimitri, T.: A closed-form solution to uncalibrated photometric stereo via diffuse maxima. In: Proceedings of the IEEE Conference on Computer Vision and Pattern Recognition, pp. 821–828 (2012)
20. Grumpe, A., Belkhir, F., Wöhler, C.: Construction of lunar DEMs based on reflectance modelling. Adv Space Res **53**(12), 1735–1767 (2014)
21. Helmsen, J., Puckett, E., Colella, P., Dorr, M.: Two new methods for simulating photolithography development in 3D. In: Optical Microlithography IX, vol. 2726, pp. 253–261. SPIE, Bellingham (1996)
22. Horn, B.K.P.: Shape from Shading: a method for obtaining the Shape of a smooth opaque object from one view. PhD thesis, Massachusetts Institute of Technology (1970)
23. Horn, B.K.P., Brooks, M.J.: The variational approach to Shape from Shading. Comput. Vis. Graph. Image Process. **33**(2), 174–208 (1986)
24. Horn, B.K.P., Brooks, M.J.: Shape from Shading. MIT, Cambridge (1989)
25. Ju, y.-C., Breuß, M., Bruhn, A., Galliani, S.: Shape from Shading for rough surfaces: analysis of the Oren-Nayar model. In: Proceedings of the British Machine Vision Conference (BMVC), pp 104.1–104.11. BMVA Press, Durham (2012)
26. Ju, Y.-C., Tozza, S., Breuß, M., Bruhn, A., Kleefeld, A.: Generalised perspective Shape from Shading with Oren-Nayar reflectance. In: Proceedings of the 24th British Machine Vision Conference (BMVC 2013), Bristol, pp. 42.1–42.11. BMVA Press (2013)
27. Kao, C.Y., Osher, S., Qian, J.: Lax-Friedrichs sweeping scheme for static Hamilton-Jacobi equations. J. Comput. Phys. **196**(1), 367–391 (2004)
28. Kao, C.Y., Osher, S., Tsai, Y.-H.: Fast sweeping methods for static Hamilton–Jacobi equations. SIAM J. Numer. Anal. **42**(6), 2612–2632 (2005)
29. Kozera, R.: Existence and uniqueness in photometric stereo. Appl. Math. Comput. **44**(1), 103 (1991)
30. Lions, P. L., Rouy, E., Tourin, A.: Shape-from-Shading, viscosity solutions and edges. Numerische Mathematik **64**(1), 323–353 (1993)
31. Lohse, V., Heipke, C., Kirk, R.L.: Derivation of planetary topography using multi-image Shape-from-Shading. Planet. Space Sci. **54**(7), 661–674 (2006)
32. Mecca, R., Falcone, M.: Uniqueness and approximation of a photometric Shape-from-Shading model. SIAM J. Imaging Sci. **6**(1), 616–659 (2013)

33. Mecca, R., Tozza, S.: Shape reconstruction of symmetric surfaces using photometric stereo. In: Breuß, M., Bruckstein, A., Maragos, P. (eds.) Innovations for Shape Analysis: Models and Algorithms. Mathematics and Visualization, pp. 219–243. Springer, Berlin/Heidelberg (2013)
34. Okatani, T., Deguchi, K.: Shape reconstruction from an endoscope image by Shape from Shading technique for a point light source at the projection center. Comput. Vis. Image Underst. **66**(2), 119–131 (1997)
35. Oren, M., Nayar, S.K.: Generalization of Lambert's reflectance model. In: Proceedings of the International Conference and Exhibition on Computer Graphics and Interactive Techniques (SIGGRAPH), pp. 239–246 (1994)
36. Oren, M., Nayar, S.K.: Generalization of the Lambertian model and implications for machine vision. Int. J. Comput. Vis. **14**(3), 227–251 (1995)
37. Phong, B.T.: Illumination for computer generated pictures. Commun. ACM **18**(6), 311–317 (1975)
38. Prados, E., Faugeras, O.: "Perspective Shape from Shading" and viscosity solutions. In: Proceedings of the Ninth IEEE International Conference on Computer Vision (ICCV), vol. 2, pp. 826–831 (2003)
39. Prados, E., Faugeras, O.: Shape from Shading: a well-posed problem? In: Proceedings of the IEEE Conference on Computer Vision and Pattern Recognition (CVPR), vol. 2, pp. 870–877 (2005)
40. Qian, J., Zhang, Y.-T., Zhao, H.-K.: A fast sweeping method for static convex Hamilton-Jacobi equations. J. Sci. Comput. **31**(1–2), 237–271 (2007)
41. Quéau, Y., Lauze, F., Durou, J.-D.: Solving uncalibrated photometric stereo using total variation. J. Math. Imaging Vis. **52**(1), 87–107 (2015)
42. Ragheb, H., Hancock, E.R.: Surface radiance correction for shape from shading. Pattern Recognit. **38**(10), 1574–1595 (2005)
43. Rouy, E., Tourin, A.: A viscosity solutions approach to Shape-from-Shading. SIAM J. Numer. Anal. **29**(3), 867–884 (1992)
44. Sethian, J.A.: Level Set Methods and Fast Marching Methods: Evolving Interfaces in Computational Geometry, Fluid Mechanics, Computer Vision, and Materials Science. Cambridge University Press, Cambridge/New York (1999)
45. Smith, W.A.P., Hancock, E.R.: Recovering facial shape and albedo using a statistical model of surface normal direction. In: Tenth IEEE International Conference on Computer Vision (ICCV), pp. 588–595, Oct 2005
46. Smith, W.A.P., Hancock, E.R.: Estimating facial albedo from a single image. IJPRAI **20**(6), 955–970 (2006)
47. Steffen, W., Koning, N., Wenger, S., Morisset, C., Magnor, M.: Shape: a 3D modeling tool for astrophysics. IEEE Trans. Vis. Comput. Graph. **17**(4), 454–465 (2011)
48. Tankus, A., Kiryati, N.: Photometric stereo under perspective projection. In: Proceedings of the IEEE International Conference on Computer Vision (ICCV), Beijing, vol. 1, pp. 611–616 (2005)
49. Tankus, A., Sochen, N., Yeshurun, Y.: Shape-from-Shading under perspective projection. Int. J. Comput. Vis. **63**(1), 21–43 (2005)
50. Tozza, S.: Analysis and Approximation of Non-Lambertian Shape-from-Shading Models. PhD thesis, Dipartimento di Matematica, Sapienza – Università di Roma, Roma, Jan 2015
51. Tozza, S., Falcone, M.: A semi-Lagrangian approximation of the Oren-Nayar PDE for the orthographic Shape–from–Shading problem. In: Battiato, S., Braz, J. (eds.) Proceedings of the 9th International Conference on Computer Vision Theory and Applications (VISAPP), vol. 3, pp. 711–716. SciTePress (2014)
52. Tozza, S., Falcone, M.: Analysis and approximation of some shape-from-shading models for non-Lambertian surfaces. J. Math. Imaging Vis. **55**(2), 153–178 (2016)
53. Tozza, S., Mecca, R., Duocastella, M., Del Bue, A.: Direct differential photometricstereo shape recovery of diffuse and specular surfaces. J. Math. Imaging Vis. **56**(1), 57–76 (2016)
54. Tsai, Y., Cheng, L., Osher, S., Zhao, H.: Fast sweeping algorithms for a class of Hamilton-Jacobi equations. SIAM J. Numer. Anal. **41**, 673–694 (2004)

55. Tsitsiklis, J.N.: Efficient algorithms for globally optimal trajectories. IEEE Trans. Autom. Control **40**(9), 1528–1538 (1995)
56. Vogel, O., Breuß, M., Weickert, J.: Perspective Shape from Shading with non-Lambertian reflectance. In: Rigoll, G. (ed.) Pattern Recognition. Volume 5096 of Lecture Notes in Computer Science, pp. 517–526. Springer, Berlin/Heidelberg (2008)
57. Vogel, O., Breuß, M., Leichtweis, T., Weickert, J.: Fast Shape from Shading for Phong-type surfaces. In: Tai, X.-C., Mørken, K., Lysaker, M., Lie, K.-A. (eds.) Scale Space and Variational Methods in Computer Vision. Volume 5567 of Lecture Notes in Computer Science, pp. 733–744. Springer, Berlin/Heidelberg (2009)
58. Woodham, R.J.: Photometric method for determining surface orientation from multiple images. Opt. Eng. **19**(1), 139–144 (1980)
59. Wu, C., Narasimhan, S., Jaramaz, B.: A multi-image Shape-from-Shading framework for near-lighting perspective endoscopes. Int. J. Comput. Vis. **86**, 211–228 (2010)
60. Yoon, K.-J., Prados, E., Sturm, P.: Joint estimation of shape and reflectance using multiple images with known illumination conditions. Int. J. Comput. Vis. **86**, 192–210 (2010)
61. Zhang, R., Tsai, P.-S., Cryer, J.E., Shah, M.: Shape from Shading: a survey. IEEE Trans. Pattern Anal. Mach. Intell. **21**(8), 690–706 (1999)
62. Zheng, Q., Chellappa, R.: Estimation of illuminant direction, Albedo, and Shape from Shading. In: Proceedings IEEE Computer Society Conference on Computer Vision and Pattern Recognition (CVPR), Lahaina, pp. 540–545, June 1991

Chapter 3
Direct Variational Perspective Shape from Shading with Cartesian Depth Parametrisation

Yong Chul Ju, Daniel Maurer, Michael Breuß and Andrés Bruhn

Abstract Most of today's state-of-the-art methods for perspective shape from shading are modelled in terms of partial differential equations (PDEs) of Hamilton-Jacobi type. To improve the robustness of such methods w.r.t. noise and missing data, first approaches have recently been proposed that seek to embed the underlying PDE into a variational framework with data and smoothness term. So far, however, such methods either make use of a radial depth parametrisation that makes the regularisation hard to interpret from a geometrical viewpoint or they consider indirect smoothness terms that require additional consistency constraints to provide valid solutions. Moreover the minimisation of such frameworks is an intricate task, since the underlying energy is typically non-convex. In this chapter we address all three of the aforementioned issues. First, we propose a novel variational model that operates directly on the Cartesian depth. In contrast to existing variational methods for perspective shape from shading this refers to both the data and the smoothness term. Moreover, we employ a direct second-order regulariser with edge-preservation property. This direct regulariser yields by construction valid solutions without requiring additional consistency constraints. Finally, we also propose a novel coarse-to-fine minimisation framework based on an alternating explicit scheme. This framework allows us to avoid local minima during the minimisation and thus to improve the accuracy of the reconstruction. Experiments show the good quality of our model as well as the usefulness of the proposed numerical scheme.

Y.C. Ju (✉) • D. Maurer • A. Bruhn
Institute for Visualization and Interactive Systems, University of Stuttgart, Stuttgart, Germany
e-mail: ju@vis.uni-stuttgart.de; maurer@vis.uni-stuttgart.de; bruhn@vis.uni-stuttgart.de

M. Breuß
Institute for Applied Mathematics and Scientific Computing, Brandenburg University of Technolog, Cottbus, Germany
e-mail: breuss@b-tu.de

M. Breuß et al. (eds.), *Perspectives in Shape Analysis*, Mathematics and Visualization, DOI 10.1007/978-3-319-24726-7_3

3.1 Introduction

Shape from Shading (SfS) is a classic task in computer vision. Given information on light reflectance and illumination in a photographed scene, the aim of SfS is to compute based on the brightness variation the 3D structure of a depicted object from a single input image. SfS has a wide variety of applications, ranging from large scale problems such as astronomy [42] or terrain reconstruction [7] to small scale tasks such as dentistry [2] or endoscopy [35, 52, 53].

Classical Methods. First approaches to SfS go back to 1951 and 1966, respectively, when Van Diggelen [19] and Rindfleisch [42] used SfS techniques to reconstruct the surface of the moon. Later on in the 1970s, Horn [24] was the first one to tackle the SfS problem by solving a partial differential equation (PDE) approach. In 1981, he and Ikeuchi were also the first ones to model the SfS problem using a variational framework [29]. The most prominent classical variational approach is given by the work of Horn and Brooks [26]. Assuming a simple *orthographic* projection model, a light source at *infinity* as well as a *Lambertian* reflectance model, they proposed to compute the normals of the unknown surface as minimiser of an energy functional.

Those first approaches, however, had several drawbacks. The model assumptions were *very simple* and mainly suitable in the context of astronomical applications. In fact, the use of an orthographic projection model with a light source located at infinity requires the distances between camera, light source and illuminated object to be huge. Also the *depth was not estimated directly* such that the SfS process required a postprocessing step that performed a numerical integration of the estimated surface normals. Thereby, *inconsistent gradient fields* turned out to be a problem, so that extensions of the original model were required that tried to enforce this consistency during or after the estimation [22, 27]. Finally, in case of variational methods, the smoothness term was restricted to a quadratic regulariser [26, 29]. While such standard smoothness terms simplify the minimisation of the underlying energy, they do not allow to preserve discontinuities in the depth and thus lead to *oversmoothed solutions* [36]. For a detailed review of most of the classical methods the reader is referred to [20, 25, 27, 54].

Perspective Shape from Shading. At the end of the 1990s research mainly focused on novel concepts for formulating orthographic SfS such as viscosity solutions [44] and level set formulations [31]. However, for most applications results were not satisfactory [54]. In the early 2000s, the situation changed completely. Inspired by the work of Okatani and Deguchi [35], independently, Prados and his co-workers [39, 40] as well as two other research groups [16, 46] proposed to consider a *perspective* camera model. Evidently, such a model is particularly appropriate for tasks that require the object to be relatively close to the camera such as e.g. in medical endoscopy. In such cases the perspective effects dominate and an orthographic projection model would cause significant systematic errors as shown in [47]. Secondly, Prados and colleagues proposed to *shift the light source location from infinity to the camera centre* which can be seen as a good approximation of a camera with photoflash. This made shape from shading attractive for a variety of photo-based applications. Finally, also a physically motivated *light attenuation term* was introduced that models a quadratic fall-off due to the inverse square law. As discussed in [9], the use of this term largely resolved the convex-concave ambiguity that was inherent to the classical orthographic model although some ambiguities are still present. Even the generalisation of such approaches to advanced reflectance models such as the Oren-Nayar [37] or the Phong reflectance model [38] have been recently investigated [4, 50].

However, this evolution of SfS models was accompanied by a different way of formulating the SfS problem. Instead of using variational methods, the perspective SfS problem was formulated in terms of hyperbolic PDEs [40]. Although such PDE formulations allow for an efficient computation of the solution using fast marching schemes [45], they suffer from two inherent drawbacks: (i) On the one hand, they are prone to *noise* and *missing data*, since they do not rely on any form of regularisation or filling-in. This can be particularly problematic in the context of real-world images. (ii) On the other hand, it is *difficult to extend* the underlying model of such PDE-based schemes by additional constraints such as smoothness terms, multiple views, or additional light sources. While there have been recently some PDE-based approaches to photometric stereo [33], one has to take care of ensuring the uniqueness of the solution if the input data from multiple images is not consistent, cf. the discussion in [34].

Variational Perspective Shape from Shading. Given the flexibility and robustness of variational methods, it is not surprising that recently researchers tried to close the evolutionary loop by integrating the perspective SfS model into a suitable variational framework. So far, however, there are only a few works in the literature that deal with this recent idea. On the one hand, there is the work of Ju et al. [30] that embeds the PDE of Prados et al. [40] as data term into a variational model and complements it with a discontinuity-preserving second order smoothness term. However, since the approach penalises deviations from the PDE directly and uses a parametrisation in terms of the radial depth, deviations in both the data and the smoothness term are difficult to interpret geometrically or photometrically. On the other hand, there is the approach of Abdelrahim et al. [3] that formulates the data term in terms of brightness differences and makes use of a Cartesian depth parametrisation. While the corresponding energy functional is thus more meaningful from a geometric and photometric viewpoint, it defines smoothness based on surface normals and thus needs an additional integrability constraint. Moreover, the corresponding smoothness term is restricted to a simple homogeneous regulariser that does not allow to preserve object edges during the reconstruction. Finally, there are the works of Zhang et al. [55] and Wu et al. [53] that also make use of a Cartesian depth parametrisation but rely on an indirect estimation using auxiliary variables. While the approach of Zhang et al. [55] resolves the resulting consistency problem by considering an integrability constraint, the method of Wu et al. [53] repeatedly integrates the surface normals during computation to ensure valid solutions. Moreover, both approaches use derivations for their surface normals that are based on the orthographic projection model of Horn and Brooks [27]. Unfortunately, the resulting models are thus only valid in case of weak perspective distortions.

A final issue that is common to all four of the aforementioned works is the difficulty of minimising the underlying energy. Since this energy is non-convex, two of the methods rely on initialisations provided by closely related PDE-based SfS approaches [3, 55]. This, however, contradicts the idea of introducing robustness into the estimation – in particular in the presence of noise or missing data. In contrast, the other two methods estimate the solution from scratch [30, 53]. However, those methods do not provide any quantitative assessment of the reconstruction quality.

Let us summarise: While from a modelling viewpoint, it would be desirable to design a variational model that directly solves for the Cartesian depth without the need of integrability constraints or repeated integrations steps, it would be helpful from an optimisation viewpoint to develop a minimisation scheme that neither depends on the solution of other SfS techniques as in [3, 55] nor requires an accurate initialisation to produce meaningful results.

Our Contributions. In this book chapter we contribute to the field of variational SfS in three ways: (i) First, we consider a variational model for perspective SfS that makes use of a Cartesian depth parametrisation and an edge-preserving Cartesian depth regularisation. By penalising deviations from the image brightness in the data term and regularising the Cartesian depth in the smoothness term directly, we obtain an approach that is geometrically and photometrically meaningful. In this context, we also point out a popular mistake in the derivation of the surface normal and show two different ways to derive the normal correctly. (ii) Our method is a direct approach to depth computation, i.e. it does not yield gradient fields that need to be integrated in a subsequent step, nor do we employ integrability constraints. (iii) Apart from the novel model, we also propose a novel minimisation strategy. By embedding an alternating explicit scheme into a coarse-to-fine scheme, we obtain an optimisation framework that allows to obtain significantly better results than a traditional explicit scheme. Experiments with synthetic and real-world images show the good quality of our reconstructions and the advantages of our numerical scheme.

Organisation of the Chapter. In Sect. 3.2 we propose a novel PDE-based model for perspective SfS that is based on a Cartesian parametrisation of the depth. In Sect. 3.3 we then embed this PDE into a variational framework with appropriate second order smoothness term. Details on the minimisation and the discretisation are provided in Sect. 3.4, while Sect. 3.5 comments on the integration of intrinsic camera parameters. Finally, a detailed evaluation of our approach is presented in Sect. 3.6. The paper concludes with a summary in Sect. 3.7.

3.2 Perspective SfS with Cartesian Depth Parametrisation

In this section, we introduce a novel PDE-based SfS model that is parametrised in terms of the *Cartesian* depth. In contrast to most existing SfS models that estimate the radial depth or multiples thereof, such a Cartesian parametrisation expresses the unknown surface directly in terms of the Euclidean distance along the z-axis, which is the axis orthogonal to the image plane.

Parametrisation of the Surface. The starting point for our new model is formed by the classical PDE-approach of Prados et al. [40] which is originally parametrised in terms of the *radial* depth. Key assumptions of this SfS model are that a point light source is located at the *optical centre* of a perspective camera and that the surface reflectance is *Lambertian* with uniform albedo that is fixed to one. The unknown surface $\mathscr{S} : \overline{\Omega}_{\mathbf{x}} \rightarrow \mathbb{R}^3$ can then be described as

$$\mathscr{S}(\mathbf{x}, u(\mathbf{x})) = \left\{ \frac{\mathtt{f}}{\sqrt{|\mathbf{x}|^2 + \mathtt{f}^2}} \, u(\mathbf{x}) \begin{bmatrix} x \\ y \\ -\mathtt{f} \end{bmatrix} \middle| \; \mathbf{x} := (x, y)^\top \in \overline{\Omega}_{\mathbf{x}} \right\}, \tag{3.1}$$

where $\mathbf{x} = (x, y)^\top \in \overline{\Omega}_{\mathbf{x}}$ is the position in the closure $\overline{\Omega}_{\mathbf{x}}$ of the rectangular image domain $\Omega_{\mathbf{x}} \subset \mathbb{R}^2$, $\mathtt{f}$ denotes the focal length of the camera and $u(\mathbf{x})$ is a multiple of $\mathtt{f}$ that describes the radial distance (depth) of the surface from the camera centre.

Since the third component in Eq. (3.1) corresponds to the negative Cartesian depth z, we can derive the following relationship to the radial depth $u \, \mathtt{f}$

$$z(\mathbf{x}) = \frac{\mathtt{f}}{\sqrt{|\mathbf{x}|^2 + \mathtt{f}^2}} \, u(\mathbf{x}) \, \mathtt{f} \overset{(3.1)}{=} Q(\mathbf{x}) \, u(\mathbf{x}) \, \mathtt{f} \,, \tag{3.2}$$

where $Q(\mathbf{x})$ denotes a spatially variant conversion factor given by

$$Q(\mathbf{x}) = \frac{\mathtt{f}}{\sqrt{|\mathbf{x}|^2 + \mathtt{f}^2}} \,. \tag{3.3}$$

This relation is illustrated in Fig. 3.1.

Plugging Eq. (3.3) into Eq. (3.1), we then obtain the parametrisation of the original surface $\mathscr{S}$ with respect to the Cartesian depth z

$$\mathscr{S}(\mathbf{x}, z(\mathbf{x})) \overset{(3.1)}{=} Q(\mathbf{x}) \, u(\mathbf{x}) \begin{bmatrix} x \\ y \\ -\mathtt{f} \end{bmatrix} \overset{(3.2)}{=} Q(\mathbf{x}) \left(\frac{z(\mathbf{x})}{\mathtt{f} \, Q(\mathbf{x})} \right) \begin{bmatrix} x \\ y \\ -\mathtt{f} \end{bmatrix} = \begin{bmatrix} \frac{z(\mathbf{x}) \, x}{\mathtt{f}} \\ \frac{z(\mathbf{x}) \, y}{\mathtt{f}} \\ -z(\mathbf{x}) \end{bmatrix} . \tag{3.4}$$

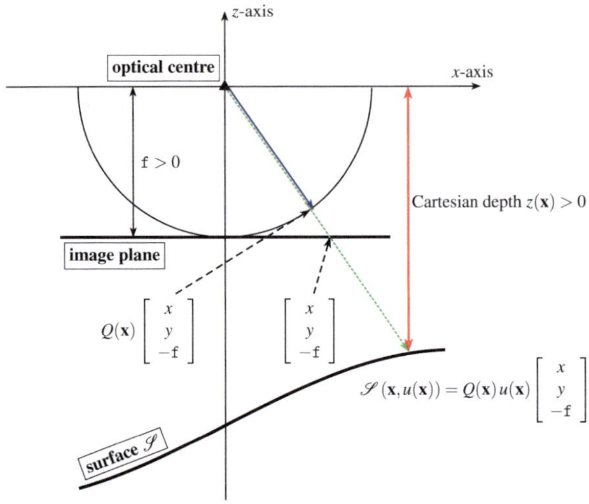

Fig. 3.1 Relation between the radial depth factor $u(\mathbf{x})$ (quotient between *green* and *blue* distance) that denotes the depth in multiples of the focal length $\mathtt{f}$ and the Cartesian depth $z(\mathbf{x})$ (*red* distance)

Brightness Equation. After we have parametrised the original surface in terms of the Cartesian depth, let us now derive the resulting brightness equation that relates the local orientation of the surface to the image brightness. Assuming a Lambertian reflectance model and a quadratic light attenuation term that follows the inverse square law, we obtain the following general brightness equation [40]:

$$I(\mathbf{x}) = \frac{1}{r(\mathbf{x})^2} \left(\frac{\mathbf{n}(\mathbf{x})}{|\mathbf{n}(\mathbf{x})|} \cdot \mathbf{L}(\mathbf{x}) \right) , \tag{3.5}$$

where I is the recorded image, $\mathbf{n}$ is the surface normal vector, $\mathbf{L}$ stands for the normalised light direction vector, and r is the (radial) distance of the light source to the surface. Knowing that $r = \mathtt{f}\, u$ and using Eq. (3.2) we can express the quadratic light attenuation term using the Cartesian depth z

$$r(\mathbf{x}) = \mathtt{f} u(\mathbf{x}) = \frac{z(\mathbf{x})}{Q(\mathbf{x})} \quad \Rightarrow \quad \frac{1}{r(\mathbf{x})^2} = \frac{Q(\mathbf{x})^2}{z(\mathbf{x})^2} . \tag{3.6}$$

What remains to be computed in terms of the Cartesian depth are the surface normal $\mathbf{n}$ and the light direction vector $\mathbf{L}$, respectively.

Surface Normal. Let us start by deriving the surface normal. Since the surface normal is the normal vector of the tangent plane, we first have to compute the partial derivatives of the surface in Eq. (3.4) in x- and y-direction, respectively

$$\mathcal{S}_x(\mathbf{x}, z) = \begin{bmatrix} \frac{z_x\, x + z}{\mathtt{f}} \\ \frac{z_x\, y}{\mathtt{f}} \\ -z_x \end{bmatrix}, \quad \mathcal{S}_y(\mathbf{x}, z) = \begin{bmatrix} \frac{z_y\, x}{\mathtt{f}} \\ \frac{z_y\, y + z}{\mathtt{f}} \\ -z_y \end{bmatrix}. \tag{3.7}$$

Here and for the whole paper we dropped the spatial dependency of z, z_x and z_y on $\mathbf{x}$ for the sake of clarity. Taking the cross-product then yields the direction of the surface normal

$$\mathbf{n}(\mathbf{x}) = \mathcal{S}_x(\mathbf{x}, z) \times \mathcal{S}_y(\mathbf{x}, z) = \begin{bmatrix} \frac{z_x\, z}{\mathtt{f}} \\ \frac{z_y\, z}{\mathtt{f}} \\ \frac{z\, [(\nabla z \cdot \mathbf{x}) + z]}{\mathtt{f}^2} \end{bmatrix}. \tag{3.8}$$

Light Direction. Let us now turn towards the computation of the light direction. Since the light source is assumed to be located in the camera centre which coincides with the origin of the coordinate system, the direction of the light rays and the direction of the optical rays coincide (up to sign). Hence, the light direction can just be read off Eq. (3.1) as

$$\mathbf{L}(\mathbf{x}) = \frac{1}{\sqrt{|\mathbf{x}|^2 + \mathtt{f}^2}} \begin{bmatrix} -x \\ -y \\ \mathtt{f} \end{bmatrix}. \tag{3.9}$$

PDE-Based Model. By plugging the surface normal from Eq. (3.8) and the light direction from Eq. (3.9) into the brightness Eq. (3.5) we finally obtain our perspective SfS model with the new Cartesian depth parametrisation

$$I - \frac{Q^3}{z\sqrt{\mathtt{f}^2\, |\nabla z|^2 + [(\nabla z \cdot \mathbf{x}) + z]^2}} = 0\,. \tag{3.10}$$

Here and for the whole paper we dropped the spatial dependency of I and Q on $\mathbf{x}$ for the sake of clarity.

The main properties of our new model (3.10) are naturally inherited from the original PDE [40]: (i) Eq. (3.10) still belongs to the class of Hamilton-Jacobi equations (HJEs) which have been intensively studied in the SfS literature. (ii) Therefore, well-posedness can be achieved in the viscosity sense [9, 14, 40]. (iii) Proper numerical discretisations must be considered when solving the HJE.

Let us note that the framework of viscosity solutions is a natural setting for HJEs such as Eq. (3.10). The basic idea behind the notion of viscosity solutions is to add a (typically, second order) regularisation term to the PDE and study the solution as this term goes to zero. This proceeding yields desirable stability properties and enables to consider even solutions with non-differentiable features like e.g. kinks. We refer the interested reader to [5, 14] for studying properties of viscosity solutions and to [12] for their use in computer vision.

Furthermore, please note that our model can be seen as a generalisation of the PDE-based approach in [56] that already makes use of the Cartesian depth parametrisation, but does not yet consider the light attenuation term from physics.

3.3 Variational Model for Perspective SfS with Cartesian Depth Parametrisation

So far we have derived a novel PDE-based model for perspective SfS with Cartesian depth parametrisation. Let us now discuss how this model can be integrated into a variational framework with smoothness term.

Variational Model. To this end, we follow the idea from [30] and use a quadratic error term based on our novel PDE as data term which is complemented with a suitable second order regulariser. More precisely, we propose to compute the Cartesian depth z as minimiser of the following energy functional

$$E(z) = \int_{\overline{\Omega}_{\mathbf{x}}} c(\mathbf{x}) \underbrace{D(\mathbf{x}, z, \nabla z)}_{\text{Data term}} + \alpha \underbrace{S(\text{Hess}(z))}_{\text{Smoothness term}} \, d\mathbf{x} \,, \tag{3.11}$$

where D is the data term, S is the smoothness term, $c : \mathbf{x} \in \overline{\Omega}_{\mathbf{x}} \subset \mathbb{R}^2 \to [0, 1]$ is a confidence function and $\alpha \in \mathbb{R}^+$ is a regularisation parameter that steers the degree of smoothness of the solution. As mentioned before our data term is based on a quadratic formulation that penalises deviations from our novel PDE. It is given by

$$D(\mathbf{x}, z, \nabla z) = \left(I(\mathbf{x}) - \frac{Q(\mathbf{x})^3}{z \, W(\mathbf{x}, z, \nabla z)} \right)^2 \tag{3.12}$$

with

$$W(\mathbf{x}, z, \nabla z) = \sqrt{f^2 \, |\nabla z|^2 + [(\nabla z \cdot \mathbf{x}) + z]^2} \,. \tag{3.13}$$

As smoothness term, we propose to use the following subquadratic and thus edge-preserving second-order regulariser based on the Frobenius norm of the Hessian

$$S(\text{Hess}(z)) = \Psi\left(\|\text{Hess}(z)\|_F^2 \right) = \Psi\left(z_{xx}^2 + 2z_{xy}^2 + z_{yy}^2 \right) \tag{3.14}$$

where Ψ is the Charbonnier function [13]

$$\Psi(s^2) = 2\lambda^2 \sqrt{1 + \frac{s^2}{\lambda^2}} \tag{3.15}$$

with contrast parameter λ. Such higher-order smoothness terms have already been successfully applied in the context of perspective SfS parametrised in terms of the radial depth [30], orthographic SfS [49], image denoising [32], optical lithography [21] and motion estimation [18]. Finally, the use of the confidence function c in the data term allows to exclude unreliable image regions which have been identified a priori, e.g. by a texture detector or by a background segmentation algorithm. Such functions are particularly useful in the context of real-world images that contain texture, noise, or missing data [17, 30].

Properties. Our variational model from Eq. (3.11) has the following distinct features:

(i) Since the data term in Eq. (3.12) is inherited from Eq. (3.10), the perspective camera projection is already taken into account. Moreover, since the reprojection error is penalised in the data term, deviations have a *photometric* interpretation.
(ii) Since the regulariser is applied directly to the Cartesian depth, also deviations from smoothness become now more meaningful than in the case of a radial depth parametrisation. In particular, they can be interpreted *geometrically*.
(iii) Moreover, in contrast to most existing approaches, the regulariser is able to *preserve edges* in the reconstruction despite of the regularisation effect.
(iv) Unreliable regions can be excluded from the data term via a *confidence function* such that the smoothness term takes over and fills in information from the neighbourhood. This can be advantageous in the context of texture, noise, or missing data. Please note that in contrast to [30], we always guarantee a fixed amount of regularisation by not restricting the smoothness term to unreliable locations.
(v) The depth of the surface is *directly* computed since we minimise for the unknown depth z in Eq. (3.11). This is in contrast to most variational methods that estimate the depth in two steps, see e.g. [10, 22, 29] where first the surface normals are computed by a variational model and then the depth is determined by integration.
(vi) The solution given by the model fulfils the *integrability constraint* per construction since we solve for z and use $z_{xy} = z_{yx}$ in the smoothness term. Otherwise, such as in [3], an additional integrability term would be needed to encourage valid solutions.
(vii) Another advantage of the new parametrisation is that it allows a straightforward *combination* with other reconstruction methods such as stereo [43] or scene flow estimation [6], since such approaches typically make use of the same Cartesian parametrisation and thus could be easily integrated into a joint framework.

Table 3.1 Comparison of the literature on variational models for perspective shape from shading

	Zhang *et al.* [55]	Wu *et al.* [52]	Abdelrahim *et al.* [3]	Ju *et al.* [29]	**Our Work**
Parametrisation	Cartesian depth	Cartesian depth	Cartesian depth	radial depth	Cartesian depth
Reprojection Error as Data Term	✓	✓	✓	–	✓
Light Attenuation Factor	–	✓	✓ [1]	✓	✓
Correct Surface Normal	– [2]	– [2]	✓ [3]	✓	✓
Regularisation	Cartesian depth	Cartesian depth	Cartesian surface normal	radial depth	Cartesian depth
Edge Preservation	–	–	–	✓	✓
No Integrability Term	–	– [4]	–	✓	✓
Direct Estimation [5]	–	–	✓	✓	✓

[1] factor not expressed in terms of the Cartesian depth
[2] see explanation in appendix
[3] no details given in the paper but derivations shown in [1]
[4] integrability constraint realised via repeated integration of surface normals
[5] depth is computed without extra variables for surface normals

(viii) A final advantage is the fact that the approach could easily be extended to *multiple views*, since transformations between the views are simpler if the approach is parametrised in terms of the Cartesian depth instead of the radial depth.

To make the difference of our model to other variational approaches from the literature explicit, the features of the different methods are compared in Table 3.1.

3.4 Minimisation

Let us now discuss the minimisation of the proposed energy. To this end, we will first derive the associated Euler-Lagrange equation and then discuss its discretisation. Finally, we will sketch a coarse-to-fine minimisation strategy with an alternating explicit scheme to solve the resulting nonlinear equations.

Euler-Lagrange Equation. The calculus of variations [15] tells us that the minimiser z of our energy in Eq. (3.11) has to fulfil the corresponding Euler-Lagrange equation. Omitting the dependencies on all variables in order to ease the readability, this equation is given by

$$
\begin{aligned}
0 &= [cD+\alpha S]_z - \frac{\partial}{\partial x}[cD+\alpha S]_{z_x} - \frac{\partial}{\partial y}[cD+\alpha S]_{z_y} \\
&\quad + \frac{\partial^2}{\partial x^2}[cD+\alpha S]_{z_{xx}} + 2\frac{\partial^2}{\partial x \partial y}[cD+\alpha S]_{z_{xy}} + \frac{\partial^2}{\partial y^2}[cD+\alpha S]_{z_{yy}} \\
&= [cD]_z - \frac{\partial}{\partial x}[cD]_{z_x} - \frac{\partial}{\partial y}[cD]_{z_y} + \frac{\partial^2}{\partial x^2}\overbrace{[cD]_{z_{xx}}}^{\equiv 0} + 2\frac{\partial^2}{\partial x \partial y}\overbrace{[cD]_{z_{xy}}}^{\equiv 0} + \frac{\partial^2}{\partial y^2}\overbrace{[cD]_{z_{yy}}}^{\equiv 0} \\
&\quad + \underbrace{[\alpha S]_z}_{\equiv 0} - \frac{\partial}{\partial x}\underbrace{[\alpha S]_{z_x}}_{\equiv 0} - \frac{\partial}{\partial y}\underbrace{[\alpha S]_{z_y}}_{\equiv 0} + \frac{\partial^2}{\partial x^2}[\alpha S]_{z_{xx}} + 2\frac{\partial^2}{\partial x \partial y}[\alpha S]_{z_{xy}} + \frac{\partial^2}{\partial y^2}[\alpha S]_{z_{yy}} \\
&= c\left([D]_z - \frac{\partial}{\partial x}[D]_{z_x} - \frac{\partial}{\partial y}[D]_{z_y}\right) + \frac{\partial^2}{\partial x^2}[\alpha S]_{z_{xx}} + 2\frac{\partial^2}{\partial x \partial y}[\alpha S]_{z_{xy}} + \frac{\partial^2}{\partial y^2}[\alpha S]_{z_{yy}} ,
\end{aligned} \tag{3.16}
$$

where we exploited the fact that

$$
\frac{\partial^2}{\partial x \partial y}[cD+\alpha S]_{z_{xy}} = \frac{\partial^2}{\partial y \partial x}[cD+\alpha S]_{z_{xy}} . \tag{3.17}
$$

On a structural level, this Euler-Lagrange equation is somewhat more complicated than its counterparts for indirect methods in [53, 55]. Such indirect methods model the surface normal using auxiliary variables $p = z_x$ and $q = z_y$ and thus do not have the additional data term contributions $\frac{\partial}{\partial x}[D]_{z_x}$ and $\frac{\partial}{\partial y}[D]_{z_y}$.

Let us now take a closer look at all the individual terms that occur in Eq. (3.17). After some computations we obtain

$$
\begin{aligned}
[D]_z &= 2\left(I - \frac{Q^3}{z\,W}\right)\left(\frac{Q^3}{z^2\,W} + \frac{Q^3}{z\,W^2}[W]_z\right) \\
&= 2\left(I - \frac{Q^3}{z\,W}\right)\frac{Q^3}{z\,W}\left(\frac{1}{z} + \frac{\nabla z \cdot \mathbf{x} + z}{W^2}\right)
\end{aligned} \tag{3.18}
$$

$$
\begin{aligned}
\frac{\partial}{\partial x}[D]_{z_x} &= \left[2\left(I - \frac{Q^3}{z\,W}\right)\frac{Q^3}{z\,W^3}\,[W]_x\right]_x \\
&= \left[2\left(I - \frac{Q^3}{z\,W}\right)\frac{Q^3}{z\,W^3}\left(\mathtt{f}^2\,z_x + [\nabla z \cdot \mathbf{x} + z]\,x\right)\right]_x
\end{aligned} \tag{3.19}
$$

$$\frac{\partial}{\partial y}\,[D]_{z_y} = \left[2\left(I - \frac{Q^3}{z\,W}\right)\frac{Q^3}{z\,W^3}\,[W]_y\right]_y \tag{3.20}$$

$$= \left[2\left(I - \frac{Q^3}{z\,W}\right)\frac{Q^3}{z\,W^3}\,\left(\mathrm{f}^2\,z_y + [\nabla z\cdot\mathbf{x} + z]\,y\right)\right]_y$$

as well as

$$\frac{\partial^2}{\partial x^2}\,[S]_{z_{xx}} = 2\,\frac{\partial^2}{\partial x^2}\left[\Psi'(\|\mathrm{Hess}(z)\|_F^2)\,z_{xx}\right], \tag{3.21}$$

$$2\frac{\partial^2}{\partial xy}\,[S]_{z_{xy}} = 4\,\frac{\partial^2}{\partial xy}\left[\Psi'(\|\mathrm{Hess}(z)\|_F^2)\,z_{xy}\right], \tag{3.22}$$

$$\frac{\partial^2}{\partial y^2}\,[S]_{z_{yy}} = 2\,\frac{\partial^2}{\partial y^2}\left[\Psi'(\|\mathrm{Hess}(z)\|_F^2)\,z_{yy}\right], \tag{3.23}$$

where the derivative of the penaliser function $\Psi(s^2)$ reads

$$\Psi'(s^2) = \frac{\partial}{\partial(s^2)}\Psi(s^2) = \frac{1}{\sqrt{1+\frac{s^2}{\lambda^2}}}\,. \tag{3.24}$$

While the contributions of the data term are related to the influence of z and ∇z on the brightness equation, the contributions of the smoothness term define an edge-preserving fourth-order diffusion process. This becomes explicit as follows: Since $\Psi'(s^2)$ becomes small for large values of s^2, this reduces the effect of the smoothing at locations with high curvature, i.e. where $\|\mathrm{Hess}(z)\|_F^2$ is large. After we have derived the resulting Euler-Lagrange equation, let us now discuss how this equation can be discretised appropriately.

Discretisation. In order to discretise the contributions of the data term given by Eqs. (3.18), (3.19) and (3.20), we employ the upwind scheme from [44] in view of the hyperbolic nature of the underlying PDE. In 1D, the corresponding upwind discretisation reads

$$\tilde{z}_x \approx \max\left(D^- z, -D^+ z, 0\right)\,, \tag{3.25}$$

with

$$D^- z = \frac{z_i - z_{i-1}}{h_x} \qquad \text{and} \qquad D^+ z = \frac{z_{i+1} - z_i}{h_x}\,, \tag{3.26}$$

where h_x denotes the grid size. Please note that in contrast to upwind schemes for eikonal equations [45] that typically approximate only the magnitude of the gradient, the sign matters in our case, such that we have to choose

$$z_x = \begin{cases} D^+ z & \text{if} \quad \tilde{z}_x = -D^+ z\,, \\ \tilde{z}_x & \text{otherwise}\,. \end{cases} \tag{3.27}$$

This selects the actual forward difference, if the second argument in (3.25) is the maximum [8, 9]. This scheme can be extended in a straightforward way to 2D. For discretising the contributions of the smoothness term, a standard central difference scheme is used.

Since it is difficult to discretise the Euler-Lagrange equation directly, we followed a *first discretise then optimise* scheme. To this end, we used the aforementioned finite difference approximations to discretise the energy in (3.11) applying the upwind scheme for the data term and a central difference approximation for the smoothness term. Then, by computing the derivatives of the discrete energy we obtain a proper discretisation for the Euler-Lagrange equation.

Finally, by using the Euler forward time discretisation method

$$z_t \approx \frac{z^{n+1} - z^n}{\tau}\,, \tag{3.28}$$

with τ being a time step size, we can reformulate the solution of Eq. (3.17) as the steady state of the corresponding evolution equation in artificial time. Thus we obtain the following explicit scheme

$$\frac{z^{n+1} - z^n}{\tau} + EL^n = 0 \qquad \Leftrightarrow \qquad z^{n+1} = z^n - \tau\, EL^n\,, \tag{3.29}$$

where EL^n is the discretisation of the Euler-Lagrange equation evaluated at time n. Please note that this discretisation may change over time, since we re-discretised the energy in each iteration by adapting the direction of the discretisation of the upwind scheme (forward, backward, no contribution) based on evaluating Eqs. (3.25), (3.26) and (3.27) for the result of the previous time step. In that sense we use a *lagged* discretisation approach, where the discretisation is updated in each iteration.

Coarse-to-Fine Approach. Since the underlying energy functional is highly non-convex, the proposed explicit scheme may get trapped in local minima. To tackle this problem, we propose to embed the estimation into a coarse-to-fine framework. Starting from a very coarse resolution, we successively refine the input image while repeatedly reconstructing the surface. Thereby, solutions from coarser levels serve as initialisation for the finer scales. Similar hierarchical schemes have already been successfully applied to many other problems in computer vision; see e.g. [8, 11].

Apart from improving the quality of the results by avoiding local minima, coarse-to-fine schemes also render the estimation more robust w.r.t. the choice of the initialisation. In fact, if sufficiently many resolution levels were used, we could hardly observe any impact of the initialisation on the quality of the final results.

Since a good initial guess can still be useful to speed up the computation, we propose to initialise the depth by pointwise solving the data term in Eq. (3.12) for $\nabla z = 0$

$$D(\mathbf{x}, z, 0) = 0 \quad \Rightarrow \quad z = \sqrt{\frac{Q(\mathbf{x})^3}{I(\mathbf{x})}} \,. \tag{3.30}$$

This can be seen as an efficient compromise between using the full model which is evidently not feasible and only considering the inverse square law, i.e. $z = 1/\sqrt{I(\mathbf{x})}$, which completely neglects the effect of the surface orientation and thus actually provides a local upper bound for the correct depth. In any case, in contrast to other variational SfS methods from the literature, our technique does not have to rely on initialisations from non-variational SfS approaches [3, 55] or surface integration methods [53] to provide meaningful results.

Let us now discuss the details of our coarse-to-fine approach. To this end, we introduce the parameter η that specifies the downsampling factor between two consecutive resolution levels and that is typically chosen in the interval $(0.5, 1)$. Then the grid size at level k of our coarse-to-fine approach can be computed as

$$h_x^k = h_x \cdot \eta^{-k} \,, \qquad h_y^k = h_y \cdot \eta^{-k} \,. \tag{3.31}$$

where $k = 0$ is the original resolution and $k = k_{\max}$ is the coarsest level. This tells us that the grid size becomes larger at coarser scales which intuitively makes sense, since the size of the image plane remains constant while the number of pixels decreases. At the same time, however, this increase of the grid size leads to a major problem: Since the contributions of the smoothness term given by Eqs. (3.21), (3.22) and (3.23) involve fourth-order derivatives that scale proportionally to $1/h^4$, the strength of the regularisation actually decreases with η^{4k} on coarser scales. In order to compensate for this effect, we thus propose to scale the smoothness weight α according to

$$\alpha^k = \eta^{-4k} \cdot \alpha \,. \tag{3.32}$$

This guarantees a similar amount of regularisation for all resolution levels.

Alternating Explicit Scheme. Finally, we observed in our experiments that the terms in Eqs. (3.19) and (3.20) that refer to the influence of the depth gradient ∇z on the brightness equation require to select the time step size τ rather small. In particular, these terms do not have a weighting parameter such as the smoothness term that can be adjusted appropriately. As a consequence, the minimisation typically needs several thousands or even millions of iterations. To counter this problem, we propose the following alternating estimation strategy at each resolution level: For a fixed number of iterations n, instead of performing n iterations using the original explicit scheme, we propose an alternating iterative scheme that first does $n/2$ iterations with a simplified explicit scheme neglecting the two terms in Eqs. (3.19) and (3.20), followed by $n/2$ iterations with the entire explicit scheme.

Since the neglected terms are based on second-order derivatives and the remaining terms did not strongly affect the convergence, we empirically found out that we can choose the time step size approximately $\min(h_x^{-2}, h_y^{-2})$ times larger for the first $n/2$ iterations (given that $h_x, h_y \ll 1$). In our experiments this leads to speed-ups of about one to four orders of magnitude. Moreover, in most cases, even the simplified scheme was sufficient to achieve excellent results. Thereby one should note that, from a numerical viewpoint, the simplified scheme can be understood as an optimisation method for a series of energy functionals of type of Eq. (3.11), where the gradient ∇z is lagging and thus has no direct influence on the optimisation.

3.5 Intrinsic Parameters

So far we have derived a variational model for perspective SfS with Cartesian depth parametrisation that is given in terms of *image coordinates*. Let us now discuss how the model and the minimisation has to be adapted if we additionally consider the intrinsic camera parameters, i.e. if we express the model in terms of *pixel coordinates*.

Coordinate Transformation. Let the corresponding calibration matrix be given by

$$K = \begin{bmatrix} \mathtt{f}/h_x & 0 & c_1 \\ 0 & \mathtt{f}/h_y & c_2 \\ 0 & 0 & 1 \end{bmatrix} . \tag{3.33}$$

where $(c_1, c_2)^\top$ denotes the location of the focal point, and h_x and h_y is the grid size in x- and y-direction, respectively [23]. Knowing this matrix allows us to reformulate the image coordinates $\mathbf{x} = (x, y)^\top$ of our original model in terms of pixel coordinates $\mathbf{a} = (a, b)^\top$. The corresponding transformation is given by

$$\begin{bmatrix} a \\ b \\ -1 \end{bmatrix} = K \, \frac{1}{\mathtt{f}} \begin{bmatrix} x \\ y \\ -\mathtt{f} \end{bmatrix} \Rightarrow \begin{bmatrix} x \\ y \\ -\mathtt{f} \end{bmatrix} = \mathtt{f} \, K^{-1} \begin{bmatrix} a \\ b \\ -1 \end{bmatrix} , \tag{3.34}$$

where one has to take care that the image plane is at distance $\mathtt{f}$ of the camera centre. Plugging Eq. (3.33) into Eq. (3.34) then yields

$$\mathbf{x}(\mathbf{a}) = \begin{bmatrix} x(a) \\ y(b) \end{bmatrix} = \begin{bmatrix} h_x & 0 \\ 0 & h_y \end{bmatrix} \begin{bmatrix} a \\ b \end{bmatrix} - \begin{bmatrix} h_x \, c_1 \\ h_y \, c_2 \end{bmatrix} . \tag{3.35}$$

Variational Model. Now we are in the position to reformulate our entire model in terms of pixel coordinates. Substituting Eq. (3.35) into our original energy and transforming the integration domain $\overline{\Omega}_{\mathbf{a}} = \mathbf{x}^{-1}(\overline{\Omega}_{\mathbf{x}})$ accordingly, we obtain the following variational model expressed in terms of pixel coordinates

$$E\,(z(\mathbf{x}(\mathbf{a}))) = \int_{\overline{\Omega}_{\mathbf{a}}} c(\mathbf{x}(\mathbf{a}))\ \underbrace{D(\mathbf{x}(\mathbf{a}), z(\mathbf{x}(\mathbf{a})), \nabla z(\mathbf{x}(\mathbf{a})))}_{\text{Data term}} + \alpha\ \underbrace{S(\text{Hess}(z)(\mathbf{x}(\mathbf{a})))}_{\text{Smoothness term}}\ \mathbf{da}\,. \tag{3.36}$$

Please note that we omitted the substitution factor given by $|\det(J(\mathbf{x}(\mathbf{a})))|$, where J is the Jacobian, since this factor is constant and thus does not change the minimiser of our energy. Let us now derive the corresponding Euler-Lagrange equation for our novel model expressed in terms of pixel coordinates.

Euler-Lagrange Equation. Analogously to Eq. (3.17) we drop the dependencies on all variables and obtain the following Euler-Lagrange equation

$$\begin{aligned} 0 \;&=\; c\left([D]_z - \frac{\partial}{\partial a}[D]_{z_a} - \frac{\partial}{\partial b}[D]_{z_b}\right) \\ &\quad + \alpha\left(\frac{\partial^2}{\partial a^2}[S]_{z_{aa}} + 2\frac{\partial^2}{\partial a \partial b}[S]_{z_{ab}} + \frac{\partial^2}{\partial b^2}[S]_{z_{bb}}\right) && (3.37) \\ &=\; c\left([D]_z - \frac{\partial}{\partial x}[D]_{z_x} - \frac{\partial}{\partial y}[D]_{z_y}\right) \\ &\quad + \alpha\left(\frac{\partial^2}{\partial x^2}[S]_{z_{xx}} + 2\frac{\partial^2}{\partial x \partial y}[S]_{z_{xy}} + \frac{\partial^2}{\partial y^2}[S]_{z_{yy}}\right), && (3.38) \end{aligned}$$

where we exploited the following relation between derivatives in pixel and image coordinates due to Eq. (3.35)

$$\frac{\partial}{\partial a} = h_x\,\frac{\partial}{\partial x}, \quad \frac{\partial}{\partial b} = h_y\,\frac{\partial}{\partial y}, \quad \frac{\partial}{\partial z_a} = \frac{1}{h_x}\frac{\partial}{\partial z_x}, \quad \frac{\partial}{\partial z_b} = \frac{1}{h_y}\frac{\partial}{\partial z_y}, \tag{3.39}$$

$$\frac{\partial}{\partial z_{aa}} = \frac{1}{h_x h_x}\frac{\partial}{\partial z_{xx}}, \quad \frac{\partial}{\partial z_{ab}} = \frac{1}{h_x h_y}\frac{\partial}{\partial z_{xy}}, \quad \frac{\partial}{\partial z_{bb}} = \frac{1}{h_y h_y}\frac{\partial}{\partial z_{yy}}. \tag{3.40}$$

The equality between Eqs. (3.37) and (3.38) shows that the Euler-Lagrange equations of our models in pixel and image coordinates are basically identical. One only has to parametrise the terms (3.18), (3.19), (3.20), (3.21), (3.22) and (3.23) that have been originally derived in image coordinates using the coordinate transform in Eq. (3.35). Apart from that, the discretisation can be performed in accordance with our explanations from the previous section. In this context, the grid size is given by the intrinsic parameters h_x and h_y. Moreover, one has to adapt the camera matrix K at each level of the coarse-to-fine scheme. This requires to scale both the grid size and the principal point $(c_1, c_2)^\top$.

3.6 Evaluation

Test Images and Error Measures. In order to evaluate our novel approach we make use of four synthetic images with ground truth that fulfil the underlying assumptions regarding reflectance and illumination. This allows us to compute two error measures: one with respect to the reconstructed surface and the other one with respect to reprojected image. The first error measure is the *relative surface error* (RSE) of a point wise computed Euclidean distance between the computed surface $\mathcal{S}$ and the ground truth surface $\mathcal{S}^{\mathrm{gt}}$. It is given by

$$\mathrm{RSE} = \frac{\sum_{\overline{\Omega}_{\mathbf{a}}} |\mathcal{S}(\mathbf{x}(\mathbf{a})) - \mathcal{S}^{\mathrm{gt}}(\mathbf{x}(\mathbf{a}))|}{\sum_{\overline{\Omega}_{\mathbf{a}}} |\mathcal{S}^{\mathrm{gt}}(\mathbf{x}(\mathbf{a}))|}, \tag{3.41}$$

where the normalisation allows to determine the reconstruction error *relative* to the ground truth shape. This in turn makes errors of differently scaled surfaces comparable. The second error measure is the *relative image error* (RIE) between the reprojected image I and the given input image I^{gt}. It is defined as follows

$$\mathrm{RIE} = \frac{\sum_{\overline{\Omega}_{\mathbf{a}}} |I(\mathbf{x}(\mathbf{a})) - I^{\mathrm{gt}}(\mathbf{x}(\mathbf{a}))|}{\sum_{\overline{\Omega}_{\mathbf{a}}} |I^{\mathrm{gt}}(\mathbf{x}(\mathbf{a}))|}. \tag{3.42}$$

This time, however, the normalisation is performed with respect to the brightness of the input image to make reprojection results for input images with different brightness scale comparable. Summarising: While the first measure reflects how well the reconstruction matches the ground truth surface, the second measure determines how well the reprojection fits the input data.

Let us now discuss the considered test images which are depicted in Fig. 3.2 in detail. The first synthetic test image *Sombrero* was generated from a known parametric surface, using the following equation

$$Z(X, Y) = 0.5\frac{\sin(r(X, Y))}{r(X, Y)} + 1.7, \quad r(X, Y) = \sqrt{(10X)^2 + (10Y)^2}. \tag{3.43}$$

The image was rendered using Eq. (3.5) at a size of 256×256 pixels, where the focal length was set $\mathtt{f} = 1$, the grid size was chosen to be $h_x = h_y = 1/200$ and

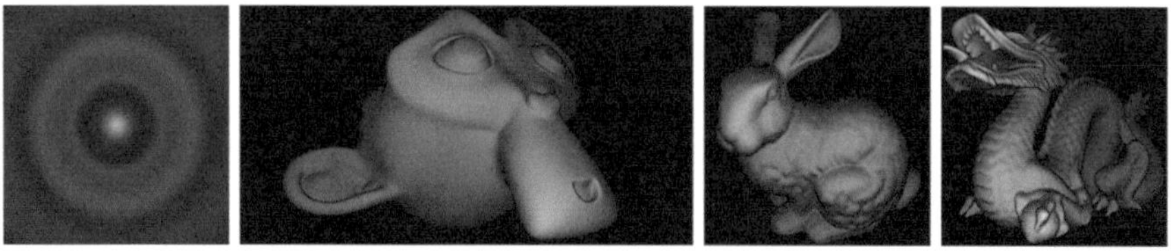

Fig. 3.2 Synthetic test images. **From left to right:** *Sombrero*, *Suzanne*, *Stanford Bunny* and *Dragon*

the principal point was fixed at $c = (128, 128)^\top$. The second test image *Suzanne* was generated using the open-source software Blender [28]. In this context, the Z-buffer of the rendering path and the corresponding intrinsic parameters ($\texttt{f} = 35$, $h_x = 1/16$, $h_y = 9/128$, $c = (256, 128)^\top$) were extracted and the final image was rendered at a size of 512×256 using Eq. (3.5) as before. The other two test images *Stanford Bunny* and *Dragon* have been computed likewise using 3-D models obtained from the Stanford 3D scanning repository [48]. For them a size of 256×256 pixels and the intrinsic parameters ($\texttt{f} = 35$, $h_x = 1/8$, $h_y = 9/128$, $c = (128, 128)^\top$) were chosen. Finally, all images were saved as 8-bit grey-value images.

Let us finally comment on the selection of the parameters in our experiments. In order to keep the number of parameters low, we choose a preferred standard set of parameters for all the following experiments, unless otherwise stated: A downsampling factor of $\eta = 0.8$ for the coarse-to-fine approach, $n = 10^6$ solver iterations on each coarse-to-fine level and a contrast parameter of $\lambda = 10^{-3}$. Moreover, the time step size τ provided in the different experiments always refers to the simplified explicit scheme. The time step size for the full explicit scheme is $\min(h_x^2, h_y^2)$ times smaller.

Results on Synthetic Test Images. In our first experiment we evaluate the reconstruction quality of our novel approach. To this end, we applied our perspective SfS algorithm to all four of the previously discussed test images and compared the reprojected image and the reconstruction to the ground truth; see Figs. 3.3 and 3.4. Herein, the depth values are colour-coded in such a way that depth increases from red via green to blue. As one can see, both the reprojected image as well as the estimated depth values coincide very well with the ground truth. This is also confirmed by the corresponding surface error maps in Fig. 3.5. Indeed, only small differences for the *Stanford bunny* (right paw) and the *Dragon* (tail tip) are visible. As a consequence both error measures which are listed in Table 3.2 are very small. Moreover, one can see that the proposed subquadratic penaliser outperforms a quadratic smoothness term in most cases. Only for the *Sombrero* which has a rather smooth surface, the reconstruction error is smaller in the quadratic case.

Influence of the Regularisation. In our second experiment we investigate the influence of the regularisation on the quality of the reconstruction and its reprojection. To this end, we consider the *Sombrero* test image and vary the regularisation parameter α while the other parameters are kept fixed ($\tau = 0.001$, $n = 10^4$). The outcome is visualised in Fig. 3.6. While the reprojection related error measure (RIE) increases for a moderate amount of regularisation but is overall very low, the surface related error measure (RSE) decreases by almost a factor 3 (from 4.4×10^{-2} to 1.7×10^{-2}). This, however, is not surprising, since the computed surface typically exhibits some form of smoothness and thus benefits from a moderate amount of regularisation. Since the actual purpose of SfS is to find the correct surface, this shows that the regularisation may have an overall positive impact on the quality of the results.

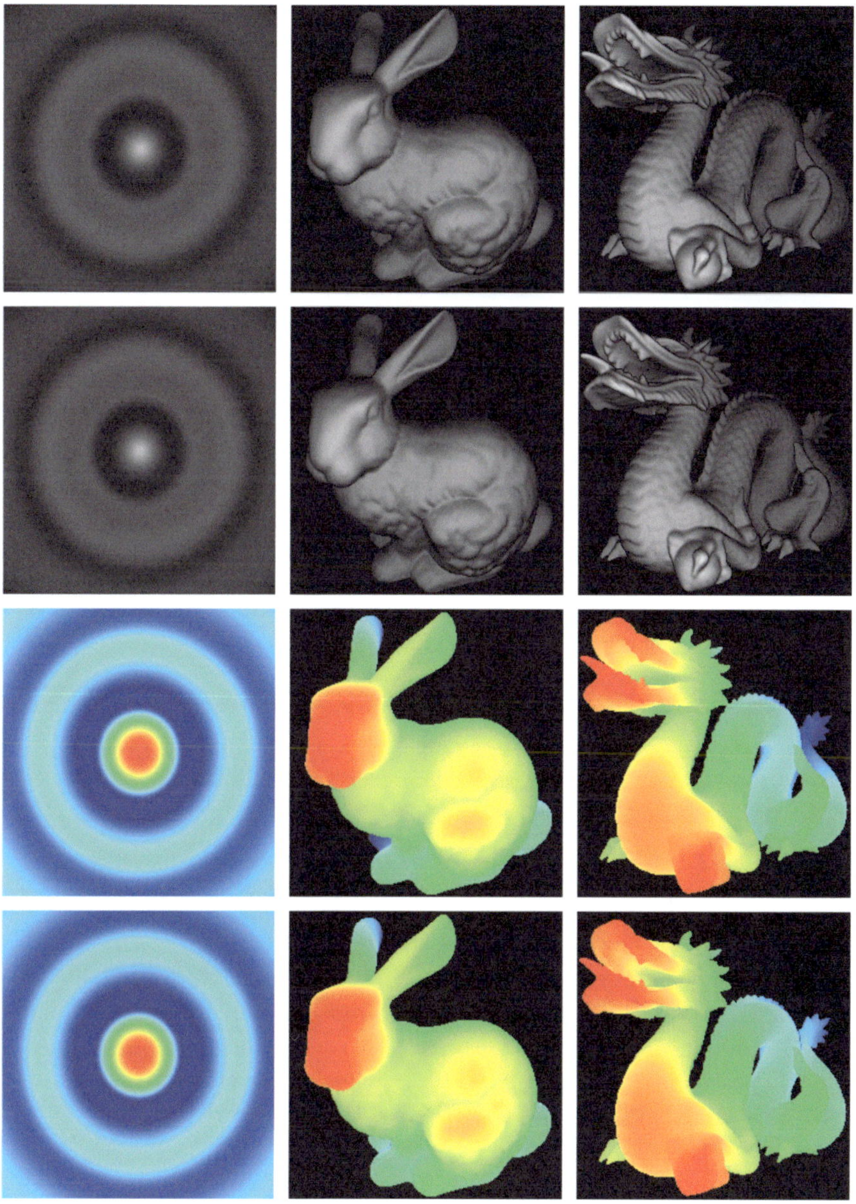

Fig. 3.3 **First column, from top to bottom:** Input image, reprojected image, ground truth depth, computed depth for the *Sombrero* test image ($\alpha = 7.5 \times 10^{-5}$, $\tau = 10^{-2}$, $n = 10^{6}$). **Second column:** Ditto for the *Stanford Bunny* test image ($\alpha = 7.5 \times 10^{-5}$, $\tau = 10^{-3}$, $n = 10^{6}$). **Third column:** Ditto for the *Dragon* test image ($\alpha = 7.5 \times 10^{-8}$, $\tau = 10^{-3}$, $n = 10^{6}$)

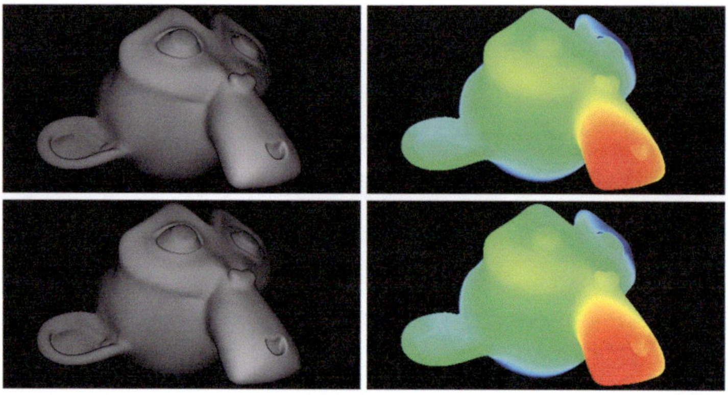

Fig. 3.4 **First row, from left to right:** Input image and ground truth depth of the *Suzanne* test image. **Second row:** Reprojected image and the computed depth ($\alpha = 10^{-7}$, $\tau = 10^{-3}$, $n = 10^6$)

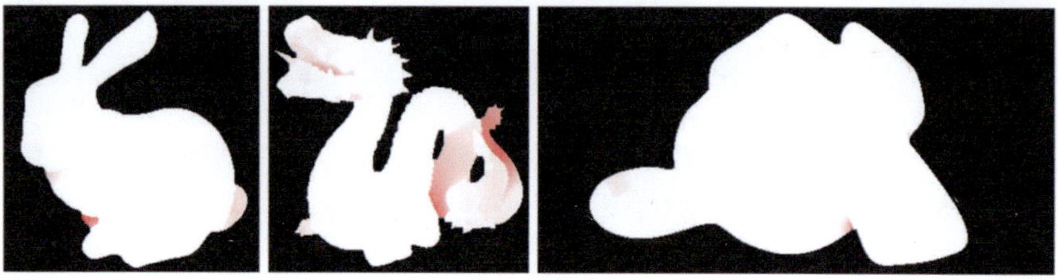

Fig. 3.5 Surface error maps. **From left to right:** *Stanford Bunny*, *Dragon* and *Suzanne*. *Red* denotes errors above 1 %, where the intensity encodes the error magnitude. *White* denotes errors below 1 %. The *Sombrero* is not shown, since the error is below 1 % everywhere

Table 3.2 Results for our approach with quadratic and subquadratic penaliser. Error measures are given in terms of the relative surface error (RSE) and the relative image error (RIE). Best results for each test image are highlighted boldface. Same parameters as in Figs. 3.3 and 3.4

	Quadratic		Subquadratic		
	RSE	RIE	RSE	RIE	Runtime
Sombrero	**0.00208**	0.00694	0.00318	**0.00209**	29,113 s
Stanford Bunny	0.00546	0.00015	**0.00439**	**0.00007**	23,969 s
Dragon	**0.01376**	**0.00028**	**0.01376**	**0.00028**	2535 s
Suzanne	0.00392	0.00011	**0.00251**	**0.00002**	48,395 s

Independence of the Initialisation. In our third experiment we analyse the dependency of our approach on the initialisation. To this end, we use the *Stanford Bunny* ($z \in [1, 2]$) and compare our initialisation on the coarsest scale of the proposed coarse-to-fine scheme (cf. Eq. 3.30) with two other initialisations based on plain surfaces ($z = 1$, $z = 10$). The initial error and the outcome after $n = 10^6$ iterations are listed in Table 3.3. While the initial error for a good guess ($z = 1$) and

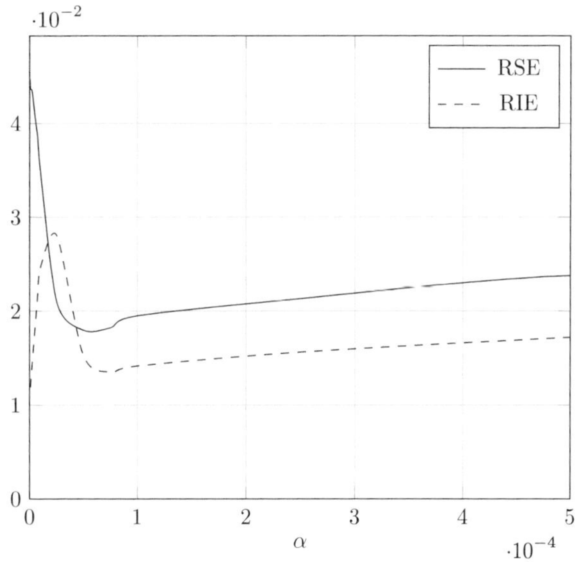

Fig. 3.6 Impact of the amount of regularisation on the reconstruction quality and the reprojection accuracy for the *Sombrero* test image

Table 3.3 Impact of different initialisations on the reconstruction quality and reprojection accuracy for the *Stanford Bunny* ($\alpha = 7.5 \times 10^{-5}$, $\tau = 10^{-3}$, $n = 10^{6}$)

	Initial error		After computation	
	RSE	RIE	RSE	RIE
Plane ($z = 1$)	0.25804	1.63174	0.00439	0.00007
Plane ($z = 10$)	6.41960	0.97373	0.00439	0.00007
Proposed	0.37712	0.74363	0.00439	0.00007

a poor initialisation ($z = 10$) differs significantly, the quality of the reconstruction and the reprojection is identical after sufficiently many iterations. This also holds for our initialisation which can be computed from the input image without requiring a specific knowledge of the depth. That all initialisations converge to the same solution, however, is not surprising since the estimation is embedded in our coarse-to-fine scheme.

Comparison of Numerical Schemes. In our fourth experiment we compare the different numerical schemes proposed in Sect. 3.4: the full explicit scheme, the simplified explicit scheme and the alternating explicit scheme. In the first part of the experiment we juxtapose the quality of the different numerical schemes for equal stopping times (iterations × time step size). As one can see from the results in Table 3.4, the full explicit scheme clearly gives the best results in terms of reconstruction quality and reprojection accuracy. However, this comes at the expense of a significantly larger runtime, since more iterations are needed due to the time step restrictions discussed in Sect. 3.4. In fact the runtime is up to four orders of magnitude larger making the approach hardly feasible for larger image sizes. In

Table 3.4 Comparison of different numerical schemes for equal stopping time $t = n \times \tau$. Results and runtimes refer to smaller versions of the four test images. Same parameters as in Figs. 3.3 and 3.4 except for n, which is given by $n = t/\tau$

	Alternating scheme		Simplified scheme		Full scheme	
Test image	RSE	RIE	RSE	RIE	RSE	RIE
Small Sombrero	0.01823	0.01920	0.01820	0.02048	**0.00785**	**0.00527**
(128 × 128)	(runtime: 30 s)		(runtime: 15 s)		(runtime: 178,021 s)	
Small Stanford Bunny	0.00659	0.00151	0.00667	0.00257	**0.00576**	**0.00097**
(128 × 128)	(runtime: 303 s)		(runtime: 150 s)		(runtime: 4278 s)	
Small Dragon	0.01667	0.00267	0.01673	0.00620	**0.01526**	**0.00205**
(128 × 128)	(runtime: 308 s)		(runtime: 149 s)		(runtime: 4304 s)	
Small Suzanne	**0.00899**	0.00514	0.01055	0.01909	0.01022	**0.00203**
(128 × 96)	(runtime: 223 s)		(runtime: 111 s)		(runtime: 2384 s)	

Table 3.5 Comparison of different numerical schemes for equal number of iterations. Results refer to the smaller versions of the four test images, see Table 3.4. The same parameters as in Figs. 3.3 and 3.4 have been used except for n, which is given by $n = 10^7$

	Alternating scheme		Simplified scheme		Full scheme	
Test image	RSE	RIE	RSE	RIE	RSE	RIE
Small Sombrero	0.02357	**0.00082**	0.02392	0.00659	**0.00358**	0.00319
Small Stanford Bunny	0.00390	**0.00001**	**0.00378**	0.00004	0.00489	0.00047
Small Dragon	0.00572	**0.00001**	**0.00562**	**0.00001**	0.00964	0.00170
Small Suzanne	**0.00319**	0.00002	0.00320	**0.00001**	0.00505	0.00056

the second part of the experiment we compared the numerical schemes for an equal number of iterations. From the results in Table 3.5 it becomes evident that in this case the simplified explicit scheme and in particular the alternating explicit scheme perform best in most cases in terms of reconstruction quality and reprojection accuracy. This demonstrates that it can be worthwhile to (partly) omit the terms that are added in the full explicit scheme since they slow down the convergence, but doing so does not necessarily compromise the quality.

Reconstruction with Inpainting. In our fifth experiment we demonstrate the inpainting capabilities of the regularisation in combination with the confidence function c embedded in the data term. For this reason we created a pair of degraded *Stanford Bunny* test images together with the corresponding confidence functions, which are both depicted in Fig. 3.7. In addition, the computed depth values and the reprojected images are shown. One can see that in both cases the missing regions in the input image can hardly deteriorate the quality of the results since the smoothness term fills in the information from the neighbourhood. This is also reflected in the error measures given in Table 3.6. In case of the perforated version the surface error even remains the same compared to the result for the original version.

Comparison with a PDE-Based Approach. In our seventh experiment we compare the results of our variational method with the PDE-based approach of Vogel

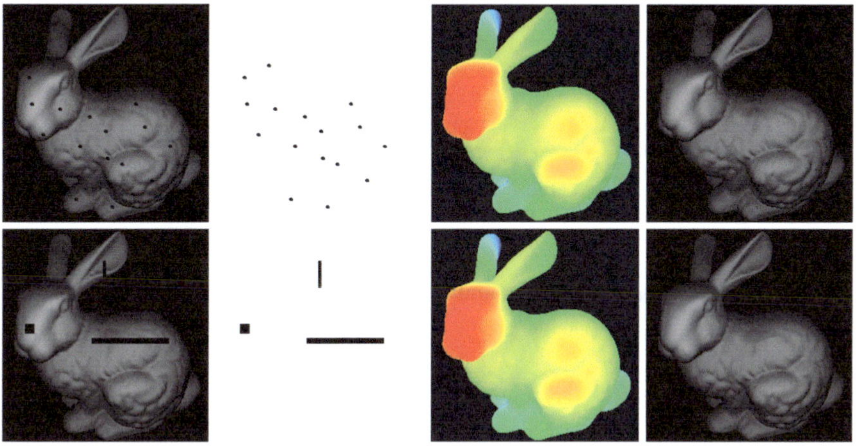

Fig. 3.7 **First row, from left to right:** Perforated version of the *Stanford Bunny* test image, corresponding confidence function c, computed depth values, reprojected image ($\alpha = 7.5 \times 10^{-5}$, $\tau = 10^{-3}$, $n = 10^{6}$). **Second row:** Ditto for the sliced version (same parameters)

Table 3.6 Evaluation of inpainting properties for degraded versions of the *Stanford Bunny* test image. Same parameters as in Fig. 3.7

	Perforated version (Fig. 3.7, top row)	Sliced version (Fig. 3.7, bottom row)	Original version (Fig. 3.2)
RSE	0.00439	0.00509	0.00439
RIE	0.00039	0.00249	0.00007

et al. [51] with Lambertian reflectance model. This essentially comes down to a comparison to the baseline method of Prados et al. [40] which is solved by Vogel et al. [51] as part of a Phong-based model using an efficient fast marching scheme [45]. In this experiment we consider two scenarios, that nicely demonstrate the advantages and shortcomings of the different types of methods: On the one hand, we use input images without noise, on the other hand, we added Gaussian noise of standard deviation $\sigma = 20$ before applying the two methods. The corresponding results are summarised in Tables 3.7 and 3.8, respectively. For the test images without noise both approaches give excellent results with errors among or below 1 % of the solution. Thereby the approach of Vogel et al. gives slightly better results in terms of the relative surface error (RSE), while the variational approach gives better results in terms of the relative image error (RIE). From the viewpoint of the variational approach this can be explained as follows: While the data term penalises deviations from the photometric reprojection error and thus gives rather small RIE values, the regulariser and the coarse-to-fine scheme yield a moderate smoothing of the surface resulting in slightly higher RSE values. In the case of the noisy input images the findings are completely different. Here, the variational method can take advantage of both the regulariser and the independence of the initialisation. While a higher smoothness weight allows to obtain a smooth surface, the hierarchical

Table 3.7 Comparison between our variational method and the PDE-based approach of Vogel et al. [51] with Lambertian reflectance model (= baseline model of Prados et al. [40]). Error measures are given in terms of the relative surface error (RSE) and the relative image error (RIE). Same parameters as in Figs. 3.3 and 3.4

	Vogel et al. [51] (PDE-based approach)		Our method (variational method)	
	RSE	RIE	RSE	RIE
Sombrero	0.00301	0.00495	0.00318	0.00209
Stanford Bunny	0.00266	0.00154	0.00439	0.00007
Dragon	0.00422	0.00255	0.01376	0.00028
Suzanne	0.00253	0.00082	0.00251	0.00002

Table 3.8 Performance under noise. Comparison between our variational method and the PDE-based approach of Vogel et al. [51] with Lambertian reflectance model (= baseline model of Prados et al. [40]). Gaussian noise of standard deviation $\sigma = 20$. Error measures are given in terms of the relative surface error (RSE) and the relative image error (RIE). The applied parameters are as follows: *Sombrero* ($\alpha = 0.1$, $\tau = 10^{-5}$, $n = 10^{6}$), *Stanford Bunny* ($\alpha = 1.0$, $\tau = 10^{-5}$, $n = 10^{6}$), *Dragon* ($\alpha = 1.0$, $\tau = 10^{-5}$, $n = 10^{6}$), *Suzanne* ($\alpha = 1.0$, $\tau = 5 \times 10^{-6}$, $n = 10^{6}$)

	Vogel et al. [51] (PDE-based approach)		Our method. (variational method)	
	RSE	RIE	RSE	RIE
Noisy Sombrero	0.19530	0.27254	0.05118	0.13239
Noisy Stanford Bunny	0.10973	0.17347	0.03235	0.15279
Noisy Dragon	0.12240	0.19409	0.05395	0.18767
Noisy Suzanne	0.12134	0.16783	0.01256	0.14302

initialisation via the coarse-to-fine scheme does not require to rely on noisy solutions at critical points as the PDE-based approach of Vogel et al. As a consequence, the resulting surface errors of 3–6 % for our variational approach are significantly lower than those of the PDE-based model (11–20 %). This can also be seen from the depth estimates for the Stanford Bunny depicted in Fig. 3.8. Not surprisingly our findings are in full accordance with the observation in [30], in which the robustness of variational methods for perspective SfS has been investigated.

Results on Real-World Images. Finally, in order to evaluate our approach on real-world images, we used two images of faces provided by Prados [41]. According to Prados, these images have been taken with a cheap digital camera in a dark place, where the scene is illuminated by the flash of the camera. The focal length is $\mathtt{f} = 5.8$ mm and the grid size is approximately $h_x = h_y = 0.018$ mm. The test images as well as additional images rendered from a new viewpoint using the computed depth are shown in Fig. 3.9. In both cases the results look quite realistic. One can also see how the depth values at the eyes have been inpainted in the reconstruction, since a

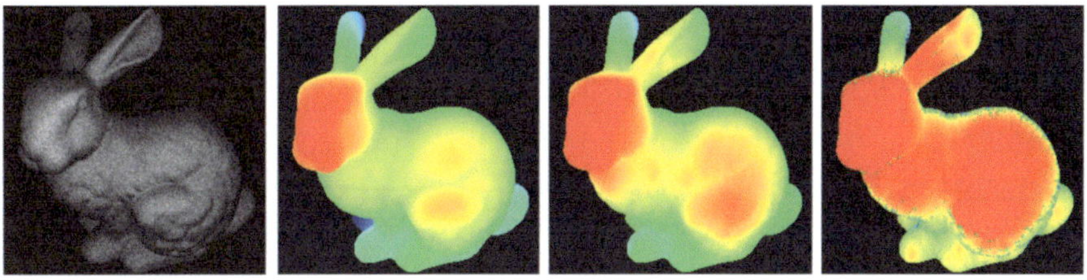

Fig. 3.8 **From left to right:** Noisy version of the *Stanford Bunny* (Gaussian noise with $\sigma = 20$), ground truth depth, computed depth using our variational approach ($\alpha = 1.0$, $\tau = 10^{-5}$, $n = 10^6$), computed depth using the PDE-based approach of Vogel et al. [51] with Lambertian model

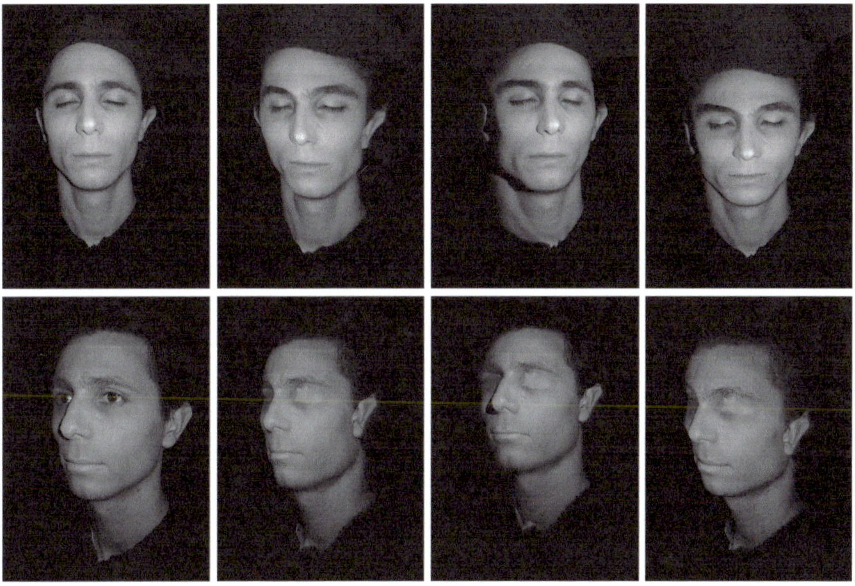

Fig. 3.9 **First row, from left to right:** Face with closed eyes, three images rendered from a new viewpoint using the estimated depth ($\alpha = 7.5 \times 10^{-5}$, $\tau = 5 \times 10^{-3}$, $n = 2 \times 10^5$). **Second row:** Ditto for the second test image ($\alpha = 7.5 \times 10^{-5}$, $\tau = 5 \times 10^{-3}$, $n = 2 \times 10^5$)

manually defined confidence function was used to mask out those regions where the assumption of a Lambertian surface is violated.

3.7 Conclusion

In this paper, we described a novel variational model for perspective shape from shading that not only has many desirable theoretical properties but also yields very convincing reconstruction results for synthetic and real-world input images, even in the presence of noise or other deteriorations in an input image. While the arising optimisation problem has turned out to be challenging, we have proposed an alternating explicit scheme embedded in a coarse-to-fine framework that is robust with respect to the initialisation and that allows reasonable computation times compared to a standard explicit scheme.

Besides the results that are documented via extensive experiments in this chapter, let us point out that we see a main contribution of our work in a different context, as we have layed the fundamental building block for a conceptually correct, working variational framework that can combine perspective shape from shading with other techniques from computer vision such as e.g. stereo vision. We aim to explore the arising possibilities in a future work.

Acknowledgements This work has been partially funded by the Deutsche Forschungsgemeinschaft (DFG) as a joint project (BR 2245/3-1, BR 4372/1-1).

Appendix

Alternative Derivation of the Surface Normal. Instead of computing the derivatives with respect to the 2-D image coordinates x and y, one can also derive the surface normal in an alternative way that is often used in the literature, see e.g. [53]. The idea is to interpret the original surface in Eq. (3.4) as a function of the 3-D coordinates X, Y and $Z(X, Y)$

$$\mathscr{S}\,(X(\mathbf{x},z), Y(\mathbf{x},z), Z(X(\mathbf{x},z), Y(\mathbf{x},z))) = \begin{bmatrix} X(\mathbf{x},z) \\ Y(\mathbf{x},z) \\ Z(X(\mathbf{x},z), Y(\mathbf{x},z)) \end{bmatrix} := \begin{bmatrix} \frac{zx}{\mathtt{f}} \\ \frac{zy}{\mathtt{f}} \\ -z \end{bmatrix}. \tag{3.44}$$

Dropping the dependency of X, Y and $Z(X, Y)$ on $\mathbf{x}$, z and computing the partial derivatives with respect to X and Y via the chain rule

$$\frac{\partial X}{\partial Y} = \frac{\partial X}{\partial x}\frac{\partial x}{\partial Y}, \qquad \frac{\partial Y}{\partial X} = \frac{\partial Y}{\partial y}\frac{\partial y}{\partial X}$$

then gives the tangent vectors to the surface

$$\mathscr{S}_X(\mathbf{x},z) = \begin{bmatrix} 1 \\ \frac{z_x y}{z + z_x x} \\ -\frac{z_x\,\mathtt{f}}{z + z_x x} \end{bmatrix}, \qquad \mathscr{S}_Y(\mathbf{x},z) = \begin{bmatrix} \frac{z_y x}{z + z_y y} \\ 1 \\ -\frac{z_y\,\mathtt{f}}{z + z_y y} \end{bmatrix}. \tag{3.45}$$

After some computations we finally obtain the corresponding normal direction

$$\hat{\mathbf{n}}(\mathbf{x}) = \mathscr{S}_X(\mathbf{x}, z) \times \mathscr{S}_Y(\mathbf{x}, z) = \frac{f^2}{(z + z_x x)(z + z_y y)} \, \mathbf{n}(\mathbf{x}) \,. \tag{3.46}$$

where $\mathbf{n}(\mathbf{x})$ is the normal direction from Eq. (3.8) As expected, both vectors only differ by scale, i.e. they have the same direction. Hence, the corresponding normalised vectors $\mathbf{n}/|\mathbf{n}|$ and $\hat{\mathbf{n}}/|\hat{\mathbf{n}}|$ are identical. While this alternative derivation was not used in our paper, it helps to clarify a common mistake in the literature that will be explained in the following.

Remark Please note that, unlike in the orthographic case, the cross derivatives $\partial X/\partial Y$ and $\partial Y/\partial X$ do not vanish for the perspective model. Hence, using the orthographic derivation of the normal direction from Horn and Brooks [27]

$$\mathbf{n}_{\text{ortho}}(\mathbf{x}) = \frac{\partial}{\partial X} \begin{bmatrix} X \\ Y \\ Z \end{bmatrix} \times \frac{\partial}{\partial Y} \begin{bmatrix} X \\ Y \\ Z \end{bmatrix} = \begin{bmatrix} 1 \\ 0 \\ Z_X \end{bmatrix} \times \begin{bmatrix} 0 \\ 1 \\ Z_Y \end{bmatrix} = \begin{bmatrix} -Z_X \\ -Z_Y \\ 1 \end{bmatrix} \tag{3.47}$$

with zero cross derivatives and simply replacing the remaining partial derivatives Z_X and Z_Y by the corresponding expressions from (3.45) is *not completely correct* for the perspective case. Such an approach has for instance been proposed in [53, 55]. It actually mixes the orthographic and the perspective model and thus typically gives worse results in the case of strong perspective distortions. Moreover, apart from not being completely correct, this strategy also yields significantly more complex models that typically require auxiliary variables to be solved, see again e.g. [53, 55].

References

1. Abdelrahim, A.S.: Three-Dimensional Modeling of the Human Jaw/Teeth Using Optics and Statistics. PhD thesis, Department of Electrical and Computer Engineering, University of Louisville, Louisville (2014)
2. Abdelrahim, A.S., Abdelrahman, M.A., Abdelmunim, H., Farag, A., Miller, M.: Novel image-based 3D reconstruction of the human jaw using shape from shading and feature descriptors. In: Proceedings of the British Machine Vision Conference, pp. 1–11 (2011)
3. Abdelrahim, A.S., Abdelmunim, H., Graham, J., Farag, A.: Novel variational approach for the perspective shape from shading problem using calibrated images. In: Proceedings of the IEEE International Conference on Image Processing, pp. 2563–2566 (2013)
4. Ahmed, A., Farag, A.: A new formulation for shape from shading for non-Lambertian surfaces. In: Proceedings of the IEEE Conference on Computer Vision and Pattern Recognition, pp. 1817–1824 (2006)
5. Barles, G.: Solutions de viscosité des équations de Hamilton-Jacobi. Mathématiques & Applications, vol. 17. Springer (1994)
6. Basha, T., Moses, Y. Kiryati, N.: Multi-view scene flow estimation: a view centered variational approach. Int. J. Comput. Vis. **101**(1), 6–21 (2012)

7. Bors, A.G., Hancock, E.R., Wilson, R.C.: Terrain analysis using radar shape-from-shading. IEEE Trans. Pattern Anal. Mach. Intell. **25**(8), 974–992 (2003)
8. Breuß, M., Cristiani, E., Durou, J.-D., Falcone, M., Vogel, O.: Numerical algorithms for perspective shape from shading. Kybernetika **46**(2), 207–225 (2010)
9. Breuß, M., Cristiani, E., Durou, J.-D., Falcone, M. , Vogel, O.: Perspective shape from shading: ambiguity analysis and numerical approximations. SIAM J. Imaging Sci. **5**(1), 311–342 (2012)
10. Brooks, M.J., Horn, B.K.P.: Shape and source from shading. In: Proceedings of the International Joint Conference in Artificial Intelligence, pp. 932–936 (1985)
11. Brox, T., Bruhn, A., Papenberg, N., Weickert, J.: High accuracy optical flow estimation based on a theory for warping. In: Proceedings of the European Conference on Computer Vision. LNCS, vol. 3024, pp. 25–36 (2004)
12. Camilli, F., Prados, E.: Viscosity solution. In: Ikeuchi, K. (ed.) The Encyclopedia of Computer Vision. Springer, New York (2014)
13. Charbonnier, P., Blanc-Féraud, L., Aubert, G., Barlaud, M.: Deterministic edge-preserving regularization in computed imaging. IEEE Trans. Image Process. **6**(2), 298–311 (1997)
14. Crandall, M.G., Lions, P.-L.: Viscosity solutions of Hamilton-Jacobi equations. Trans. Am. Math. Soc. **277**(1), 1–42 (1983)
15. Courant, R., Hilbert, D.: Methods of Mathematical Physics. Interscience Publishers, Inc., New York (1953)
16. Courteille, F., Crouzil, A., Durou, J.-D., Gurdjos, P.: Towards shape from shading under realistic photographic conditions. In: Proceedings of the IEEE International Conference on Pattern Recognition, pp. 277–280 (2004)
17. Courteille, F., Crouzil, A., Durou, J.-D., Gurdjos, P.: 3D-spline reconstruction using shape from shading: spline from shading. Image Vis. Comput. **26**(4), 466–479 (2008)
18. Demetz, O., Stoll, M., Volz, S., Weickert, J., Bruhn, A.: Learning brightness transfer functions for the joint recovery of illumination changes and optical flow. In: Proceedings of the European Conference on Computer Vision. LNCS, vol. 8689, pp. 455–471 (2014)
19. Diggelen, J.V.: A photometric investigation of the slopes and heights of the ranges of hills in the maria of the moon. Bull. Astron. Inst. Neth. **XI**(423), 283–289 (1951)
20. Durou, J.-D., Falcone, M., Sagona, M.: Numerical methods for shape-from-shading: a new survey with benchmarks. Comput. Vis. Image Underst. **109**(1), 22–43 (2008)
21. Estellers, V., Thiran, J.-P., Gabrani, M.: Surface reconstruction from microscopic images in optical lithography. IEEE Trans. Image Process. **23**(8), 3560–3573 (2014)
22. Frankot, R.T., Chellappa, R.: A method for enforcing integrability in shape from shading algorithms. IEEE Trans. Pattern Anal. Mach. Intell. **10**(4), 439–451 (1988)
23. Hartley, R., Zisserman, A.: Multiple View Geometry in Computer Vision. Cambridge University Press, Cambridge (2004)
24. Horn, B.K.P.: Shape from Shading: A Method for Obtaining the Shape of a Smooth Opaque Object from One View. PhD thesis, Department of Electrical Engineering, MIT, Cambridge (1970)
25. Horn, B.K.P.: Robot Vision. MIT, Cambridge (1986)
26. Horn, B.K.P., Brooks, M.J.: The variational approach to shape from shading. Comput. Vis. Graph. Image Process. **33**, 174–208 (1986)
27. Horn, B.K.P., Brooks, M.J.: Shape from Shading. Artificial Intelligence Series. MIT, Cambridge (1989)
28. http://www.blender.org. Last visited on 05 May 2015
29. Ikeuchi, K., Horn, B.K.P.: Numerical shape from shading and occluding boundaries. Artif. Intell. **17**(1–3), 141–184 (1981)
30. Ju, Y.C., Breuß, M., Bruhn, A.: Variational perspective shape from shading. In: Proceedings of the International Conference on Scale Space and Variational Methods in Computer Vision. LNCS, vol. 9087, pp. 538–550 (2015)
31. Kimmel, R., Siddiqi, K, Kimia, B.B., Bruckstein, A.M.: Shape from shading: level set propagation and viscosity solutions. Int. J. Comput. Vis. **16**, 107–133 (1995)

32. Lysaker, M., Lundervold, A., Tai, X.-C.: Noise removal using fourth-order partial differential equation with applications to medical magnetic resonance images in space and time. IEEE Trans. Image Process. **12**(12), 1057–1590 (2003)
33. Mecca, R., Falcone, M.: Uniqueness and approximation of a photometric shape-from-shading model. SIAM J. Imaging Sci. **6**, 616–659 (2013)
34. Mecca, R., Wetzler, A., Bruckstein, A.M., Kimmel, R.: Near field photometric stereo with point light sources. SIAM J. Imaging Sci. **7**(4), 2732–2770 (2014)
35. Okatani, T., Deguchi, K.: Shape reconstruction from an endoscope image by shape from shading technique for a point light source at the projection center. Comput. Vis. Image Underst. **66**, 119–131 (1997)
36. Oliensis, J.: Shape from shading as a partially well-constrained problem. Comput. Vis. Graph. Image Process: Image Underst. **54**(2), 163–183 (1991)
37. Oren, M., Nayar, S.: Generalization of the Lambertian model and implications for machine vision. Int. J. Comput. Vis. **14**(3), 227–251 (1995)
38. Phong, B.T.: Illumination for computer-generated pictures. Commun. ACM **18**(6), 311–317 (1975)
39. Prados, E., Faugeras, O.: "Perspective shape from shading" and viscosity solutions. In: Proceedings of the IEEE International Conference Computer Vision, pp. 826–831 (2003)
40. Prados, E., Faugeras, O.: Shape from shading: a well-posed problem? In: Proceedings of the IEEE Conference on Computer Vision and Pattern Recognition, pp. 870–877 (2005)
41. Prados, E., Camilli, F., Faugeras, O.: A unifying and rigorous shape from shading method adapted to realistic data and applications. J. Math. Imaging Vis. **25**(3), 307–328 (2006)
42. Rindfleisch, T.: Photometric method for lunar topography. Photogramm. Eng. **32**(2), 262–277 (1966)
43. Robert, L., Deriche, R.: Dense depth map reconstruction: a minimization and regularization approach which preserves discontinuities. In: Proceedings of the European Conference on Computer Vision. LNCS, vol. 1064, pp. 439–451 (1996)
44. Rouy, E., Tourin, A.: A viscosity solution approach to shape-from-shading. SIAM J. Numer. Anal. **29**(3), 867–884 (1992)
45. Sethian, J.: Level Set Methods and Fast Marching Methods. Cambridge University Press, Cambridge (1999)
46. Tankus, A., Sochen, N, Yeshurun, Y.: A new perspective [on] shape-from-shading. In: Proceedings of the IEEE International Conference on Computer Vision, pp. 862–869 (2003)
47. Tankus, A., Sochen, N., Yeshurun, Y.: Shape-from-shading under perspective projection. Int. J. Comput. Vis. **63**(1), 21–43 (2005)
48. The Stanford 3D Scanning Repository, http://graphics.stanford.edu/data/3Dscanrep/. Last visited on 05 May 2015
49. Vogel, O., Bruhn, A., Weickert, J., Didas, S.: Direct shape-from-shading with adaptive higher order regularisation. In: Proceedings of the International Conference on Scale Space and Variational Methods in Computer Vision. LNCS, vol. 4485, pp. 871–882 (2007)
50. Vogel, O., Breuß, M., Weickert, J.: Perspective shape from shading with non-Lambertian reflectance. In: Proceedings of the German Conference on Pattern Recognition. LNCS, vol. 5096, pp. 517–526 (2008)
51. Vogel, O., Breuß, M., Leichtweis, T., Weickert, J.: Fast shape from shading for Phong-type surfaces. In: Proceedings of the International Conference on Scale Space and Variational Methods in Computer Vision. LNCS, vol. 5567, pp. 733–744 (2009)
52. Wang, G.H., Han, J.Q., Zhang, X.M.: Three-dimensional reconstruction of endoscope images by a fast shape from shading method. Meas. Sci. Technol. **20**(12) (2009)
53. Wu, C., Narasimhan, S., Jaramaz, B.: A multi-image shape-from-shading framework for near-lighting perspective endoscopes. Int. J. Comput. Vis. **86**, 211–228 (2010). http://iopscience.iop.org/article/10.1088/0957-0233/20/12/125801/meta;jsessionid=2362840D33B53BD14BC41A3CE06C16D8.c5.iopscience.cld.iop.org
54. Zhang, R., Tsai, P.-S., Cryer, J.E., Shah, M.: Shape from shading: a survey. IEEE Trans. Pattern Anal. Mach. Intell. **21**(8), 690–706 (1999)

55. Zhang, L., Yip, A.M., Tan, C.T.: Shape from shading based on Lax-Friedrichs fast sweeping and regularization techniques with applications to document image restoration. In: Proceedings of the IEEE Conference on Computer Vision and Pattern Recognition, pp. 1–8 (2007)
56. Zhang, L., Yip, A.M., Tan, C.T.: A restoration framework for correcting photometric and geometric distortions in camera-based document images. In: Proceedings of the IEEE International Conference on Computer Vision, pp. 1–8 (2007)

Chapter 4
Amoeba Techniques for Shape and Texture Analysis

Martin Welk

Abstract Morphological amoebas are image-adaptive structuring elements for morphological and other local image filters introduced by Lerallut et al. Their construction is based on combining spatial distance with contrast information into an image-dependent metric. Amoeba filters show interesting parallels to image filtering methods based on partial differential equations (PDEs), which can be confirmed by asymptotic equivalence results. In computing amoebas, graph structures are generated that hold information about local image texture. This chapter reviews and summarises the work of the author and his coauthors on morphological amoebas, particularly their relations to PDE filters and texture analysis. It presents some extensions and points out directions for future investigation on the subject.

4.1 Introduction

Mathematical morphology [38, 45, 46] has developed since the 1960s as a powerful theoretical framework from which versatile instruments for shape analysis in images can be derived, such as for structure-preserving denoising or shape simplification [23]. The fundamental building blocks of classical mathematical morphology are non-linear local image filters like dilation, erosion, and median filters. They rely on aggregating intensities within a neighbourhood of any given pixel by e.g. maximum, minimum, and median operations. The selection of neighbourhoods for processing is classically done by shifting a sliding window of fixed size and shape across the image. In the context of morphology, this sliding window is known as structuring element.

M. Welk (✉)
Biomedical Image Analysis Division, Department of Biomedical Computer Science and Mechatronics, University for Health Sciences, Medical Informatics and Technology, Eduard-Wallnöfer-Zentrum 1, 6060 Hall/Tyrol, Austria
e-mail: martin.welk@umit.at

M. Breuß et al. (eds.), *Perspectives in Shape Analysis*, Mathematics and Visualization, DOI 10.1007/978-3-319-24726-7_4

More recently, concepts for adaptivity have been developed generally in image filtering and also specifically in morphology [5, 51]. One recent concept for adaptive morphology are morphological amoebas introduced by Lerallut et al. [32]. These are space-variant structuring elements constructed from a combination of spatial distance measurement with local contrast measurement via an amoeba metric.

In earlier work by the author of the present chapter and his coauthors, properties of amoeba filters and their relations to image filters based on partial differential equations (PDEs) were investigated [54, 57, 58]. As an application to image segmentation, an amoeba-based active contour method was designed [53, 54, 56]. Recently, a combination of edge-weighted graphs generated in the computation of amoebas with graph indices was used to introduce a new class of texture descriptors [55] which are currently under further investigation. This chapter reviews and summarises the results from these works. Directions of ongoing research on this topic are sketched.

With focus on giving a comprehensive overview of the theory that has been developed in various earlier publications, the (mostly lengthy) proofs of the results are omitted here and referred to the respective original sources. Nevertheless, the main principles underlying the proofs are shortly outlined. Although amoeba filtering of multi-channel images has been addressed to some extent in [57], this aspect of the topic presents itself in a stage too early for a summarised presentation, and is therefore not included in the present chapter.

In the following the structure of the chapter is detailed, highlighting contributions that are novel in this presentation.

Section 4.2 introduces the concept of morphological amoebas as image-adaptive structuring elements in the space-discrete as well as the space-continuous setting. To ease bridging to the graph techniques discussed later in Sect. 4.6, the presentation in the discrete case emphasises the modelling of discrete images by neighbourhood graphs and uses standard terminology from graph theory, thereby following [55]. The presentation of the space-continuous case is similar to that e.g. in [57].

The application of amoebas in image filtering is the topic of Sect. 4.3. Median filters, morphological dilation and erosion are presented together with their relationship to PDE image filters, reproducing herein results from [54, 56–58]. Regarding the association between amoeba metrics on the discrete filtering side and edge-stopping functions occurring in the corresponding PDEs, the current work adds to the previously considered exemplary L^1 and L^2 (Euclidean) amoeba metrics as a third simple case the L^∞ (maximum) amoeba metric and states explicitly the corresponding edge-stopping function. Moreover, the amoeba variants of morphological opening and closing are included in the description for the first time. For dilation, erosion, opening and closing filters, the presentation here emphasises the algebraic background including max-plus/min-plus convolution and conjugacy of structure elements.

Section 4.4 considers the application of amoeba techniques to devise basic algorithms for unsupervised segmentation of grey-value images, namely the *amoeba active contours (AAC)* first introduced in [53] and further investigated in [54, 56]. Results from [56] on the relation between AAC and geodesic active contours are reported.

In image filtering by nonlinear PDEs, one often computes the nonlinearities not from the input images themselves but from Gaussian pre-smoothed versions of these, in order to reduce noise sensitivity of filters and to improve numerical stability. This is also the case with self-snakes and active contour PDEs; note that the self-snakes PDE is even ill-posed without such pre-smoothing. Section 4.5 investigates the effect of pre-smoothing in the self-snakes PDE using perturbation analysis on a synthetic example; furthermore, it discusses how a comparable stabilisation can be achieved in the amoeba median filter framework. The analysis presented in this section relies on previous work in [54, 57] in which oscillatory perturbations aligned with the gradient direction were studied, and extends it by including also perturbations aligned with the level line direction.

Section 4.6 is devoted to a different direction of application of amoeba ideas. Noticing that the computation of discrete amoeba structuring elements is intimately related with graph structures – a weighted neighbourhood graph, weighted and unweighted Dijkstra search trees – in the neighbourhood of each pixel, one can try to extract local texture information from these graphs. Quantitative graph theory [13] offers a variety of graph indices for generating quantitative information from graph structures. The presentation of the construction of texture descriptors from amoebas and graph indices in this section follows [55]. Compared to the large set of descriptors covered in [55], only a few representatives are shown here, complementing their mathematical description by a visualised example. Extending the previous work on texture discrimination in [55], the present chapter also shows a first example of the new texture descriptors in texture segmentation by using the descriptors as components of an input image for multi-channel GAC segmentation.

4.2 Morphological Amoebas

Well-known local image filters such as the mean filter, median filter, morphological dilation or erosion consist of two steps: a sliding-window *selection* step, and the *aggregation* of selected input values by taking e.g. the arithmetic mean, median, maximum or minimum. A strategy to improve the sensitivity of such filters to important image structures is to modify the selection step by using spatially adaptive neighbourhoods instead of a fixed sliding window. The general idea is to give preference in the selection to neighbouring image locations with similar intensities, and thus to reduce the flow of grey-value information across high contrast steps or slopes in the filter process.

First introduced by Lerallut et al. [32, 33] as structuring elements for adaptive morphology, morphological amoebas are a specific type of such spatially adaptive neighbourhoods. Their construction relies on the combination of spatial distance in the image domain with grey-value contrast into a modified metric on the image.

4.2.1 Edge-Weighted Neighbourhood Graph

To define morphological amoebas on discrete images, we start by considering edge-weighted graphs based on the image grid.

Definition 4.1 Let f be a discrete image. Construct an edge-weighted graph $G_w(f) := (V, E, w)$ with vertex set V, edge set E and weights w as follows. The vertex set V is formed by all pixels of f. Two vertices i, j are connected, $\{i,j\} \in E$, if and only if pixels i, j are neighbours under a suitably chosen neighbourhood notion. To define the edge weights $w_{i,j}$ for an edge $\{i,j\} \in E$, consider the corresponding pixel locations $\boldsymbol{p}_i$ and $\boldsymbol{p}_j$ as well as the intensities f_i and f_j, and set $w_{i,j}$ to

$$w_{ij} := \varphi\big(\|\boldsymbol{p}_i - \boldsymbol{p}_j\|_2, \beta\,|f_i - f_j|\big) \tag{4.1}$$

where $\|\boldsymbol{p}_i - \boldsymbol{p}_j\|_2$ denotes Euclidean distance in the image plane, $\beta > 0$ is a contrast scale parameter weighting between spatial and tonal distances, and φ is a norm on $\mathbb{R}^2$ which can be rewritten as

$$\varphi(s,t) = \begin{cases} |t| \cdot \nu(|s/t|)\,, & t > 0\,, \\ |s|\,, & t = 0 \end{cases} \tag{4.2}$$

with a monotonically increasing function $\nu : \mathbb{R}_0^+ \to \mathbb{R}^+$ (by continuity, $\nu(0) = 1$).

The edge-weighted graph $G_w(f)$ is called *neighbourhood graph* of f.

In this definition, neighbourhood can be understood as a 4-neighbourhood, as done in [32], or as an 8-neighbourhood as in [55, 57, 58]. The latter choice gets somewhat closer to a Euclidean measurement of spatial distances in the image plane and is therefore also considered the default in the present work.

As to the norm function ν, the setting $\nu(z) \equiv \nu_1(z) = 1 + z$ corresponds to the L^1 metric also used in [32] that gives

$$w_{ij} = \|\boldsymbol{p}_i - \boldsymbol{p}_j\|_2 + \beta\,|f_i - f_j|\,, \tag{4.3}$$

whereas $\nu(z) \equiv \nu_2(z) = \sqrt{1 + z^2}$ entails a Euclidean (L^2) metric in which the edge weights are obtained by the Pythagorean sum

$$w_{ij} = \sqrt{\|\boldsymbol{p}_i - \boldsymbol{p}_j\|_2^2 + \beta^2 |f_i - f_j|^2} \tag{4.4}$$

A straightforward generalisation is

$$\nu_p(z) = (1 + z^p)^{1/p} \quad \text{for } p \geq 1 , \tag{4.5}$$

which in the limit $p \to +\infty$ also includes $\nu_\infty(z) = \max\{1, z\}$ and the corresponding edge weight

$$w_{ij} = \max\left\{\|\boldsymbol{p}_i - \boldsymbol{p}_j\|_2, \beta\,|f_i - f_j|\right\} . \tag{4.6}$$

4.2.2 Discrete Amoeba Metric

We use the edge-weighted neighbourhood graph to define the discrete amoeba metric on image f.

Definition 4.2 Let a discrete image f be given. Let $G_w(f)$ be its neighbourhood graph with edge weights given by (4.1). Define for two pixels i and j their distance $d(i,j)$ as the minimal total weight (length) among all paths between i and j in $G_w(f)$. Then d is called *(discrete) amoeba metric* on f.

The metric d is called L^p amoeba metric, $1 \leq p < \infty$, if it is derived from (4.5), or L^∞ amoeba metric if it is obtained from $\nu(z) = \max\{1, z\}$. The L^2 amoeba metric is also called Euclidean amoeba metric.

Definition 4.3 In a discrete image f with amoeba metric d, an *amoeba structuring element* (short: *amoeba*) $\mathscr{A}_\varrho(i) \equiv \mathscr{A}_\varrho(f;i)$ with amoeba radius ϱ and reference point at pixel i is a discrete ϱ-ball around pixel i in the amoeba metric, i.e. the set of all vertices within a distance ϱ from i,

$$\mathscr{A}_\varrho(i) := \{j \mid d(i,j) \leq \varrho\} . \tag{4.7}$$

The derivation of amoebas from a metric with a global radius parameter ϱ has an interesting consequence: for two pixels i, j, one has

$$i \in \mathscr{A}_\varrho(j) \quad \Leftrightarrow \quad j \in \mathscr{A}_\varrho(i) , \tag{4.8}$$

which is helpful in the design of some morphological filters.

4.2.3 Computation of Discrete Amoebas

To compute amoebas in a discrete image, one has to search the neighbourhood of each given reference pixel i in order to identify the pixels j with amoeba distance $d(i,j) \leq \varrho$. Given that the edge weights $w_{i,j}$ in $G_w(f)$ are nonnegative, this can be achieved by running Dijkstra's shortest path algorithm [16] on $G_w(f)$ starting at pixel i. As this algorithm enumerates neighbour pixels in order of increasing path weight, it can be stopped as soon as a pixel j with $d(i,j) > \varrho$ is visited.

Moreover, by the construction of the amoeba distance it is clear that the Euclidean distance in the image domain is a lower bound for the amoeba distance between pixels. Therefore the Dijkstra algorithm for the start vertex i can be run on the subgraph of $G_f(w)$ that contains just the pixels from the Euclidean ϱ-neighbourhood of i.

4.2.4 Amoebas on Continuous Domains

Even superficial inspection of results obtained by some amoeba filters indicates that they have striking similarities to image processing methods based on partial differential equations (PDEs). This observation has been substantiated in [56–58] by studying space-continuous versions of amoeba filters; the results proven there allow to interpret amoeba filters as time steps of explicit discretisations for suitable PDEs.

To devise space-continuous versions of amoeba filters, one has to translate first the notion of amoeba metric to the space-continuous setting. Once this is done, the definition of an amoeba as a ϱ-ball around a reference point is straightforward.

The amoeba metric for a space-continuous greyvalue image – a real-valued function f over a connected compact image domain $\Omega \subset \mathbb{R}^n$ – can be stated by assigning to each two given points $\boldsymbol{p}, \boldsymbol{q} \in \Omega$ as their distance the minimum of a path integral between $\boldsymbol{p}$ and $\boldsymbol{q}$. Just like the edge weights in the discrete amoeba construction, the integrand of the path integral is obtained by applying a suitable norm φ to the spatial metric (the Euclidean curve element of the path) and the greyvalue metric (the standard metric on the real domain), such that the amoeba distance reads as

$$\begin{aligned} d(\boldsymbol{p},\boldsymbol{q}) &= \min_{\boldsymbol{c}} \int_0^1 \varphi\big(\|\boldsymbol{c}'(t)\|_2, \beta\,|(f\circ\boldsymbol{c})'(t)|\big)\,\mathrm{d}t \\ &= \min_{\boldsymbol{c}} \int_0^1 \varphi\big(\|\boldsymbol{c}'(t)\|_2, \beta\,|\nabla f^{\mathrm{T}}\boldsymbol{c}'(t)|\big)\,\mathrm{d}t \end{aligned} \tag{4.9}$$

where $\boldsymbol{c}$ runs over all regular curves $\boldsymbol{c} : [0,1] \to \Omega$ with $\boldsymbol{c}(0) = \boldsymbol{p}$, $\boldsymbol{c}(1) = \boldsymbol{q}$, and φ can be chosen as in the discrete case.

Let us associate to the function $f : \mathbb{R}^2 \supset \Omega \to \mathbb{R}$ its (vertically rescaled) *graph*, the manifold $\Gamma := \{(x, y, \beta f(x, y)) \mid (x, y) \in \Omega\} \subset \mathbb{R}^3$. Then we see that the amoeba distance $d(\pm p, \pm q)$ between two points $\pm p$, $\pm q$ in the image domain Ω can be interpreted as a distance $\hat{d}(\boldsymbol{p}', \boldsymbol{q}')$ on Γ. The points $\boldsymbol{p}', \boldsymbol{q}' \in \Gamma$ herein are

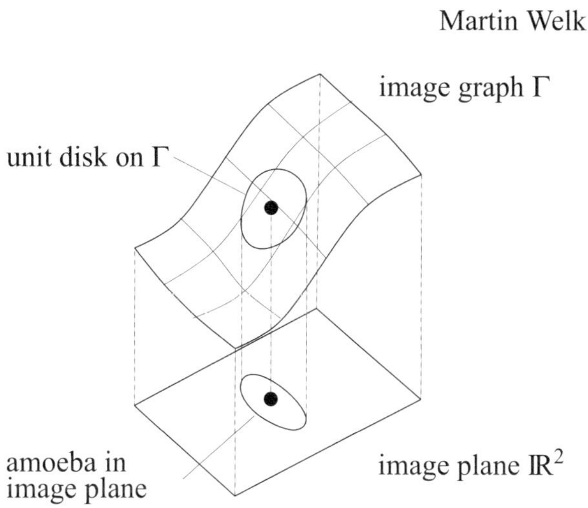

Fig. 4.1 Amoeba as projection of the unit disk on the image graph to the image plane

given by $\boldsymbol{p}' := (\boldsymbol{p}, f(\boldsymbol{p}))$, $\boldsymbol{q}' := (\boldsymbol{q}, f(\boldsymbol{q}))$. To define the metric $\hat{d}$ on Γ, consider a metric $\tilde{d}$ in the surrounding space $\mathbb{R}^3$ that combines Euclidean metric in the x-y-plane with the standard metric in z-direction via the function φ from (4.2) that appeared already in the original construction of the amoeba metric. Using $\tilde{d}$ in $\mathbb{R}^3$, the metric $\hat{d}$ is obtained as its induced metric on the submanifold $\Gamma \subset \mathbb{R}^3$. Figure 4.1 illustrates that the amoeba structuring element is then the projection of a unit disk on Γ back to the image plane.

Figure 4.2 shows typical amoeba shapes in smooth image regions for the three exemplary amoeba metrics exposed in Sect. 4.2.2.

4.3 Amoeba-Based Image Filters

To obtain applicable image filters, the amoeba procedure described above is used as a selection step and needs to be complemented by some aggregation step. We consider here standard choices of aggregation operators from classical local filters; introducing also modifications into this part of the filtering procedure is left as a possible direction for future research. Moreover, keeping close to the original context in which amoebas were developed, we focus on morphological operators. Here, morphological operators are characterised by their invariance under arbitrary monotonically increasing transformations of the intensities, see e.g. [37], which means that also median and quantiles belong to this class.

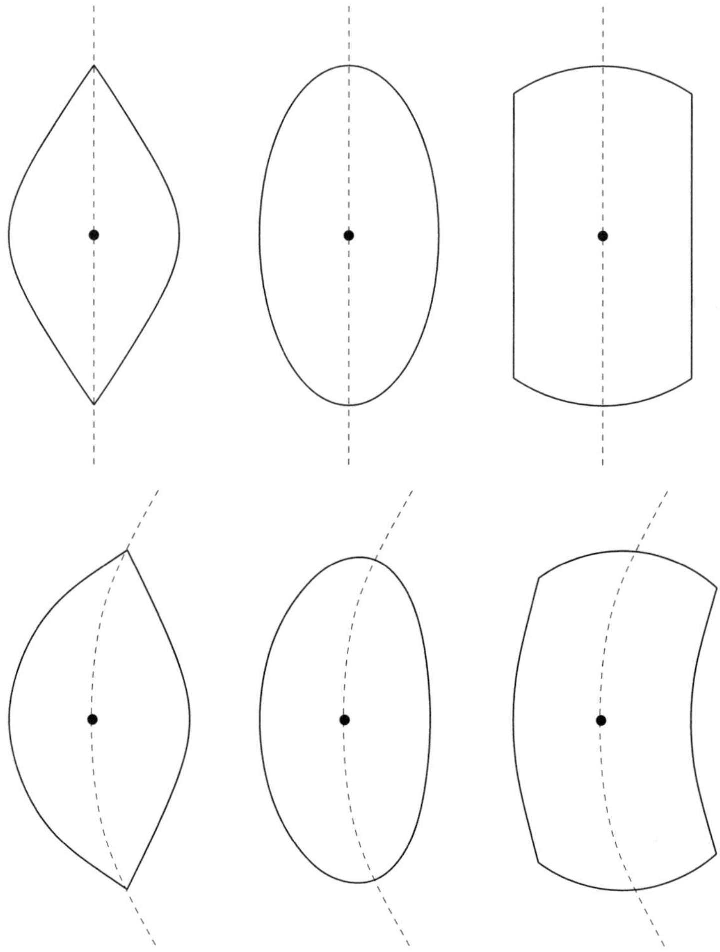

Fig. 4.2 Typical shapes of amoebas in the continuous domain for different amoeba metrics. *Top row* shows amoebas on an image with equidistant straight level lines, *bottom row* shows amoebas on *curved level lines* (schematic). *Left column* shows L^1 amoeba metric, middle column Euclidean amoeba metric, and the *right column* shows the maximum (L^∞) amoeba metric. Each amoeba is shown with its reference point (*bold*) and *level line* through the reference point (*dashed*)

4.3.1 Median

A median filter aggregates the intensity values of the selected pixels by taking their median. In the non-adaptive, sliding-window setting this filter can be traced back to Tukey [50], and since then it has gained high popularity as a simple denoising filter that preserves discontinuities (edges) and its robustness with respect to some types of noise. Median filtering can be iterated. Unlike average filters, the median filter on a discrete image possesses non-trivial steady states, so-called root signals [17], that depend on the filter window. The smaller the filter window, the faster the iterated median filtering process locks in at a root signal.

Despite the nice preservation of edges, the non-adaptive median filter involves a displacement of curved edges in inward direction and rounding of corners that is often undesired. Amoeba median filtering greatly reduces this effect. Figure 4.3 demonstrates this by an example.

4.3.1.1 PDE Approximation

As noticed already in 1997 by Guichard and Morel [20], the overall robust denoising effect and the characteristic corner-rounding behaviour of standard median filtering resemble the properties of the well-known (mean) curvature motion PDE [1]. Further analysis confirmed this observation by proving an asymptotic relationship between the two filters, as set forth in the following proposition.

Proposition 4.1 (Guichard and Morel [20]) *For a smooth function* $u : \Omega \to \mathbb{R}$, *one iteration of median filtering with a* ϱ*-ball as structuring element approximates for* $\varrho \to 0$ *a time step of size* $\tau = \varrho^2/6$ *of the curvature motion PDE [1]*

$$u_t = |\nabla u| \operatorname{div}\left(\frac{\nabla u}{|\nabla u|}\right) . \tag{4.10}$$

This seminal result motivates the investigation of relations between amoeba and PDE filters whose results are reviewed in the further course of the present paper.

Just like amoeba median filtering differs from standard median filtering by an adaptation procedure that suppresses smoothing across edges, the curvature motion equation (4.10) has a counterpart in which also the flow across edges is suppressed. This so-called self-snakes filter [44] allows curvature-based image smoothing and simplification, preserves and even enhances edges, while at the same time avoiding

Fig. 4.3 Non-adaptive and amoeba median filtering. (**a**) Original image. (**b**) Filtered by 5 iterations of standard median filtering with a discrete disk of radius 2 as structuring element. (**c**) Filtered by 5 iterations of amoeba median filtering with Euclidean amoeba metric, $\beta = 0.2$, $\varrho = 7$

to shift them, as curvature motion does. It turns out that indeed amoeba median filtering is connected to self-snakes by a similar asymptotic relationship as that of Proposition 4.1, as follows.

Theorem 4.1 ([57, 58]) *For a smooth function $u : \Omega \to \mathbb{R}$, one iteration of amoeba median filtering with amoeba radius ϱ approximates for $\varrho \to 0$ a time step of size $\tau = \varrho^2/6$ of the self-snakes PDE [44]*

$$u_t = |\nabla u| \operatorname{div} \left(g\big(|\nabla u|\big) \frac{\nabla u}{|\nabla u|} \right) \tag{4.11}$$

where $g : \mathbb{R}_0^+ \to \mathbb{R}_0^+$ is a decreasing edge-stopping function that depends on the amoeba metric being used.

Proofs for Theorem 4.1 have been given in [57, 58]. While these proofs are not reproduced in detail here, it is of interest to describe the two different strategies that are used in these proofs. These approaches form also the basis for the further amoeba–PDE asymptotics results presented in Sect. 4.3.2.

4.3.1.2 Proof Strategies

The crucial observation for all median filter–PDE equivalence results since Guichard and Morel's proof of Proposition 4.1 in [20] is that the median of a smooth function u within a given compact structuring element $\mathscr{A}$ is the function value whose corresponding level line divides the structuring element into two parts of equal area. Herein it is assumed that each value of u within the structuring element is associated with a unique level line segment inside $\mathscr{A}$, which is satisfied for sufficiently small fixed or amoeba structuring elements whose reference point $\boldsymbol{x}_0$ is not an extremum of u, and therefore acceptable when studying the limit $\varrho \to 0$.

The amount by which a single median filtering step changes the function value at the reference point $\boldsymbol{x}_0$ of the structuring element then corresponds, up to multiplication with $|\nabla u|$, to the distance between the area-bisecting level line and the level line through $\boldsymbol{x}_0$, see the illustration in Fig. 4.4a. The two approaches discussed in the following differ in the way how they measure the area of the structuring elements and parts thereof.

Proof Strategy I

The first strategy has been followed in [58] to prove Theorem 4.1 for the entire class of amoeba metrics discussed in Sect. 4.2 above, see also the more detailed version in [57, Section 4.1.1]. It is close to the approach from [20] in that it develops the smooth function u around the reference point $\boldsymbol{x}_0$ into a Taylor expansion up to second order. The Taylor expansion is then used to approximate, for an amoeba

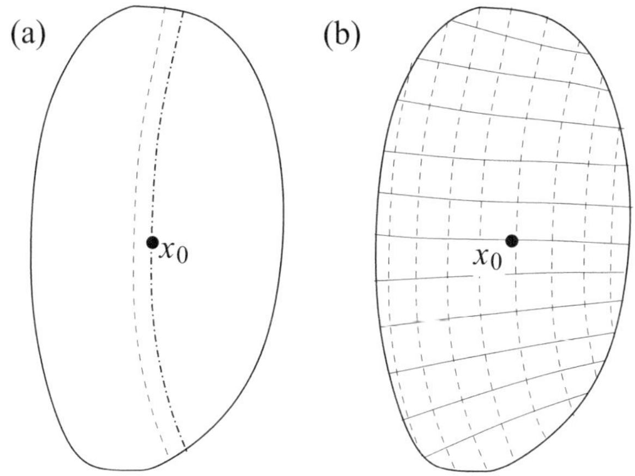

Fig. 4.4 (**a**) Amoeba with reference point $\boldsymbol{x}_0$, level line through $\boldsymbol{x}_0$ (*dot-dashed*) and bisecting level line (*dashed*), schematic. (**b**) Amoeba with curvilinear coordinate system formed by level lines (*dashed*) and gradient flow lines (*solid*)

$\mathscr{A} = \mathscr{A}_\varrho(\boldsymbol{x}_0)$, three items: first, the range of function values occurring within $\mathscr{A}$, i.e. the minimum $\min\limits_{\mathscr{A}} u$ and maximum $\max_{\mathscr{A}} u$, second, the length $L(z)$ of the level line segment for each $z \in [\min_{\mathscr{A}} u, \max_{\mathscr{A}} u]$, and third, the density $\delta(z)$ of level lines around each z, which equals the steepness of the slope of u near the level line of z.

Integrating the lengths of level lines over function values, weighted with their reciprocal densities, yields the area of $\mathscr{A}$, i.e.

$$\text{Area}(\mathscr{A}) = \int\limits_{\min_{\mathscr{A}} u}^{\max_{\mathscr{A}} u} \frac{L(z)}{\delta(z)}\, \mathrm{d}z\,. \tag{4.12}$$

As this integral effectively runs over level lines, splitting the integration interval exactly corresponds to cutting $\mathscr{A}$ at some level line. The calculation of the desired median of u within $\mathscr{A}$ is then achieved by determining a suitable splitting point in the integration interval so that the integrals on both sub-intervals become equal.

Summarising, this strategy describes the amoeba shape in terms of a curvilinear coordinate system aligned with the gradient and level line directions at $\boldsymbol{x}_0$, in which the level lines take the role of coordinate lines, compare Fig. 4.4b.

Proof Strategy II

The second strategy abandons the consideration of the individual level lines within $\mathscr{A}$; the only level line that is explicitly studied is the one through $\boldsymbol{x}_0$ itself. Instead of the distorted Cartesian coordinate system one uses polar coordinates to describe the shape of the amoeba. This approach has first been used in [54] in the context of amoeba active contours (see Sect. 4.4.1), and again in [57, Section 4.1.2], both times restricted to the Euclidean amoeba metric. It has been extended to cover the full generality of amoeba metrics under consideration in [56], again for amoeba active contours.

Writing the outline of $\mathscr{A}$ as a function $a(\alpha)$ of the polar angle $\alpha \in [0, 2\pi]$, the amoeba's area is stated by the standard integral for areas enclosed by function graphs in polar coordinates as

$$\mathrm{Area}(\mathscr{A}) = \frac{1}{2}\int_0^{2\pi} a(\alpha)^2 \,\mathrm{d}\alpha \,. \tag{4.13}$$

Unlike for (4.12), splitting this integral yields areas of sectors instead of segments; however, if the level line through $\boldsymbol{x}_0$ happens to be a straight line, splitting up the integral (4.13) at the pair of opposite angles corresponding to the level line direction yields the areas of two segments into which $\mathscr{A}$ is cut by that level line, compare Fig. 4.5a.

Provided that $\mathscr{A}$ is symmetric (w.r.t. point reflection at the reference point), the two segments are of equal area, making in this case the median equal to $u(\boldsymbol{x}_0)$. Deviations from this situation that make the median differ from $u(\boldsymbol{x}_0)$ can be separated into two contributions: first, the asymmetry of the amoeba; second, the curvature of the level lines. Cross-effects of the two contributions influence only higher order terms that can be neglected in the asymptotic analysis; thus the two sources can be studied independently. In approximating the area difference Δ_1 caused by the asymmetric amoeba shape, one can assume that the level lines are straight, see Fig. 4.5b, while the level line curvature effect Δ_2 can be studied under the assumption that $\mathscr{A}$ has symmetric shape, see Fig. 4.5c.

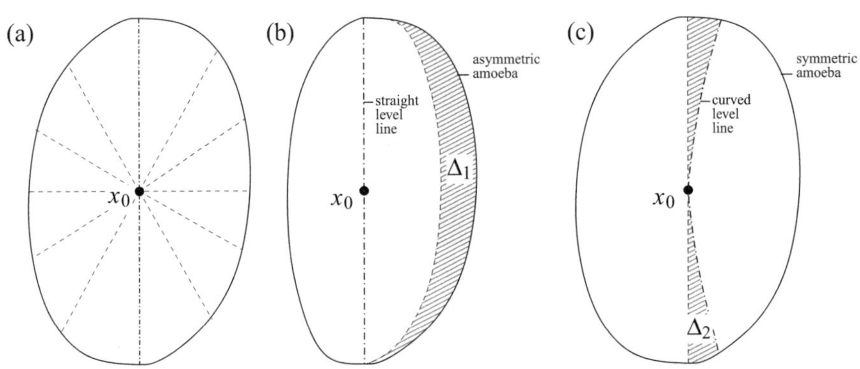

Fig. 4.5 (**a**) Amoeba with straight level line (*dot-dashed*) through its reference point $\boldsymbol{x}_0$ and further radial lines (*dashed*) of a polar coordinate system centred at $\boldsymbol{x}_0$. (**b**) Area difference Δ_1 in an asymmetric amoeba with straight level lines. The hashed region is enclosed between the right arc of the amoeba contour and the point-mirrored copy of its left arc. (**c**) Area difference Δ_2 in a symmetric amoeba with curved level lines. (**b**), (**c**) from [54]

Finally, the combined effect $\Delta_1 + \Delta_2$ must be compensated by a parallel shift of the level line through $\boldsymbol{x}_0$, compare again Fig. 4.4a. From the shift the median, and thus the right-hand side of the PDE approximated by the amoeba filter, can be derived.

4.3.1.3 Amoeba Metrics and Edge-Stopping Functions

It remains to specify the relation between amoeba metric and edge-stopping function mentioned in Theorem 4.1. In [57, 58], the following representation of g in terms of the function ν defining the amoeba metric has been proven.

$$g(z) = \frac{3}{\beta^2 s^2 \nu^3(1/(\beta z))} \int_0^1 \xi^2 \sqrt{\nu^{-2}\left(\frac{1}{\xi}\,\nu\left(\frac{1}{\beta z}\right)\right) - \frac{1}{\beta^2 z^2}}\,\mathrm{d}\xi\;, \tag{4.14}$$

where $\nu^{-2}(z)$ is short for $(\nu^{-1}(z))^2$, i.e. the square of the inverse of ν, and $\nu^3(z)$ for the cube $(\nu(z))^3$.

In the case of the Euclidean amoeba metric, $\nu(z) = \sqrt{1+z^2}$, the expression (4.14) simplifies to

$$g(z) \equiv g_2(z) = \frac{1}{1+\beta^2 z^2}\;, \tag{4.15}$$

which is, up to the substitution $\lambda = 1/\beta$, the Perona-Malik diffusivity [39] that is also one of the common choices for g in the self-snakes equation.

When using the L^1 amoeba metric, $\nu(z) = 1 + z$, the integral in (4.14) can be numerically evaluated, and one obtains an edge-stopping function $g(s) \equiv g_1(s)$ that differs from (4.15) in that it decreases away from $g(0) = 1$ already with nonvanishing negative slope, thus reacting more sensitive to even small image contrasts.

Finally, for the L^∞ amoeba metric, $\nu(z) = \max\{1, z\}$, it is again possible to state g in closed form,

$$g(z) \equiv g_\infty(z) = \begin{cases} 1\,, & \beta z \le 1\,,\\ 1 - \left(1 - \dfrac{1}{\beta^2 z^2}\right)^{3/2}\,, & \beta z > 1 \end{cases} \tag{4.16}$$

which shows that g_∞ is completely insensitive to image contrasts up to $z = 1/\beta$ and then starts decreasing with a kink. All three edge-stopping functions are depicted in Fig. 4.6.

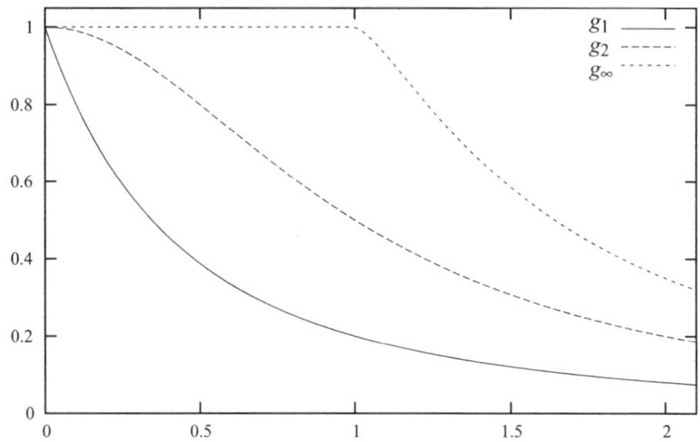

Fig. 4.6 Edge-stopping functions g_1, g_2 and g_∞ associated to L^1, Euclidean and L^∞ amoeba metrics, respectively. Throughout these metrics, the contrast scale β has been set to 1

Fig. 4.7 Grey-scale image (256 × 256 pixels) used to demonstrate non-adaptive and amoeba-based morphological filters

4.3.2 Dilation and Erosion

The two most fundamental operations of mathematical morphology, dilation and erosion, use as aggregation step the maximum and minimum of intensities, respectively. This can naturally be done also in combination with an amoeba-based pixel selection step (Fig. 4.7).

We point out that the standard dilation of an image u with fixed structuring element S can be written as

$$(u \oplus S)(i) = \max_{j \in i+S} u(j) = \max_{j \in \Omega} \big(u(j) + \omega_S^-(i-j)\big) , \tag{4.17}$$

where ω_S^- denotes the function

$$\omega_S^-(k) = \begin{cases} 0 , & -k \in S , \\ -\infty , & \text{else.} \end{cases} \tag{4.18}$$

The last term in (4.17) allows an interesting interpretation in terms of the max-plus algebra [3, 42], an algebraic structure on $\mathbb{R} \cup \{+\infty, -\infty\}$ in which the maximum operation takes the role of addition in the usual algebra of real numbers, while addition takes the role of multiplication. It is evident that (4.17) is nothing else but a convolution of u and ω_S^- in the max-plus algebra, see [36].

In writing erosion in an analogous way, we follow a convention frequently used in the literature by using instead of the structuring element S the conjugate structuring element S^*, which comes down geometrically to a point reflection on the origin, $S^* = -S$. The advantage of this convention is that subsequent definitions like those for opening and closing become simpler [24], compare Sect. 4.3.3.
Defining then $\omega_{S^*}^+$ as zero on S, but $+\infty$ outside, erosion is stated as

$$(u \ominus S)(i) = \min_{j \in j+S^*} u(j) = \min_{j \in \Omega}\big(u(j) + \omega_{S^*}^+(i-j)\big) = \min_{j \in \Omega}\big(u(j) + \omega_S^+(j-i)\big)\;, \tag{4.19}$$

which can be interpreted again as a convolution of u and $\omega_{S^*}^+$ in the min-plus algebra [36].

Abandoning the fixed window and using a family $\mathscr{S} := \{i \mapsto S(i) \mid i \in \Omega\}$ of structuring elements $S(i)$ located at pixel i, one can write amoeba dilation as

$$(u \oplus \mathscr{S})(i) = \max_{j \in \Omega}\big(u(j) + \omega_{\mathscr{S}}^-(i,j)\big)\;, \tag{4.20}$$

$$\omega_{\mathscr{S}}^-(i,j) = \begin{cases} 0\;, & j \in S(i)\;, \\ -\infty\;, & \text{else.} \end{cases} \tag{4.21}$$

Just as the last term in (4.17) is a max-plus convolution, the right-hand side (4.20) is the max-plus analogon of a (discretised) integral operator. Herein, $\omega_{\mathscr{S}}^-(i,j)$ acts as the max-plus counterpart of just the same type of integral kernel that appears as point-spread function in space-variant image deconvolution models.

Similarly, amoeba erosion becomes a min-plus integral operator with a min-plus kernel $\omega_{\mathscr{S}^*}^+(i,j) \equiv \omega_{\mathscr{S}}^+(j,i)$. Generally, conjugate structuring elements in the space-variant case are given by

$$S^*(i) = \{j \in \Omega \mid i \in S(j)\}\;. \tag{4.22}$$

Interestingly, if $\mathscr{S}$ is made up by amoebas $S(i) \equiv \mathscr{A}_\varrho(i)$, there is no difference whether the conjugate structuring elements $\mathscr{S}^*$ or standard structuring elements $\mathscr{S}$ are used in erosion: property (4.8) of the amoebas entails $\omega_{\mathscr{S}}^\pm(j,i) = \omega_{\mathscr{S}}^\pm(i,j)$ for all $i, j \in \Omega$, or equivalently

$$\mathscr{A}_\varrho^*(i) \equiv \mathscr{A}_\varrho(i)\;. \tag{4.23}$$

We will denote this property as *self-conjugacy* of amoebas.

Figure 4.7 shows the results of non-adaptive and amoeba dilation and erosion of an example image depicted in Fig. 4.8. Non-adaptive as well as amoeba-based

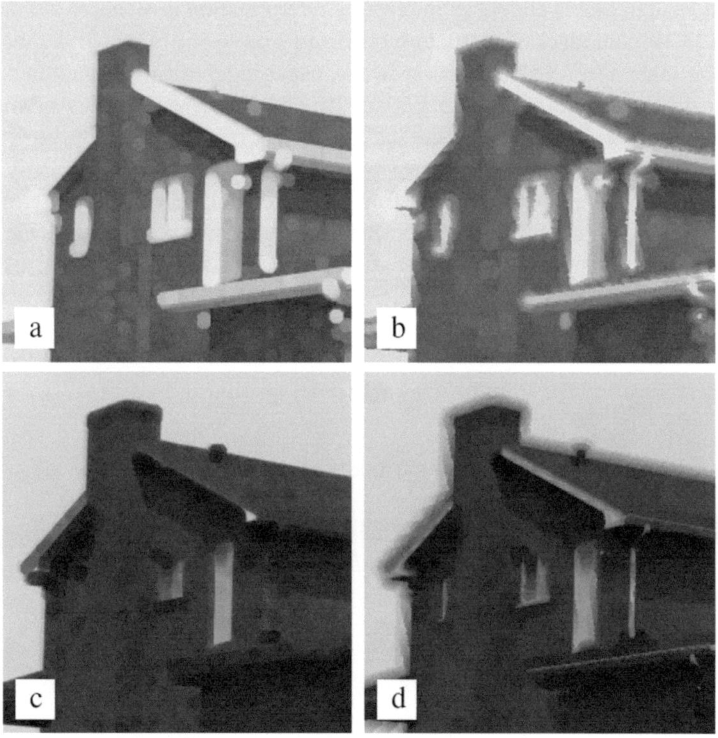

Fig. 4.8 Morphological dilation and erosion, non-adaptive and amoeba-based, of the test image from Fig. 4.7. (**a**) Non-adaptive morphological dilation with disk of radius $\varrho = 5$ as structuring element. (**b**) Amoeba dilation with Euclidean amoeba metric, $\beta = 0.1$, $\varrho = 10$. (**c**) Non-adaptive morphological erosion with structuring element as in (**a**). (**d**) Amoeba erosion with amoeba parameters as in (**b**)

dilation extend bright image details, but it can be seen that the spreading of bright image parts is stopped at strong edges; similarly for the propagation of dark details by erosion.

4.3.2.1 PDE Approximation

It is a well-known fact that Hamilton-Jacobi PDEs

$$u_t = \pm|\nabla u| \tag{4.24}$$

describe dilation ("+" case) and erosion ("−") of continuous-scale images or level-set functions u in the sense that evolution of an initial image $u(t = 0) = f$ by (4.24) up to time $T = \varrho$ yields the dilation or erosion of f with a Euclidean ball-shaped structuring element of radius ϱ. It can therefore be expected that amoeba dilation and erosion, too, should be related to hyperbolic PDEs resembling (4.24). The following result from [57] confirms this intuition.

Theorem 4.2 ([57]) *For a smooth function $u : \Omega \to \mathbb{R}$, one step of amoeba dilation or amoeba erosion with amoeba radius ϱ and Euclidean amoeba metric approximates for $\varrho \to 0$ a time step of size $\tau = \varrho$ of an explicit time discretisation of the Hamilton-Jacobi-type PDE*

$$u_t = \pm \frac{|\nabla u|}{\sqrt{1 + \beta^2 |\nabla u|^2}} , \tag{4.25}$$

where the "+" sign applies for dilation, and "−" for erosion.

The proof of this result can be found in [57]; it is based on Proof Strategy I from Sect. 4.3.1.2.

Note that unlike in Theorem 4.1 the time step size here depends linearly, not quadratically, on ϱ. In [57] the theorem is formulated slightly more general to cover also amoeba α-quantile filters that interpolate in a natural way between median filtering ($\alpha = 1/2$), dilation ($\alpha = 1$) and erosion ($\alpha = 0$). As a result of the different order of decay of τ for $\varrho \to 0$, it comes as no surprise that for $\alpha \neq 1/2$ always the advection behaviour of the Hamilton-Jacobi equation (4.25) dominates over the parabolic equation (4.10), thus turning quantile filters into "slower" approximations to the same PDE.

4.3.3 Opening and Closing

In mathematical morphology, the opening of an image f with (fixed) structuring element S is defined as the concatenation of an erosion followed by a dilation with S. In case S is not point-symmetric it is essential that, as mentioned in Sect. 4.3.2, the conjugate structuring element S^* is used in the erosion step. Opening therefore reads as

$$(f \circ S)(i) = \big((f \ominus S) \oplus S\big)(i) = \max_{j\in\Omega} \min_{k\in\Omega} \big(f(k) + \omega_{S^*}^{+}(j-k) + \omega_S^{-}(i-j)\big) . \tag{4.26}$$

Analogously, closing is defined as dilation followed by erosion,

$$(f \bullet S)(i) = \big((f \oplus S) \ominus S\big)(i) = \min_{j\in\Omega} \max_{k\in\Omega} \big(f(k) + \omega_S^{-}(j-k) + \omega_{S^*}^{+}(i-j)\big) . \tag{4.27}$$

Again, it is straightforward to turn these operations into adaptive variants by using amoeba structuring elements. Amoeba opening and closing of image f with amoebas of radius ϱ are given as

$$f \circ \mathscr{S}_\varrho(f) = \big(f \ominus \mathscr{S}_\varrho(f)\big) \oplus \mathscr{S}_\varrho(f) , \tag{4.28}$$

$$f \bullet \mathscr{S}_\varrho(f) = \bigl(f \oplus \mathscr{S}_\varrho(f)\bigr) \ominus \mathscr{S}_\varrho(f) \tag{4.29}$$

where $\mathscr{S}_\varrho(f) = \{i \mapsto \mathscr{A}_\varrho(f;i) \mid i \in \Omega\}$.

It is worth noticing that the difficulty about using the conjugate set of structuring elements for erosion disappears here due to the self-conjugacy (4.23) of the amoeba structuring element set.

As it is essential to use the same set of structuring elements in the dilation and erosion step, both steps must be carried out with the amoebas obtained from the original image. The underlying principle is that in the second step (dilation for opening or erosion for closing) each pixel should influence exactly those pixels which have influenced it in the first step before. As a consequence, e.g. amoeba opening is not exactly the same as amoeba erosion followed by amoeba dilation – this sequence would be understood by default as recalculating amoebas after the erosion step, i.e.

$$\bigl(f \ominus \mathscr{S}_\varrho(f)\bigr) \oplus \mathscr{S}_\varrho\bigl(f \ominus \mathscr{S}_\varrho(f)\bigr)\,, \tag{4.30}$$

which is inappropriate for an opening operation.

In Fig. 4.9 exemplary results of non-adaptive and amoeba-based closing and opening of the test image from Fig. 4.7 are shown. Like its non-adaptive counterparts, amoeba-based closing and opening remove small-scale dark or bright details, respectively. However, the amoeba versions do this in a less aggressive way. Extended narrow structures that are often removed partially by the non-adaptive filters are more often preserved as a whole, with reduced contrast, or removed completely by the amoeba filters, see e.g. the roof front edge descending to the right from the chimney, and the acute roof corner separating it from the sky.

4.3.3.1 Opening and Closing Scale Spaces and PDEs

The association between median, dilation and erosion filters and PDEs is inherently related to the scale space structures of these filters, compare [25]. All of these filters form an additive semi-group in the sense that iterative application of the same filter yields an increasing filter effect that naturally adds up over iteration numbers. In the case of dilation and erosion iteration numbers are also in linear relation with increasing structuring element size, as dilating an initial image n times with (non-adaptive) structuring element radius ϱ is equivalent to dilating once with radius $n\varrho$. Such an additive semi-group structure perfectly matches initial value problems for PDEs in which, too, evolution times add up.

While opening and closing, too, have a scale space structure, their semi-group operation is not additive but *supremal* as it is based on taking the maximum of parameters. For example, repeating the same opening or closing operation on a given image just reproduces the result of the first application of the filter (i.e., opening and closing operators are idempotent); and concatenating two openings or

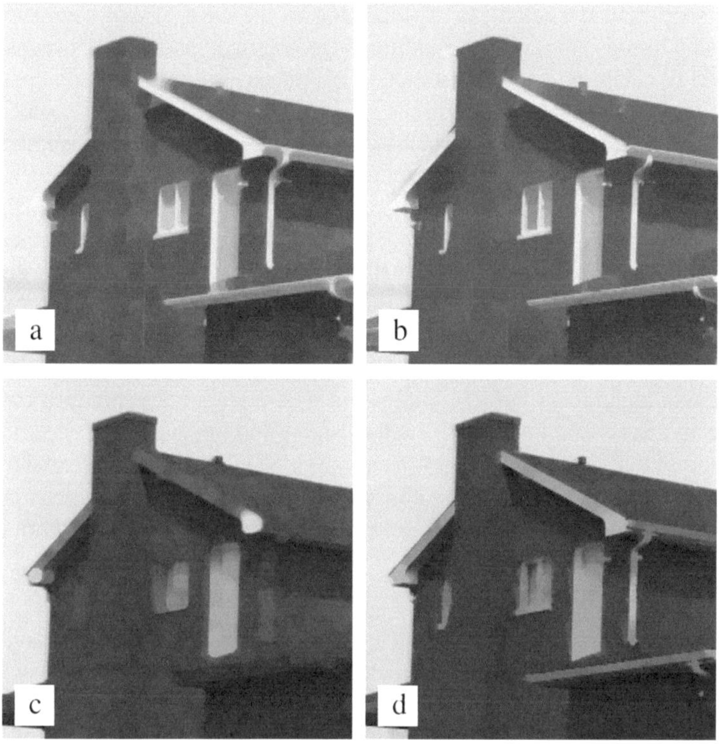

Fig. 4.9 Non-adaptive and amoeba-based morphological closing and opening applied to the test image from Fig. 4.8. (**a**) Non-adaptive closing with disk-shaped structuring element of radius $\varrho = 5$. (**b**) Amoeba closing with Euclidean amoeba metric, $\beta = 0.1$, $\varrho = 10$. (**c**) Non-adaptive opening with structuring element as in (**a**). (**d**) Amoeba opening with amoeba parameters as in (**b**)

two closings with structuring element radii ϱ_1, ϱ_2 gives an opening or closing with radius $\max\{\varrho_1, \varrho_2\}$.

For this reason, also amoeba opening and closing are not associated with PDE evolutions in the same way as the previous filters. Possible relations to PDE-based filters may be considered in future research.

4.4 Grey-Scale Segmentation

Following established terminology, image segmentation denotes the task to decompose a given image into regions that are in the one or other way homogeneous in themselves but different from each other, with the intention that these regions are meaningful in that they are associated to objects being depicted. Intensity-based segmentation uses intensity as the main criterion of homogeneity within and dissimilarity between segments. Specialising to the case of two segments (foreground and background) with the additional geometric hypothesis that segments are separated by sharp and smooth contours, contour-based segmentation approaches based on curve or level set evolutions lend themselves as tools for segmentation, with active contours as an important representative. In this section we show how amoeba algorithms can be made useful in this context.

Despite the fact that experiments on magnetic resonance data are used to illustrate the concepts in this section, this is not meant to make a claim that neither active contour nor the related active region methods (which are not discussed further here) in their pure form could serve as a state-of-the-art segmentation method for medical images. In fact, competitive results in medical image segmentation are nowadays achieved by complex frameworks that often include active contours and/or active regions as a component but in combination with additional techniques that allow to bring in anatomical knowledge such as shape and appearance models [11]. An early representative of these frameworks is [34], which has been followed by many more since then. Like classical geodesic active contours, the amoeba active contours presented in the following could be integrated into this type of framework but this has not been done so far.

4.4.1 Amoeba Active Contours

The standard procedure of an *active contour,* or *snake,* method starts with some initial contour which may be obtained automatically from some previous knowledge or heuristics regarding the position of a sought structure, or from human operator input. Representing this contour either by a sampled curve or by a level-set function, it is then evolved up to a given evolution time or up to a steady state by the action of some parabolic PDE, which is often derived as a gradient descent of a segmentation energy in the image plane. An important representative are geodesic active contours (GAC) [9, 30]. Their segmentation energy is essentially a curve length measure of the contour in a modified metric on the image plane that favours placing the contour in high-contrast locations. The PDE for GAC in level-set representation reads

$$u_t = |\nabla u| \operatorname{div}\left(g\big(|\nabla f|\big) \frac{\nabla u}{|\nabla u|} \right) . \tag{4.31}$$

Herein, u is the evolving level-set function in the plane that represents the actual evolving contour as one of its level sets (by default, the zero-level set), and f is the invariable image being segmented. The similarity of (4.31) to self-snakes (4.11) (which were actually inspired from active contours, thus the name) together with the link between amoeba median filtering and self snakes established by Theorem 4.1 suggest that an amoeba median approach could be used to evolve the level set function u instead of equation (4.31).

Introduced in [53], the resulting *amoeba active contour (AAC)* algorithm proceeds as follows:

1. Compute amoeba structuring elements based on the input image f.
2. Initialise the evolving level-set function u to represent the initial contour.
3. Evolve the image u by median filtering with the amoebas from Step 1 as structuring elements.

Results from this algorithm look qualitatively fairly similar to those from GAC, as will also be demonstrated later in this section.

4.4.2 PDE Approximation

In order to study the relation between AAC and GAC, it makes sense again to consider a space-continuous model and to investigate the PDE approximated by AAC in the case of vanishing amoeba radius. The following result was proven in [56]. Note that in this theorem the contrast scale parameter β is fixed to 1 for simplicity, which, however, is no restriction of the result because in the active contour setting in question, the case $\beta \neq 1$ is easily mapped to $\beta = 1$ by just scaling the intensities of image f by β.

Theorem 4.3 ([56]) *Let a smooth level-set function u be filtered by amoeba median filtering, where the amoebas are generated from a smooth image f. Assume that the amoeba metric is given by (4.9), (4.2) with $\beta = 1$. One step of this filter for u then approximates for $\varrho \to 0$ a time step of size $\tau = \varrho^2/6$ of an explicit time discretisation of the PDE*

$$u_t = G\,u_{\xi\xi} - |\nabla u| \cdot \bigl(H_1 f_{\chi\chi} + 2H_2 f_{\chi\eta} + H_3 f_{\eta\eta}\bigr) \tag{4.32}$$

with the coefficients given by

$$G \equiv G\bigl(|\nabla f|, \alpha\bigr) = \frac{1}{\nu\bigl(|\nabla f| \sin\alpha\bigr)^2}\,, \tag{4.33}$$

$$\begin{pmatrix} H_1 & H_2 \\ H_2 & H_3 \end{pmatrix} \equiv \begin{pmatrix} H_1\bigl(|\nabla f|, \alpha\bigr) & H_2\bigl(|\nabla f|, \alpha\bigr) \\ H_2\bigl(|\nabla f|, \alpha\bigr) & H_3\bigl(|\nabla f|, \alpha\bigr) \end{pmatrix}$$

$$= \frac{3}{2}\,\nu\bigl(|\nabla f| \sin\alpha\bigr) \int\limits_{\alpha-\pi/2}^{\alpha+\pi/2} \frac{\nu'\bigl(|\nabla f| \sin\vartheta\bigr)}{\nu\bigl(|\nabla f| \sin\vartheta\bigr)^4} \begin{pmatrix} \cos^2\vartheta & \sin\vartheta\cos\vartheta \\ \sin\vartheta\cos\vartheta & \cos^2\vartheta \end{pmatrix} \mathrm{d}\vartheta\,. \tag{4.34}$$

Here, $\eta = \nabla u/|\nabla u|$ and $\boldsymbol{\xi} \perp \eta$ are unit vectors in gradient and level line direction, respectively, for u, whereas $\boldsymbol{\chi} = \nabla f/|\nabla f|$ and $\boldsymbol{\zeta} \perp \boldsymbol{\chi}$ are the corresponding unit vectors for f, and $\alpha = \angle(\boldsymbol{\eta}, \boldsymbol{\chi})$ is the angle between both gradient directions.

The proof of this result is found for the case of the Euclidean amoeba metric in [54], and for general amoeba metric in [56]. It relies on Proof Strategy II from Sect. 4.3.1.2.

An attempt to analyse AAC using Proof Strategy I had been made in [53], where, however, only a special case was successfully treated: The theorem proven in [53] states that AAC approximates the GAC equation (4.31) if image f and level set function u are rotationally symmetric about the same centre.

In fact, the rotational symmetry hypothesis can be weakened; what is needed for (4.32), (4.33) and (4.34) to reduce to the exact GAC equation is actually, whenever $\alpha = 0$ (thus, $\eta = \chi$, $\xi = \zeta$), $u_{\xi\eta} = f_{\xi\eta} = 0$ and $u_{\xi\xi}/|\nabla u| = f_{\xi\xi}/|\nabla f|$ hold, (4.32), (4.33) and (4.34) boil down to the GAC equation (4.31).

At first glance, this is still a very artificial choice; however, looking at the geometrical implications of this setting, one sees that it means that the level lines of u are aligned to those of f, have the same curvature, and the image contrast in both f and u does not change along these level lines. Thereby the hypothesis of this special case is well approximated in the near-convergence stage of a segmentation process when the object–background contrast is more or less uniform along the contour.

As a consequence, the coincidence of AAC and GAC in this case justifies that both approaches can expected to yield very similar types of segmentations. The convergence behaviour towards these segmentations may differ more; a closer comparison of both PDEs in [54, 56] based on typical geometric configurations indicates that the amoeba active contour PDE drives contours toward image contours in a more pronounced way.

Figure 4.10 presents an example that confirms the overall similarity between amoeba and geodesic active contours but also the tendency of AAC to adapt more precise to very small-scaled edge details. Frame (a) shows the original image with

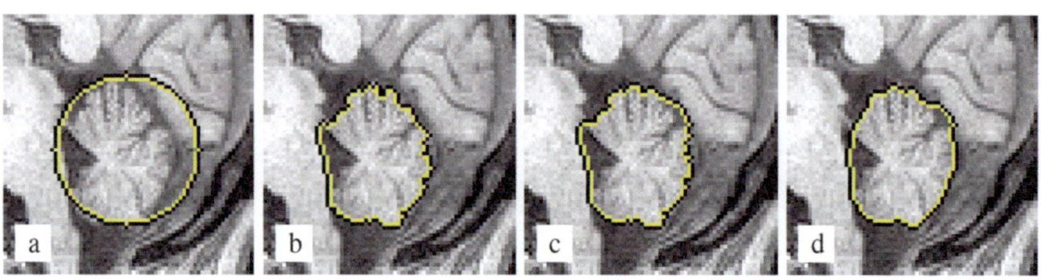

Fig. 4.10 Amoeba and geodesic active contour segmentation. (**a**) Detail (70 × 70 pixels) from an MR slice of a human brain with initial contour enclosing the cerebellum. (**b**) Amoeba active contours with Euclidean amoeba metric, $\beta = 0.1$, $\varrho = 12$, 10 iterations. (**c**) Amoeba active contours with L^1 amoeba metric, $\beta = 0.1$, $\varrho = 12$, 60 iterations. (**d**) Geodesic active contours with Perona-Malik edge-stopping function, $\lambda = 10$, 960 iterations of explicit scheme with time step size $\tau = 0.25$ (From [53, 56])

an initial contour roughly enclosing the cerebellum. Frames (b) and (c) demonstrate segmentation by AAC with Euclidean and L^1 amoeba metrics, respectively, while Frame (d) shows a GAC result for comparison.

4.4.3 *Force Terms*

Geodesic active contours in their basic form (4.31) suffer from some limitations. First of all, when initialised with a contour enclosing a large area with one or several small objects inside, the active contour process spends plenty of evolution time to slowly move the contour inwards until it hits an object boundary, due to the initially small curvature of the contour. Secondly, for pronounced concave object geometries, the process tends to lock in at undesired local minima that detect well some convex contour parts but short-cut concave parts via straight line segments. Similar problems can occur when segmenting multiple objects within one initial contour, see the examples in [31]. Thirdly, as the basic curvature motion process involves only inward movement of contours, it is generally not possible with (4.31) to segment objects from initial contours inside the object, which is sometimes desirable in applications. Due to their similarity to GAC, amoeba active contours share these problems.

A common remedy for these problems in the literature on active contour segmentation is the introduction of a *force term*. Its typical form is $\pm\gamma|\nabla u|$, i.e. essentially the right-hand side of a Hamilton-Jacobi PDE for dilation or erosion, compare (4.24). An erosion force accelerates the inward motion of the contour; it allows to get past homogeneous areas faster, and helps the contour to find concave object boundaries and to separate multiple objects. By a dilation force it is possible to push the contour evolution in outward direction, which makes it possible to use initial contours inside objects.

In both cases, however, the force strength needs careful adjustment because dilation or erosion may also push the contour evolution across object boundaries, thereby preventing their detection.

In [10] where this modification was proposed first (by the name of "balloon force"), γ was chosen as constant, but the possibility to steer it contrast-dependent, was mentioned. This has been done in [9, 31, 35] by modulating the force term in a geodesic active contour model with the same edge-stopping function g, such that the entire force term reads as $\pm\gamma\, g(|\nabla f|)\, |\nabla u|$ with constant γ.

The relation between amoeba quantile filters and Hamilton-Jacobi PDEs mentioned in Sect. 4.3.2 indicates how to achieve a similar modification in the amoeba active contour algorithm: the median filter step should be biased, basically by replacing the median with some quantile. The most obvious way to do this is to use the α-quantile with a fixed $\alpha \neq 1/2$. Within a discrete amoeba containing p pixels, this means to choose the value ranked αp in the ordered sequence of intensities. However, taking into account that the amoeba size p (or the amoeba area in the continuous setting) varies even for fixed ϱ with local image contrast, it is not less natural to think of α as varying with the amoeba size. If one chooses $\alpha - 1/2$ inversely proportional to the amoeba size, this comes down to modify the median with a fixed rank offset b, such that in an amoeba of p pixels one would choose the intensity value with rank $p/2 + b$. These two variants of the AAC algorithm have been proposed in [53]. In [56] a third variant ("quadratic bias") was introduced which chooses from the rank order the element with index $p/2 + rp^2$ with fixed r. For these three scenarios, further analysis was provided in [56], based on the Euclidean amoeba metric. We summarise the results here.

Fixed Offset Bias

Choosing the entry at position $p/2+b$ from the rank order approximates a force term $+\gamma_b\,|\nabla u|\,\nu(|\nabla f|\,\sin\alpha)$ with $\gamma_b \sim b$. Note that in the symmetric case in which the PDE approximated by AAC coincides with the GAC equation this becomes exactly the "balloon force" term with constant dilation/erosion weight from [10].

Quantile Bias

Choosing the element with index $p/2 + qp$ from the rank order within each amoeba approximates a force term $+\gamma_q\,|\nabla u|\,\sqrt{(1 + |\nabla f|^2 \sin^2\alpha)/(1 + |\nabla f|^2)}$ with $\gamma_q \sim q$. In the rotationally symmetric case this term lies between the constant weight of [10] and the g-weight from [31].

Quadratic Bias

Choosing the entry at index $p/2 + rp^2$ from the rank order of intensities yields an approximated force term $+\gamma_r\,|\nabla u|\,\nu(|\nabla f|)/\nu(|\nabla f|)^2$. In the rotationally symmetric case this corresponds to the g-weight from [31].

To illustrate amoeba active contours with bias, Fig. 4.11 presents an example (shortened from [56]). Frame (a) is a test image with initial contour inside a mostly homogeneous object (the corpus callosum). Figure 4.11b, c then show contours computed by amoeba active contours with fixed offset bias for two different evolution times, one intermediate, one displaying the final segmentation. For comparison, a segmentation with geodesic active contours is shown in (d).

We remark that in the AAC examples, a few pixels within the corpus callosum region are excluded from the segment, see the small isolated contour loops there. This is not a numerical artifact but a result from the precise adaption of amoebas to image structures even up to the resolution limit (pixel precision) of the image – the pixels not included in the segment are noise pixels with intensities significantly deviating from the neighbourhood, which are simply not included in any amoeba of outside pixels. Modifications like presmoothing input images can be applied to

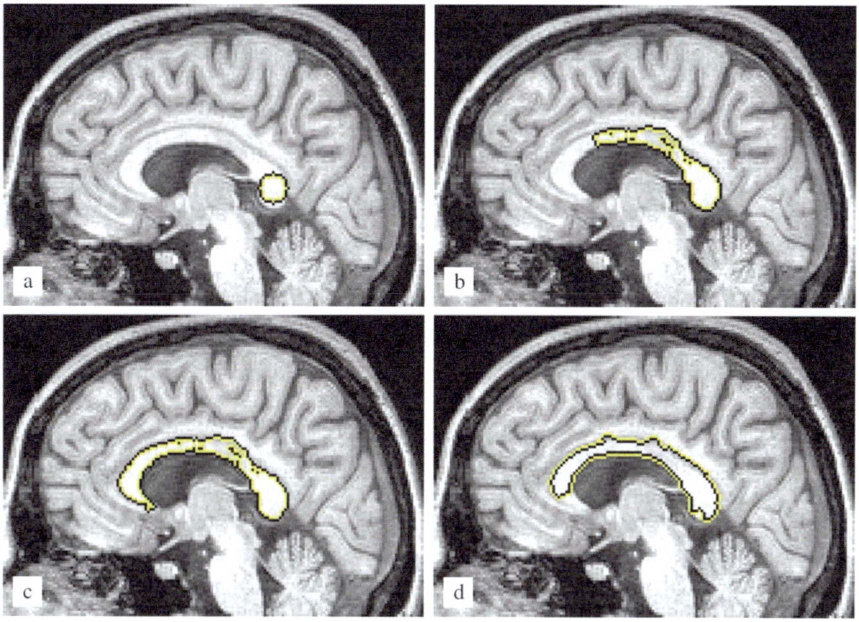

Fig. 4.11 Segmentation with initialisation inside the sought object by amoeba and geodesic active contours with dilation force. (**a**) Detail (164×114 pixels) from an MR slice of human brain with initial contour placed inside the corpus callosum. (**b**) Amoeba active contour evolution with Euclidean amoeba metric, $\beta = 2$, $\varrho = 20$, fixed offset bias $b = 10$, and 20 iterations. (**c**) Same as in (**b**) but 35 iterations. (**d**) Geodesic active contours with Perona-Malik edge-stopping function, $\lambda = 0.5$, dilation force $\gamma = -0.16$ (multiplied with the edge-stopping function) and erosion force $\gamma_c = 5 \times 10^{-4}$ (independent of the edge-stopping function), explicit scheme with time step size $\tau = 0.25$, 18,960,000 iterations (From [56])

avoid this. On the contrary, the absence of such difficulties in the GAC example is a beneficial effect of the otherwise often undesirable numerical blurring effect of the finite-difference scheme.

4.5 Pre-smoothing in Self-Snakes and Amoeba Filters

The approximation result of Theorem 4.1 associates iterated amoeba median filtering with the self-snakes equation (4.11). Unlike (mean) curvature motion (4.10), self-snakes possess edge-enhancing properties. Rewriting (4.11) by the product rule, one can state the self-snakes process as

$$u_t = g\big(|\nabla u|\big)\,|\nabla u|\,\mathrm{div}\left(\frac{\nabla u}{|\nabla u|}\right) + \langle \nabla g, \nabla u\rangle \tag{4.35}$$

in which the first summand is just a curvature motion process modulated by g, whereas the second, advective, term is responsible for the edge-enhancing behaviour. Unfortunately, this term has a shock-filter property which makes not only its numerical treatment difficult – in finite difference schemes usually an upwind discretisation will be required to approximate it – but even entails ill-posedness of the PDE itself that is reflected in a noticeable staircasing behaviour. Indeed, as demonstrated by an experiment in [58], the result of a numerical computation of a self-snakes evolution differs significantly if the underlying grid resolution is changed.

A common remedy to this ill-posed behaviour is to use pre-smoothing in the argument of the edge-stopping function, i.e. to replace $g(|\nabla u|)$ in (4.11) or (4.35) by $g(|\nabla u_\sigma|)$ where u_σ is the result of convolving u with a Gaussian of standard deviation σ. Thereby, the ill-posedness of self-snakes is removed, and a stable filtering achieved, at the cost of the additional smoothing-scale parameter σ.

In this section, we deal with the question whether this staircasing phenomenon has also an analogue in the amoeba median filtering context, and what is an appropriate counterpart for the pre-smoothing modification on the amoeba side. This is done by quantitative analysis of a synthetic example, the first part of which has been published before in [54, 57].

4.5.1 Pre-smoothing in Amoeba Median Filtering, and Amoeba Radius

First of all, notice that a straightforward translation of the pre-smoothing procedure to the amoeba median filtering context is to use u_σ in place of u when computing the structuring elements in an amoeba median filtering step. This is actually an instance of the generalised amoeba median filtering procedure of the amoeba active contour setting, Sects. 4.4.1 and 4.4.2, such that the PDE approximation result from Theorem 4.3 can be applied to see that it would approximate a PDE which is not identical to the standard self-snakes with pre-smoothing, but closely related to it.

At second glance, however, it can be questioned whether the introduction of the smoothing-scale parameter σ into the amoeba median filter is necessary. Unlike finite-difference schemes for self-snakes, amoeba filtering by construction already involves a very similar smoothing-scale parameter, namely, the amoeba radius ϱ. One can conjecture that the positive ϱ necessarily used in any amoeba computation could already provide a pre-smoothing effect similar to the Gaussian convolution in the PDE setting. This conjecture will be investigated in the following.

4.5.2 *Perturbation Analysis of Test Cases*

The starting point for constructing the test cases is a simple slope function that would be stationary under both self-snakes and amoeba median filter evolutions, see Fig. 4.12a. From this slope, described by the function $u_0 : \mathbb{R}^2 \to \mathbb{R}$, $u_0(x, y) = x$, test cases are derived by adding small single-frequency oscillations such as $\varepsilon \cos\langle \boldsymbol{k}, \boldsymbol{x} \rangle$ with frequency vectors $\boldsymbol{k}$.

Given the nonlinear nature of the filters under investigation, there is no superposition property for the effects of different perturbations of u_0. Nevertheless, interactions between u_0 and the perturbations are always of higher order $\mathcal{O}(\varepsilon^2)$, such that the analysis of the first-order effects of perturbations still gives a useful intuition about the behaviour of the filters.

4.5.2.1 Test Case 1: Gradient-Aligned Oscillation

For the first test case, see [54, 57], the perturbation frequency is aligned with the gradient direction, $\boldsymbol{k} = (k, 0)$, yielding the input signal schematically depicted in Fig. 4.12b,

$$u(x, y) = x + \varepsilon \cos(kx) , \quad \varepsilon \ll 1 . \tag{4.36}$$

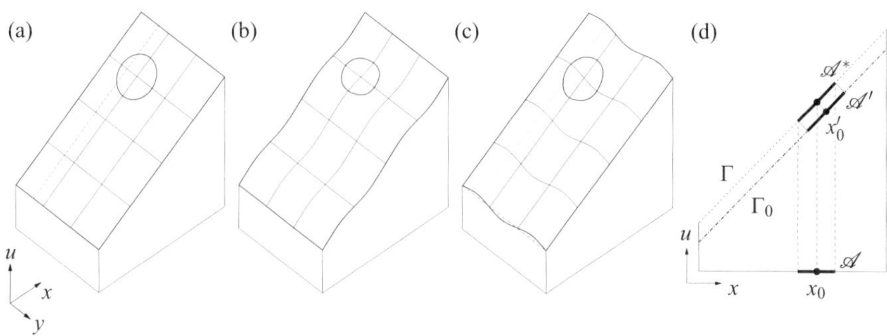

Fig. 4.12 Schematic representation of example functions used in the perturbation analysis, Sect. 4.5.2. (**a**) Graph Γ_0 of unperturbed function $u_0 = x$, with a Euclidean ϱ-disk whose projection to the x-y plane yields an amoeba. (**b**) Graph Γ of a function u of type (4.36) including a gradient-aligned perturbation. (**c**) Graph Γ of a function u of type (4.39) including a level-line-aligned perturbation. (**d**) Cut in x direction through the graph Γ from (**c**) and the unperturbed graph Γ_0 from (**a**). The sketch includes further the amoeba $\mathcal{A}$ around (x_0, y_0), the corresponding Euclidean disk $\mathcal{A}^*$ on Γ and the projection $\mathcal{A}'$ of $\mathcal{A}^*$ to Γ_0 which is centred at (x_0', y_0)

Self-Snakes Analysis

To determine the response of the self-snakes evolution (4.35) to the perturbed signal (4.36), notice first that level lines of (4.36) are straight and parallel, such that one has $\operatorname{div}(\nabla u/|\nabla u|) \equiv 0$ and $\langle \nabla g, \nabla u\rangle = g_x u_x$. Further, one has $u_x = 1 - \varepsilon\, k \sin(kx)$ and $g_x = \varepsilon\, k^2 \cos(kx)/2 + \mathcal{O}(\varepsilon^2)$, finally turning (4.35) into

$$u_t = g_x u_x = \frac{1}{2} k^2 \varepsilon \cos(kx) + \mathcal{O}(\varepsilon^2) \; . \tag{4.37}$$

From this it can be read off that a frequency response factor $k^2/2$ occurs that grows indefinitely for high frequencies. Since the nonlinearity of (4.35) instantaneously spreads out the single perturbation frequency k to higher harmonics, arbitrarily high amplification appears already within short evolution time, and the regularity of the evolving function is lost. This explains the stair-casing behaviour of self-snakes without pre-smoothing.

Using pre-smoothed u_σ in the edge-stopping function argument, one has instead $\partial_x u_\sigma = x + \varepsilon \exp(-k^2\sigma^2/2) \cos(kx)$, $g_x = k^2 \varepsilon \exp(-k^2\sigma^2/2) \cos(kx)/2$ and therefore

$$u_t = \frac{1}{2} k^2 \varepsilon \exp\left(-\frac{k^2\sigma^2}{2}\right) \cos(kx) + \mathcal{O}(\varepsilon^2) \; , \tag{4.38}$$

with the frequency response factor $k^2 \exp(-k^2\sigma^2/2)/2$ that is globally bounded with its maximum at $k = \sqrt{2}/\sigma$. Therefore, pre-smoothing ensures that the regularity of the evolving function is maintained.

Amoeba Filter Analysis

To analyse the effect of amoeba median filtering (with Euclidean amoeba metric) on the function (4.36), consider an amoeba of amoeba radius ϱ around (x_0, y_0), and assume that the contrast scale is chosen as $\beta = 1$. The median of u within that amoeba can be expressed via an integral formula, see [54, 57], which can be numerically evaluated to be approximately equal to $u(x_0, y_0) + \delta(k) \cdot \varepsilon \cos(kx_0)$ with a frequency response factor $\delta(k)$. In other words, one amoeba median filter step amplifies the perturbation $u - u_0$ of (4.36) versus $u_0(x, y) = x$ by the amplification factor $\lambda(k) := 1 + \delta(k)$.

Figure 4.13 shows results of numerical approximation of one amoeba median filtering step with $\beta = 1$, $\varrho = 1$, on test images of type (4.36) with two different frequencies k. The numerical computation was carried out on a discrete grid with mesh size $h = 0.0025$. For best approximation to the space-continuous case, amoeba distances between pixels were computed by numerical integration instead of the Dijkstra search on the pixel graph. Denoting the filtered image by v, numerical amplification factors can be computed as $\langle v - u_0, u - u_0\rangle / \langle u - u_0, u - u_0\rangle$ (with the

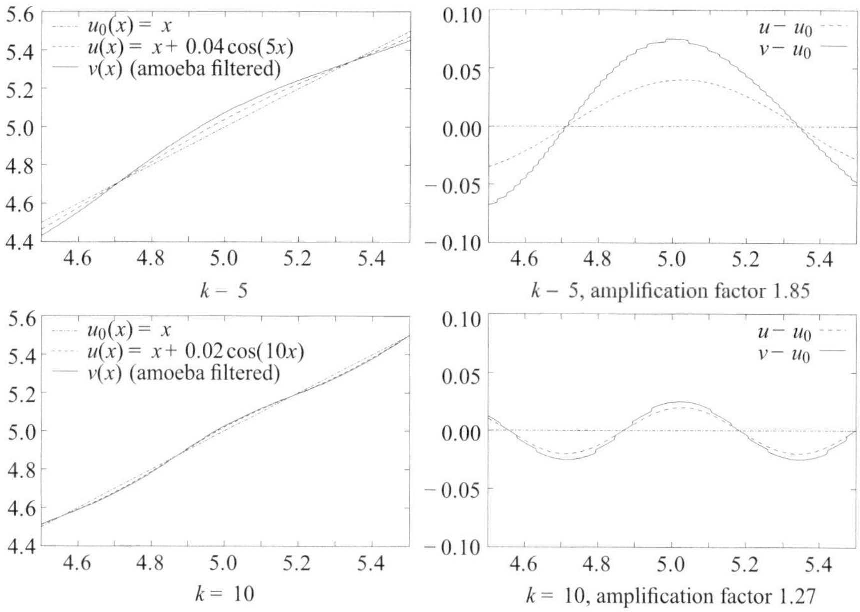

Fig. 4.13 Numerical computation results for the amplification of a gradient-aligned perturbation of a linear slope function by one amoeba median filtering step. *Top row* shows $k = 5$, *bottom row* $k = 10$. Graphs in *left column* show unperturbed function u_0, perturbed input function u, and filter result v; graphs in *right column* show perturbations $u - u_0$ and $v - u_0$. Horizontal axes represent x, vertical axes represent function values. Computations were carried out on a grid with mesh size 0.0025

usual scalar product of functions on a suitable bounded interval); these are in good accordance with the theoretical result.

Figure 4.14 shows the amplification function $\lambda(k)$ for $\varrho = 1$ together with its counterpart $\lambda_s(k) := 1 + 1/6 \cdot k^2 \exp(-k^2\sigma^2/2)/2$ for one time step of self-snakes with pre-smoothing, with the time step size $\varrho^2/6 = 1/6$ matching the amoeba radius according to Theorem 4.1. The figure also includes numerical amplification factors for amoeba median filtering with the same parameters for frequencies $k = 1, 2, \ldots, 30$. The parameter $\sigma = 0.268$ in the self-snakes case has been chosen for a good match to the first wave of $\delta(k)$. With this parameter, the amplification behaviour for frequencies up to approx. 10 is very similar for the pre-smoothed self-snakes equation and amoeba median filtering. However, for higher frequencies the amplification factor of pre-smoothed self-snakes rapidly approaches one (no amplification) whereas it oscillates around $3/2$ for the amoeba filter.

As a result, oscillations with sufficiently high frequency are just almost not amplified in the pre-smoothed self-snakes evolution. With amoeba median filtering, they are amplified by the globally bounded factor $\lambda(k)$ in each iteration step. Whatever ε was in the initial image u from (4.36), after a finite number of iterations the oscillations grow to a level for which the hypothesis $\varepsilon \ll 1$ of our analysis is no

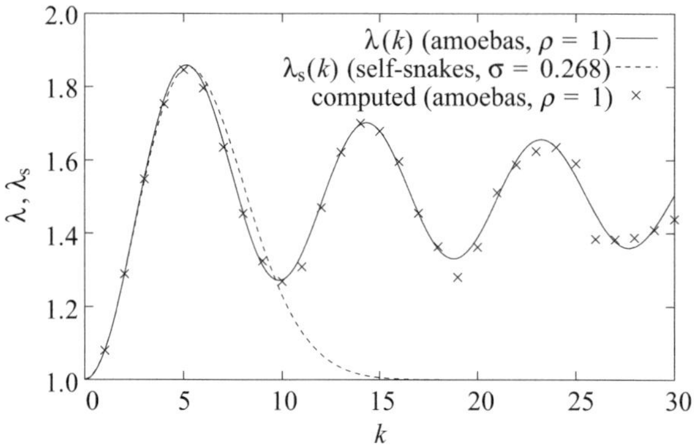

Fig. 4.14 Amplification of a gradient-aligned perturbation of a linear slope function by one amoeba median filtering step (theoretical and numerical values) and a corresponding time step of an explicit scheme for self-snakes with pre-smoothing (Adapted and extended from [54])

longer valid. Even in the space-continuous setting under consideration, oscillations cannot actually grow infinitely because the median operation obeys the maximum–minimum principle.

In practice, amoeba filters are computed in a space-discrete setting such that the effective range of spatial frequencies (parametrised by the angular frequency k of oscillations) is limited by the sampling theorem. For fixed amoeba radius $\varrho = 1$ as in Fig. 4.14, the relevant range of frequencies is determined by the mesh size of the pixel grid. If this mesh size is not below approx. $\pi/10$, the higher lobes of the amplification function $\lambda(k)$ that make up the difference to self-snakes with pre-smoothing do not take effect at all. Translating this to a grid with mesh size 1, as common in image processing, this means that for amoeba radius ϱ up to approx. $10/\pi \approx 3$ the frequency response of amoeba median filtering does almost not differ from that of self-snakes with pre-smoothing.

4.5.2.2 Test Case 2: Level-Line-Aligned Oscillation

To complement the perturbation analysis of gradient-aligned oscillations, a second test case is considered in which the perturbation frequency is aligned with the level line direction, $\boldsymbol{k} = (0, k)$. The resulting input signal, compare the schematic representation in Fig. 4.12c, reads

$$u(x, y) = x + \varepsilon \cos(ky) , \quad \varepsilon \ll 1 . \tag{4.39}$$

This test case was not presented in [54, 57]. Given that self-snakes act smoothing along level line direction, it can be expected that this kind of perturbation is dampened by their evolution. This will be confirmed by the analysis, and the corresponding behaviour of the amoeba median filter will be stated.

Self-Snakes Analysis

Unlike for the first test case, gradient directions of u now vary across the image range, combining constant $u_x = 1$ with $u_y = -k\varepsilon\sin(ky)$. Accordingly, the edge-stopping function takes the values

$$g(x,y) = \frac{1}{2+k^2\varepsilon^2\sin^2(ky)} = \frac{1}{2}\left(1-\frac{k^2\varepsilon^2}{2}\sin^2(ky)\right) + \mathcal{O}(\varepsilon^3) \tag{4.40}$$

and thereby $g_x(x,y) = \mathcal{O}(\varepsilon^3)$, $g_y(x,y) = -k^3\varepsilon^2\sin(ky)\cos(ky)/2 + \mathcal{O}(\varepsilon^3)$.

This leads further to

$$|\nabla u| = 1 + \frac{k^2\varepsilon^2}{2}\sin^2(ky) + \mathcal{O}(\varepsilon^4)\ , \tag{4.41}$$

$$\begin{aligned}\operatorname{div}\left(\frac{\nabla u}{|\nabla u|}\right) &= \partial_x\left(1-\frac{k^2\varepsilon^2}{2}\sin^2(ky)\right) + \partial_y\bigl(-k\varepsilon\sin(ky)\bigr) + \mathcal{O}(\varepsilon^3)\\ &= -k^2\varepsilon\cos(ky) + \mathcal{O}(\varepsilon^2)\ ,\end{aligned} \tag{4.42}$$

$$\langle\nabla g, \nabla u\rangle = \mathcal{O}(\varepsilon^3)\ , \tag{4.43}$$

thus after inserting into (4.35)

$$u_t = -\frac{1}{2}k^2\varepsilon\cos(ky) + \mathcal{O}(\varepsilon^2) \tag{4.44}$$

which confirms by the negative sign of the frequency response factor $-k^2/2$ that the perturbation is smoothed out by the self-snakes process.

Pre-smoothing here leads to

$$g(x,y) = \frac{1}{2}\left(1-\frac{k^2\varepsilon^2}{2}\exp(-k^2\sigma^2)\sin^2(ky)\right) + \mathcal{O}(\varepsilon^3)\ , \tag{4.45}$$

which in the further course of the calculation only influences higher-order terms, such that (4.44) is replicated.

Remark on explicit time discretisations. A difference to the first test case to be noted here is that the negative amplification factor does not depend on σ. This implies a time step size limit for explicit time discretisations of pre-smoothed self-snakes: With k denoting the highest perturbation frequency that can occur in the discretised image, given by the Nyquist frequency of the grid ($k = \pi$ for spatial mesh size $h = 1$), the amplification factor $\lambda_s(k) := 1 - \tau k^2/2$ within a single time step of size τ must not become -1 or lower, thus $\tau < 4/k^2$ must be observed.

Amoeba Filter Analysis

To determine the response of an amoeba median filter step to the perturbation (4.39), we consider again Euclidean amoeba metric and $\beta = 1$. The image graph $\Gamma = \{(x, y, u(x, y)) \mid (x, y) \in \mathbb{R}^2\}$ of (4.39), compare Sect. 4.2.4, is a developable surface. The amoeba structuring element $\mathscr{A}$ around (x_0, y_0) then is the projection of a bent Euclidean ϱ-disk $\mathscr{A}^*$ affixed to Γ to the image plane, compare Fig. 4.12c.

The orthogonal projection $\mathscr{A}'$ of the same bent ϱ-disk $\mathscr{A}^*$ not to the image plane but to the unperturbed image graph $\Gamma_0 = \{(x, y, x) \mid (x, y) \in \mathbb{R}^2\}$, compare Fig. 4.12d, is symmetric w.r.t. the line $x = x_0' := x_0 + \varepsilon \cos(kx_0)/\sqrt{2}$; note that the point $(x_0, y_0, u(x_0, y_0))$ projects to (x_0', y_0, x_0'). Moreover, the projection from Γ to Γ_0 changes areas only by a factor $1 + \mathscr{O}(\varepsilon^2)$. Similarly, projection from Γ to the image plane changes areas by a factor $\sqrt{2}/2 + \mathscr{O}(\varepsilon^2)$.

The amoeba median can therefore be computed up to $\mathscr{O}(\varepsilon^2)$ from an area difference within $\mathscr{A}'$ that solely results from the deviation of the projected level line on Γ from the line $x = x_0'$.

The level line of u corresponding to (x_0, y_0) is given by $u(x, y) = u(x_0, y_0)$, thus $x(y) = x_0 + \varepsilon \cos(ky_0) - \varepsilon \cos(ky)$; it projects on Γ_0 as

$$x(y) = x_0' + \frac{1}{2}\bigl(\varepsilon \cos(ky_0) - \varepsilon \cos(ky)\bigr) + \mathscr{O}(\varepsilon^2)\,. \tag{4.46}$$

As the level line extends in y direction from $y_0 - \varrho + \mathscr{O}(\varepsilon^2)$ to $y_0 + \varrho + \mathscr{O}(\varepsilon^2)$, the resulting area difference on Γ_0 is compensated by a level line shift of

$$\begin{aligned} \Delta x &= \frac{-2}{2\,\varrho} \int\limits_{y_0-\varrho}^{y_0+\varrho} \frac{\varepsilon}{2}\bigl(\cos(ky_0) - \cos(ky)\bigr)\,\mathrm{d}y + \mathscr{O}(\varepsilon^2) \\ &= \left(\frac{\sin(k\varrho)}{k\varrho} - 1\right) \varepsilon \cos(ky_0) + \mathscr{O}(\varepsilon^2)\,, \end{aligned} \tag{4.47}$$

making $x_0 + \Delta x$ the sought median, and leading to a frequency reponse factor $\delta(k) := \operatorname{sinc}(k\varrho) - 1$ for the increment of the perturbation.

As before, one amoeba median filter step changes the initial perturbation $u - u_0$ of (4.39) versus $u_0(x, y) = x$ by the amplification factor $\lambda(k) = 1 + \delta(k)$, i.e. $\lambda(k) = \operatorname{sinc}(k\varrho)$. Since $\lambda(k)$ is within $(-1, 1)$ for all $k > 0$, perturbations of all frequencies are dampened.

Figure 4.15 shows the graphs of both amplification functions, $\lambda(k)$ for amoeba median filtering with $\varrho = 1$, and $\lambda_s(k) = 1 + 1/6 \cdot (-k^2/2)$ for the corresponding time step of (4.44) with time step size $\varrho^2/6 = 1/6$, along with numerically computed amplification factors for amoeba median filtering with the same parameters for $k = 1, 2, \ldots, 30$.

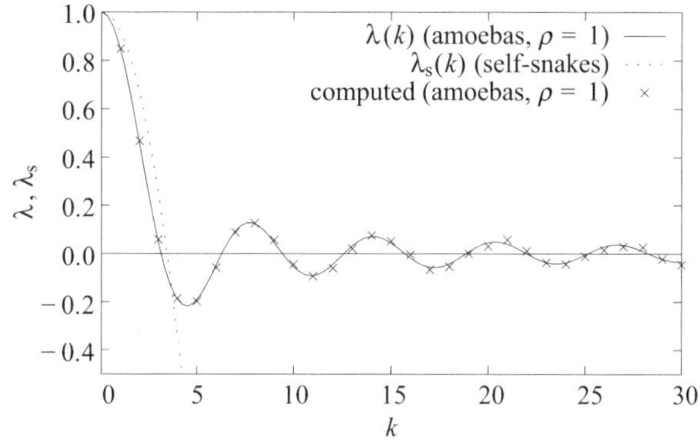

Fig. 4.15 Amplification of a level-line-aligned perturbation of a linear slope function by one amoeba median filtering step (theoretical and numerical values) and a corresponding time step of an explicit scheme for self-snakes (with or without pre-smoothing)

4.6 Amoebas and Texture

As mentioned before, Dijkstra's shortest path algorithm on the neighbourhood graph $G_w(f)$ or a subgraph thereof is used to compute amoeba structuring elements. Whereas in image filtering, only the resulting pixel set $\mathscr{A}_\varrho(i)$ around pixel i is of interest, the search tree created by Dijkstra's algorithm bears valuable information in itself: its structure depends sensitively on the local structure of contrasts in the image, thus, on its texture. Building on work first presented in [55], this section discusses an approach directed at exploiting this information for texture analysis.

4.6.1 Six Graph Structures for Local Texture Analysis

Looking at the amoeba construction in more detail, information about local image texture is distributed to several features. The first aspect are the amoeba distances between adjacent pixels themselves, i.e. the edge weights of $G_w(f)$. A second source of information is the selected pixel set of the amoeba $\mathscr{A}_\varrho(i)$. The third one is the connectivity of the Dijkstra search tree. This leads to six setups for graphs that encode these information cues in different combinations. Figure 4.16 illustrates these setups.

For the first group of three graphs, the pixels within $\mathscr{A}_\varrho(i)$ serve as vertices. For these, one can consider either the full weighted subgraph of $G_w(f)$, which will be denoted by G_w^{A}, the superscript A referring to the use of the amoeba patch. Next, one can consider just the weighted Dijkstra tree, T_w^{A}. Third, deleting the edge weights from this tree yields an unweighted tree, T_u^{A}. Despite suspending the direct use

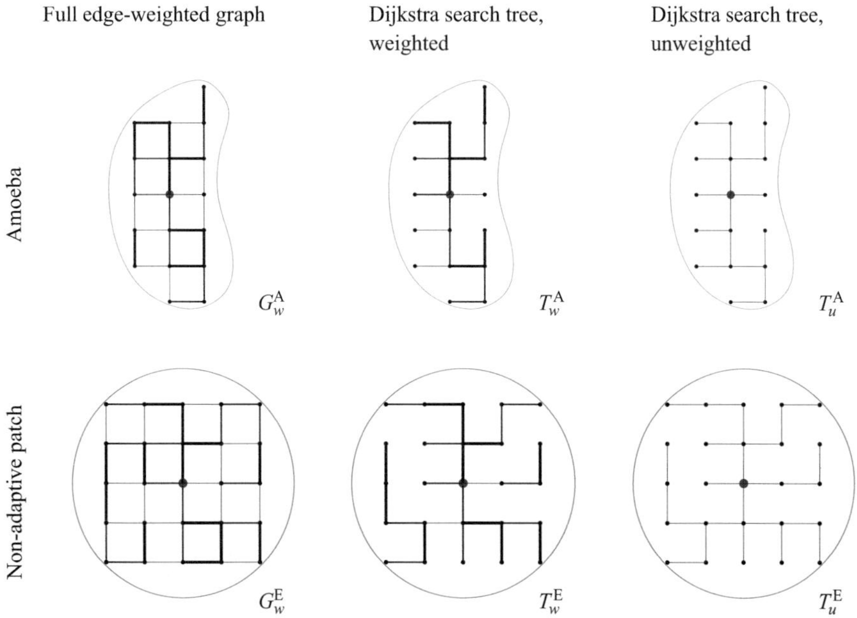

Fig. 4.16 Six graph setups for texture feature construction from amoebas (schematic). For simplicity, graphs are drawn based on 4-neighbourhood connectivity here. In the weighted graphs, different line thicknesses symbolise edge weights

of edge weights in this setting, the connectivity structure derived thereof remains present.

The second group of three graphs is analogous to the first one but chooses the pixels of a fixed window of Euclidean radius ϱ around pixel i. Again, one has the corresponding weighted subgraph of $G_w(f)$, which will be denoted as G_w^E, with the superscript E referring to the Euclidean patch, the weighted Dijkstra tree T_w^E and the unweighted Dijkstra tree T_u^E.

4.6.2 Quantitative Graph Theory: Graph Indices

We turn now to introduce exemplary graph descriptors that can be computed from the previously mentioned graphs in order to obtain quantitative texture descriptors. A larger set of graph descriptors is discussed in the same context in [55].

These graph descriptors are just samples from a tremendous variety of more than 900 concepts [14] that have been established over almost 70 years of research, motivated from applications like the analysis of molecule connectivity in computational chemistry, see e.g. [4, 26, 29, 41, 61], inexact graph matching [19, 43], or the quantitative analysis of (for instance, metabolistic) networks, see e.g. [12, 18].

In the recent decade, the systematic study of these measures has been bundled in the field of *quantitative graph theory*, see e.g. [13, 15, 18].

4.6.2.1 Distance-Based Indices

The historically first class of graph indices are computed directly from the vertex distances within a graph.

Originally introduced for unweighted graphs G, the Wiener index [61] is obtained by just summing up the distances (path lengths) between all pairs $\{i,j\}$ of vertices,

$$W(G) := \sum_{\{i,j\}} d(i,j) \,. \tag{4.48}$$

A modification is the Harary index introduced by Plavšić et al. [41] that sums the reciprocals instead of the distances themselves,

$$H(G) := \sum_{\{i,j\}} \frac{1}{d(i,j)} \,. \tag{4.49}$$

It is straightforward to apply both indices also for weighted graphs, replacing path lengths as distances by total path weights just as in the amoeba definition.

4.6.2.2 Information-Theoretic Indices

Another important class of graph indices is based on entropy concepts. Since Shannon's work [47], the entropy

$$H(p) := -\sum_{k=1}^{n} p(k) \log_2 p(k) \tag{4.50}$$

has been established as the fundamental measure of the information content of a discrete probability measure p on $\{1,\ldots,n\}$.

Bonchev-Trinajstić Information Indices

In [4], entropy has been applied in several ways to the distribution of distances within unweighted graphs to characterise graph connectivity. We pick here two of them. We consider a graph G with vertices $1,\ldots,n$ and denote by $D(G)$ its diameter, i.e. the largest path distance between two of its vertices. By k_d we denote for $d = 1,\ldots,D(G)$ the number of vertex pairs of exact distance d,

$$k_d := \#\{(i,j) \mid 1 \le i < j \le n,\ d(i,j) = d\}\ . \tag{4.51}$$

In [4], the *mean information on distances* $\bar{I}_{\mathrm{D}}^{\mathrm{E}}$ and the *total information on the realised distances* $I_{\mathrm{D}}^{\mathrm{W}}$ of G are defined, which (with a slight rewrite for $I_{\mathrm{D}}^{\mathrm{W}}$) read as

$$\bar{I}_{\mathrm{D}}^{\mathrm{E}}(G) := -\sum_{d=1}^{D(G)} \frac{k_d}{\binom{n}{2}} \log_2 \frac{k_d}{\binom{n}{2}}\ , \tag{4.52}$$

$$I_{\mathrm{D}}^{\mathrm{W}}(G) := W(G)\log_2 W(G) - \sum_{1\le i<j\le n} d(i,j)\log_2 d(i,j)\ , \tag{4.53}$$

where $W(G)$ is the Wiener index (4.48). Again, both definitions can formally be applied to weighted graphs by performing the summation over the weighted path lengths d occurring in G; however, in non-degenerate cases all k_d will equal 1, turning the mean information on distances $\bar{I}_{\mathrm{D}}^{\mathrm{E}}$ into a quantity that depends essentially only on n, and does therefore not reveal much information about the graph. In our texture analysis framework, $\bar{I}_{\mathrm{D}}^{\mathrm{E}}$ makes therefore sense only for the unweighted graphs T_u^{A} and T_u^{E}. In contrast, the total information measure $I_{\mathrm{D}}^{\mathrm{W}}$ makes perfect sense for weighted graphs and thus for all six graph setups under consideration.

Dehmer Entropies

While the Bonchev-Trinajstić indices are based on entropies on the set of distances in a graph, a class of entropy indices defined in [12] works with distributions on the vertex set. An arbitrary positive-valued function f *(information functional)* on the vertices $1,\ldots,n$ of a graph G is converted into a probability density by normalising the sum of all values to 1, such that the individual probabilities $p(i)$ read as

$$p(i) := \frac{f(i)}{\sum_{j=1}^{n} f(j)}\ . \tag{4.54}$$

The entropy

$$I_f(G) := H(p) \tag{4.55}$$

is then a graph index based on the information functional f.

In [12], two choices for f have been considered in the case of unweighted graphs, named f^V and f^P. For each of them, $f(i)$ is obtained from considering the set of neighbourhoods of increasing radius around vertex i in the path metric of the graph. While $f^V(i)$ is the exponential of a weighted sum over the cardinalities of such neighbourhoods, $f^P(i)$ is the exponential of a weighted sum over the distance sums within these neighbourhoods (i.e. the Wiener indices of the corresponding subgraphs). The weight factors assigned to increasing neighbourhoods in both f^P and f^V can be chosen in different ways. Using what is called *exponential weighting scheme* in [15] and measuring distances d by total edge weights along paths in edge-weighted graphs, the resulting information functionals can be stated as

$$f^V(i) := \exp\left(M \sum_{j=1}^{n} q^{d(i,j)}\right), \tag{4.56}$$

$$f^P(i) := \exp\left(M \sum_{j=1}^{n} q^{d(i,j)} d(i,j)\right) \tag{4.57}$$

with parameters $M > 0$ and $q \in (0, 1)$, see [55] where it is also detailed how these expressions are derived from the original definitions from [12].

For the resulting entropy indices I_{f^P} and I_{f^V} as well as for a third one, I_{f^Δ}, which is not discussed here, [15] demonstrated excellent discriminative power for unweighted graphs, i.e. they are able to uniquely distinguish large sets of different unweighted graphs. This finding lets appear I_{f^P} and I_{f^V} also as outstanding candidates for texture analysis tasks.

4.6.3 *Texture Discrimination*

As a first, yet simple, application of the framework that combines amoebas and graph indices, texture discrimination is considered. In [55], a total of 42 candidate texture descriptors was considered. These descriptors resulted from applying nine graph indices, including those described in Sect. 4.6.2 above, to the six graph setups introduced in Sect. 4.6.1, using only those combinations that made sense (as e.g. some graph indices cannot be used for weighted graphs). These graph indices were compared to Haralick features [21, 22], a set of region-based texture descriptors derived from several statistics of co-occurrence matrices of intensities. Despite their long history of more than 40 years, Haralick features are still prominent in texture analysis; together with some more recent modifications they continue to yield competitive results [27, 28, 49].

For the texture discrimination task, the experimental setup in [55] was built to suit the region-based Haralick features by aggregating the, actually local, amoeba-graph features regionwise.

Amoeba-graph descriptors as well as Haralick features were computed for a set of nine texture images from the *VisTex* database, [40]. Figure 4.17 shows a composite image made up of the nine textures used in [55]. Figure 4.18 visualises selected amoeba-graph features on this test image. It can be seen that the different features respond with different degrees of sensitivity and locality to the local structure of the textures.

Fig. 4.17 Composite image containing patches of nine different textures; *top left* to *bottom right* in rows: *brick, fabric, flowers, food, leaves, metal, stone, water, wood.* Texture patches originate from the *VisTex* database, [40]; they have been converted to greyscale, downsampled and clipped (VisTex database ©1995 Massachusetts Institute of Technology. Developed by Rosalind Picard, Chris Graczyk, Steve Mann, Josh Wachman, Len Picard, and Lee Campbell at the Media Laboratory, MIT, Cambridge, Massachusetts)

For each descriptor and texture pair, a statistical discrepancy measure $u := |\mu_1 - \mu_2|/\sigma$ was computed from the mean values μ_1, μ_2 of the texture descriptor on both textures and the joint standard deviation σ. Due to the variability of each descriptor even within the same texture, thresholds for discrimination were gauged from the measured discrepancies for different patches of the same textures: A higher threshold, T_1, was chosen as double the maximum of the nine intra-texture discrepancies measured, and a lower threshold, T_2, as the third-highest of the nine intra-texture values. Texture pairs with discrepancy at least T_1 were considered as "certainly different", and those with discrepancy at least T_2 as "probably different".

While not each texture descriptor could equally well distinguish each pair of textures, it turns out that almost all texture pairs can be told apart by at least some descriptors, with the overall discrimination capability being well comparable with that achieved by the Haralick feature set under consideration. Indeed, the pair *water*/*wood* (the last two patches in the bottom row of Fig. 4.17 was the only one

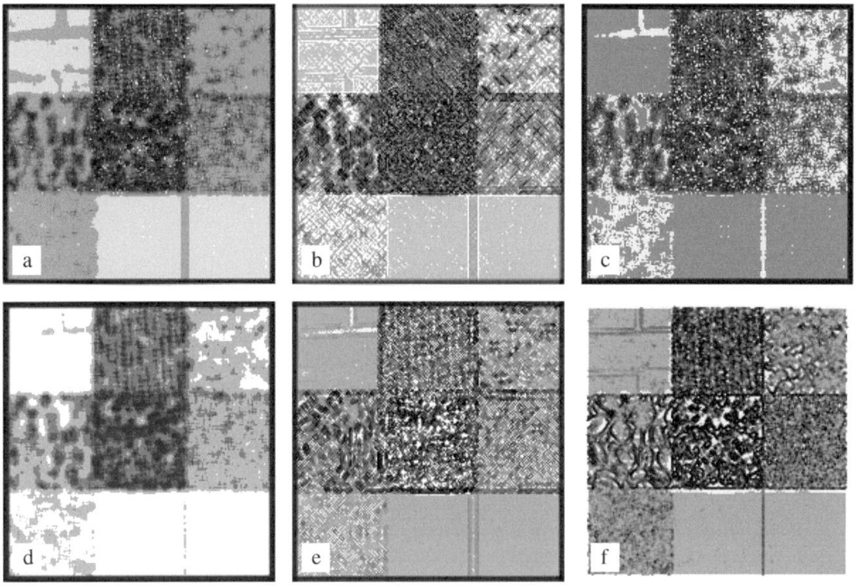

Fig. 4.18 Examples of graph-index-based feature descriptors computed on the test image shown in Fig. 4.17. Graph indices have been computed from amoebas with Euclidean amoeba metric, $\beta = 0.1$ and $\varrho = 5$. All graph index images shown here are histogram equalised. (**a**) Harary index on the weighted amoeba tree T_w^{A}. (**b**) Dehmer entropy I_{f^P} on T_w^{A}. (**c**) $I_{\mathrm{D}}^{\mathrm{W}}$ on T_w^{A}. (**d**) Harary index on the weighted tree in the Euclidean neighbourhood T_w^{E}. (**e**) Dehmer entropy I_{f^P} on T_w^{E}. (**f**) Dehmer entropy I_{f^V} on T_w^{E}

that could not be distinguished with sufficient certainty, neither by the Haralick nor the amoeba-graph feature set. The difficulty to distinguish these two textures can also be seen in Fig. 4.18.

Given that different texture pairs are distinguished best with different descriptors, it is of interest to study the similarity and dissimilarity of different amoeba-graph texture descriptors with regard to what texture pairs they can distinguish. In [55] a metric on the set of texture descriptors has been established in this way. In the further perspective, this is intended to guide the selection of a subset of just a few descriptors that complement each other well, which could therefore be a well-manageable feature set for practical applications.

4.6.4 Texture Segmentation

Finally, we show a simple example that demonstrates the applicability of amoeba-graph indices for texture segmentation. Here graph descriptors have been used as input to a standard geodesic active contour method with an outward force term $\gamma\, g\, |\nabla u|$.

Figure 4.19a shows a test image displaying a striped ring in front of a noisy background. Figure 4.19b shows the field of graph indices I_{fP} computed on weighted Dijkstra trees in Euclidean patches, T_w^{E}, while Fig. 4.19c shows $\bar{I}_{\mathrm{D}}^{\mathrm{E}}$ on T_u^{A}. It is evident from these examples that amoeba-graph indices can turn the textured foreground object into a more homogeneous region. Using just the two graph descriptors as input channels for geodesic active contours one obtains a reasonable

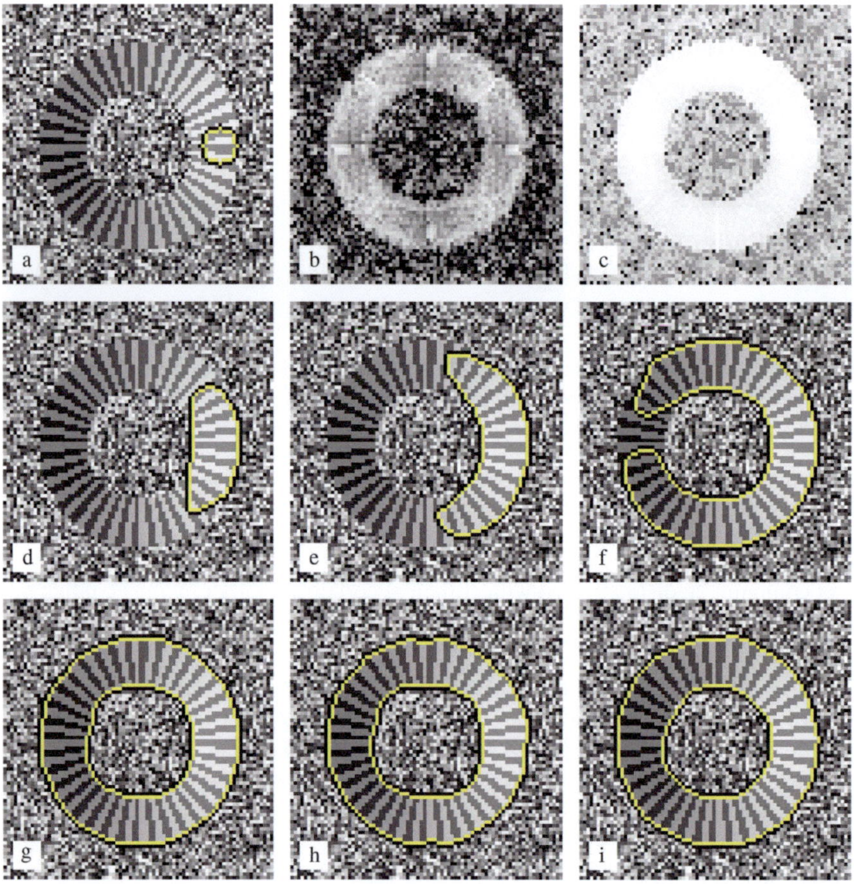

Fig. 4.19 Texture segmentation by geodesic active contour evolution based on amoeba/graph index texture features, pre-smoothing $\sigma = 3$, force term $\gamma = -2$, time step size $\tau = 0.1$. (**a**) Original image with initial contour. (**b**) Graph index I_{fP} on weighted tree T_w^{E} (normalised from $[0, 3.72]$ to $[0, 255]$). (**c**) Graph index $\bar{I}_{\mathrm{D}}^{\mathrm{E}}$ on unweighted tree T_u^{A} (normalised from $[0, 2.93]$ to $[0, 255]$). (**d**) Contour after 500 iterations of GAC evolution using I_{fP} on T_w^{E} and $\bar{I}_{\mathrm{D}}^{\mathrm{E}}$ on T_u^{A} each weighted 0.5, Perona-Malik threshold $\lambda = 0.036$. (**e**) Same as (**d**) but 1000 iterations. (**f**) Same as (**d**) but 2500 iterations. (**g**) Steady state of the segmentation process from (**d**)–(**f**) reached after 3300 iterations. (**h**) Segmentation using only I_{fP} on T_w^{E}, Perona-Malik threshold 0.48, steady state reached after 7500 iterations. (**i**) Segmentation using only $\bar{I}_{\mathrm{D}}^{\mathrm{E}}$ on T_u^{A}, Perona-Malik threshold 0.4, steady state reached after 1200 iterations

segmentation, see Fig. 4.19g. One might ask whether one graph index alone does the job, too. In the present example, this is indeed true; however, the results in Fig. 4.19h, i are visibly less precise in locating the contour separating foreground and background.

Note that this example is only a first proof of concept. A deeper investigation of the potential of this approach to texture segmentation as well as the study of parameter choice and comparison to other texture segmentation methods are topics for future research.

4.7 Outlook

From the results reviewed in this chapter it can be seen that morphological amoebas provide a powerful framework for adaptive image filtering with interesting cross-relations to other classes of filters. They can also be applied fruitfully to related tasks such as image segmentation. Combining amoeba procedures with ideas from quantitative graph theory even allows to construct a new class of texture descriptors.

At the same time, there remain many questions for future research. So far, the amoeba framework introduces adaptivity into local image filters solely by modifying the first step of the filter procedure, i.e. the selection stage. The aggregation step like median, maximum, or minimum is left unchanged. Could further improvements of adaptivity be achieved by envisioning also image-dependent modifications to the aggregation step? How do modifications of selection and aggregation step interact?

Addressing the selection step itself, it would be possible to weaken the dichotomy of including or not including neighbour locations, and to consider unsharp or weighted neighbourhoods.

No amoeba filter for multi-channel (such as colour) images have been studied in the present chapter. In principle, there is little to prevent one from applying amoeba procedures to multi-channel data. The amoeba computation step generalises straightforwardly. There are also generalisations of median filters [2, 48, 52, 59, 60] and supremum/infimum operations to multi-channel data [6–8] at hand. The theoretical understanding of multi-channel amoeba filters, however, lags behind that in the single-channel case. A result in [57] indicates that the median–PDE relation even in its non-adaptive form, see Proposition 4.1, has no equally simple multi-channel counterpart, thus leaving little hope to derive manageable PDE equivalents of multi-channel amoeba filters. New approaches to a deeper understanding of the properties of multi-channel amoeba filters will have to be sought.

The field of texture analysis addressed in Sect. 4.6 still is at an early stage of research. Ongoing research is directed at extending the experimental evaluation of the newly introduced amoeba-graph texture descriptors for texture discrimination to a broader body of data. Another goal is the selection of a powerful set of a few amoeba-graph descriptors with a high combined discrimination rate across multiple textures. Tuning of the parameters of the descriptors has not been studied extensively so far and will therefore be addressed in the future. Attempts are also underway to analyse the effect of the amoeba-graph descriptors theoretically.

In the field of texture segmentation the combination of amoeba-graph descriptors with other segmentation frameworks than the GAC considered in Sect. 4.6.4 will be investigated. An integration with an amoeba active contour procedure could lead to a texture segmentation framework that uses the same sort of theoretically founded procedure for both texture feature extraction and the actual segmentation step. In many existing approaches, and also in the preliminary example from Sect. 4.6.4, these two steps are based on rather unrelated approaches. With regard to the graph-theoretical roots of the texture features under consideration, also graph-cut approaches for the segmentation stage could be a candidate for further investigation.

References

1. Alvarez, L., Lions, P.-L., Morel, J.-M.: Image selective smoothing and edge detection by nonlinear diffusion. II. SIAM J. Numer. Anal. **29**, 845–866 (1992)
2. Austin, T.L.: An approximation to the point of minimum aggregate distance. Metron **19**, 10–21 (1959)
3. Baccelli, F., Cohen, G., Olsder, G.J., Quadrat, J.: Synchronization and Linearity: An Algebra for Discrete Event Systems. Wiley, Chichester (1992)
4. Bonchev, D., Trinajstić, N.: Information theory, distance matrix, and molecular branching. J. Chem. Phys. **67**(10), 4517–4533 (1977)
5. Braga-Neto, U.M.: Alternating sequential filters by adaptive neighborhood structuring functions. In: Maragos, P., Schafer, R.W., Butt, M.A. (eds.) Mathematical Morphology and Its Applications to Image and Signal Processing. Volume 5 of Computational Imaging and Vision, pp. 139–146. Kluwer, Dordrecht (1996)
6. Burgeth, B., Kleefeld, A.: Morphology for color images via Loewner order for matrix fields. In: Luengo Hendriks, C.L., Borgefors, G., Strand, R. (eds.) Mathematical Morphology and Its Applications to Signal and Image Processing. Volume 7883 of Lecture Notes in Computer Science, pp. 243–254. Springer, Berlin (2013)
7. Burgeth, B., Bruhn, A., Papenberg, N., Welk, M., Weickert, J.: Mathematical morphology for matrix fields induced by the Loewner ordering in higher dimensions. Signal Process. **87**(2), 277–290 (2007)
8. Burgeth, B., Welk, M., Feddern, C., Weickert, J.: Morphological operations on matrix-valued images. In: Pajdla, T., Matas, J. (eds.) Computer Vision – ECCV 2004, Part IV. Volume 3024 of Lecture Notes in Computer Science, pp. 155–167. Springer, Berlin (2004)
9. Caselles, V., Kimmel, R., Sapiro, G.: Geodesic active contours. In: Proceedings of the Fifth International Conference on Computer Vision, Cambridge, June 1995, pp. 694–699. IEEE Computer Society Press.
10. Cohen, L.D.: On active contour models and balloons. CVGIP: Image Underst. **53**(2), 211–218 (1991)
11. Cootes, T.F., Taylor, C.J.: Statistical models of appearance for computer vision. Technical report, University of Manchester, Oct 2001
12. Dehmer, M.: Information processing in complex networks: graph entropy and information functionals. Appl. Math. Comput. **201**, 82–94 (2008)
13. Dehmer, M., Emmert-Streib, F., Mehler, A. (eds.): Towards an Information Theory of Complex Networks: Statistical Methods and Applications. Birkhäuser Publishing, Basel (2012)
14. Dehmer, M., Emmert-Streib, F., Tripathi, S.: Large-scale evaluation of molecular descriptors by means of clustering. PloS ONE **8**(12), e83956 (2013)

15. Dehmer, M., Sivakumar, L.: Recent developments in quantitative graph theory: information inequalities for networks. PLoS ONE **7**(2), e31395 (2012)
16. Dijkstra, E.: A note on two problems in connexion with graphs. Numer. Math. **1**, 269–271 (1959)
17. Eckhardt, U.: Root images of median filters. J. Math. Imaging Vis. **19**, 63–70 (2003)
18. Emmert-Streib, F., Dehmer, M.: Information theoretic measures of UHG graphs with low computational complexity. Appl. Math. Comput. **190**, 1783–1794 (2007)
19. Ferrer, M., Bunke, H.: Graph edit distance–theory, algorithms, and applications. In: Lezoray, O., Grady, L. (eds.) Image Processing and Analysis with Graphs: Theory and Practice, chapter 13, pp. 383–422. CRC Press, Boca Raton (2012)
20. Guichard, F., Morel, J.-M.: Partial differential equations and image iterative filtering. In: Duff, I.S., Watson, G.A. (eds.) The State of the Art in Numerical Analysis. Number 63 in IMA Conference Series (New Series), pp. 525–562. Clarendon Press, Oxford (1997)
21. Haralick, R.: Statistical and structural approaches to texture. Proc. IEEE **67**(5), 786–804 (1979)
22. Haralick, R., Shanmugam, K., Dinstein I.: Textural features for image classification. IEEE Trans. Syst. Man Cybern. **3**(6), 610–621 (1973)
23. Heijmans, H.J.A.M.: Morphological Image Operators. Academic, Boston (1994)
24. Heijmans, H.J.A.M., Ronse, C.: The algebraic basis of mathematical morphology. I: dilations and erosions. Comput. Vis. Graph. Image Process. **50**, 245–295 (1990)
25. Heijmans, H.J.A.M., van den Boomgaard, R.: Algebraic framework for linear and morphological scale-spaces. J. Vis. Commun. Image Represent. **13**(1/2), 269–301 (2002)
26. Hosoya, H.: Topological index: a newly proposed quantity characterizing the topological nature of structural isomers of saturated hydrocarbons. Bull. Chem. Soc. Jpn **44**(9), 2332–2339 (1971)
27. Howarth, P., Rüger, S.: Evaluation of texture features for content-based image retrieval. In: Enser, P., Kompatsiaris, Y., O'Connor, N., Smeaton, A., Smeulders, A. (eds.) Image and Video Retrieval. Volume 3115 of Lecture Notes in Computer Science, pp. 326–334. Springer, Berlin (2004)
28. Huang, K., Murphy, R.: Automated classification of subcellular patterns in multicell images without segmentation into single cells. In: Proceedings of the 2004 IEEE International Symposium on Biomedical Imaging, Apr 2004, vol. 2, pp. 1139–1142
29. Ivanciuc, O., Balaban, T.-S., Balaban, A.: Design of topological indices. Part 4. Reciprocal distance matrix, related local vertex invariants and topological indices. J. Math. Chem. **12**(1), 309–318 (1993)
30. Kichenassamy, S., Kumar, A., Olver, P., Tannenbaum, A., Yezzi, A.: Gradient flows and geometric active contour models. In: Proceedings of the Fifth International Conference on Computer Vision, Cambridge, June 1995, pp. 810–815. IEEE Computer Society Press
31. Kichenassamy, S., Kumar, A., Olver, P., Tannenbaum, A., Yezzi, A.: Conformal curvature flows: from phase transitions to active vision. Arch. Ration. Mech. Anal. **134**, 275–301 (1996)
32. Lerallut, R., Decencière, É., Meyer, F.: Image processing using morphological amoebas. In: Ronse, C., Najman, L., Decencière, E. (eds.) Mathematical Morphology: 40 Years on. Volume 30 of Computational Imaging and Vision, pp. 13–22. Springer, Dordrecht (2005)
33. Lerallut, R., Decencière, É., Meyer, F.: Image filtering using morphological amoebas. Image Vis. Comput. **25**(4), 395–404 (2007)
34. Leventon, M.E., Grimson, W.E.L., Faugeras, O.: Statistical shape influence in geodesic active contours. In: Proceedings of the 2000 IEEE International Conference on Computer Vision and Pattern Recognition, vol. 1, pp. 316–323, Hilton Head Island, June 2000
35. Malladi, R., Sethian, J., Vemuri, B.: Shape modeling with front propagation: a level set approach. IEEE Trans. Pattern Anal. Mach. Intell. **17**, 158–175 (1995)
36. Maragos, P.: Lattice image processing: a unification of morphological and fuzzy algebraic systems. J. Math. Imaging Vis. **22**, 333–353 (2005)
37. Maragos, P., Vachier, C.: Overview of adaptive morphology: trends and perspectives. In: Proceedings of the 2009 IEEE International Conference on Image Processing, Cairo, Nov 2009, pp. 2241–2244
38. Matheron, G.: Eléments pour une théorie des milieux poreux. Masson, Paris (1967)

39. Perona, P., Malik, J.: Scale space and edge detection using anisotropic diffusion. IEEE Trans. Pattern Anal. Mach. Intell. **12**, 629–639 (1990)
40. Picard, R., Graczyk, C., Mann, S., Wachman, J., Picard, L., Campbell, L.: VisTex database. Online ressource, http://vismod.media.mit.edu/vismod/imagery/VisionTexture/vistex.html (1995). Retrieved 20 Nov 2013
41. Plavšić, D., Nikolić, S., Trinajstić, N.: On the Harary index for the characterization of chemical graphs. J. Math. Chem. **12**(1), 235–250 (1993)
42. Quadrat, J.-P.: Max-plus algebra and applications to system theory and optimal control. In: Chatterji, S.D. (ed.) Proceedings of the International Congress of Mathematicians, pp. 1511–1522. Birkhäuser, Basel (1995)
43. Sanfeliu, A., Fu, K.-S.: A distance measure between attributed relational graphs for pattern recognition. IEEE Trans. Syst. Man Cybern. **13**(3), 353–362 (1983)
44. Sapiro, G.: Vector (self) snakes: a geometric framework for color, texture and multiscale image segmentation. In: Proceedings of the 1996 IEEE International Conference on Image Processing, Lausanne, Sept 1996, vol. 1, pp. 817–820
45. Serra, J.: Image Analysis and Mathematical Morphology, vol. 1. Academic, London (1982)
46. Serra, J.: Image Analysis and Mathematical Morphology, vol. 2. Academic, London (1988)
47. Shannon, C.: A mathematical theory of communication. Bell Syst. Tech. J. **27**, 379–423; 623–656 (1948)
48. Spence, C., Fancourt, C.: An iterative method for vector median filtering. In: Proceedings of the 2007 IEEE International Conference on Image Processing, vol. 5, pp. 265–268 (2007)
49. Tesař, L., Shimizu, A., Smutek, D., Kobatake, H., Nawano, S.: Medical image analysis of 3D CT images based on extension of Haralick texture features. Comput. Med. Imaging Graph. **32**(6), 513–520 (2008)
50. Tukey, J.W.: Exploratory Data Analysis. Addison–Wesley, Menlo Park (1971)
51. Verly, J.G., Delanoy, R.L.: Adaptive mathematical morphology for range imagery. IEEE Trans. Image Process. **2**(2), 272–275 (1993)
52. Weiszfeld, E.: Sur le point pour lequel la somme des distances de *n* points donnés est minimum. Tôhoku Mathematics Journal **43**, 355–386 (1937)
53. Welk, M.: Amoeba active contours. In: Bruckstein, A.M., ter Haar Romeny, B., Bronstein, A.M., Bronstein, M.M. (eds.) Scale Space and Variational Methods in Computer Vision. Volume 6667 of Lecture Notes in Computer Science, pp. 374–385. Springer, Berlin (2012)
54. Welk, M.: Relations between amoeba median algorithms and curvature-based PDEs. In: Kuijper, A., Pock, T., Bredies, K., Bischof, H. (eds.) Scale Space and Variational Methods in Computer Vision. Volume 7893 of Lecture Notes in Computer Science, pp. 392–403. Springer, Berlin (2013)
55. Welk, M.: Discrimination of image textures using graph indices. In: Dehmer, M., Emmert-Streib, F. (eds.) Quantitative Graph Theory: Mathematical Foundations and Applications, chapter 12, pp. 355–386. CRC Press, Boca Raton (2014)
56. Welk, M.: Analysis of amoeba active contours. J. Math. Imaging Vis. **52**, 37–54 (2015)
57. Welk, M., Breuß, M.: Morphological amoebas and partial differential equations. In: Hawkes, P.W. (ed.) Advances in Imaging and Electron Physics, vol. 185, pp. 139–212. Elsevier/Academic, Amsterdam (2014)
58. Welk, M., Breuß, M., Vogel, O.: Morphological amoebas are self-snakes. J. Math. Imaging Vis. **39**, 87–99 (2011)
59. Welk, M., Feddern, C., Burgeth, B., Weickert, J.: Median filtering of tensor-valued images. In: Michaelis, B., Krell, G. (eds.) Pattern Recognition. Volume 2781 of Lecture Notes in Computer Science, pp. 17–24. Springer, Berlin (2003)
60. Welk, M., Weickert, J., Becker, F., Schnörr, C., Feddern, C., Burgeth, B.: Median and related local filters for tensor-valued images. Signal Process. **87**, 291–308 (2007)
61. Wiener, H.: Structural determination of paraffin boiling points. J. Am. Chem. Soc. **69**(1), 17–20 (1947)

Chapter 5
Increasing the Power of Shape Descriptor Based Object Analysis Techniques

Joviša Žunić, Paul L. Rosin, and Mehmet Ali Aktaş

Abstract An advantage of shape based techniques, for object analysis tasks, is that shape allows a large number of numerical characterizations. Some of these have an intuitively clear meaning, while others do not, but they are still very useful because they satisfy some desirable properties (e.g. invariance with respect to a set of certain transformations). In this chapter we focus on numerical shape characteristics that have a clear intuitive interpretation – i.e. based on such numerical values, we can predict, to some extent, what the considered object looks like. This is beneficial, since it enables a priori appraisal of whether certain shape characteristics have suitable discriminative potential that make them appropriate for the intended task. By their nature, the number of such methods cannot be as large as the number of methods to allocate shape/object characteristics based on some formalism (algebraic, geometric, probabilistic, etc.). Because of that, some other possibilities to increase the discriminative capacity of the methods based on numerical shape characteristics, with an intuitively predictable meaning, are considered. Herein, we observe two such possibilities: the use of tuning parameters to obtain a family of shape characteristics, and the use of multiple shapes derived from the objects analyzed.

5.1 Introduction

Shape is an important component of the human visual system, and is also widely used in computer vision to provide a means of describing objects as a precursor to identifying them. If object boundaries can be reliably extracted (which of course

J. Žunić (✉)
Mathematical Institute, Serbian Academy of Sciences and Arts, Belgrade, Serbia
e-mail: jovisa_zunic@mi.sanu.ac.rs

P.L. Rosin
School of Computer Science & Informatics, Cardiff University, Cardiff, CF24 3AA, UK
e-mail: Paul.Rosin@cs.cardiff.ac.uk

M.A. Aktaş
Computer Science, Toros University, Mersin, Turkey
e-mail: mehmet.aktas@toros.edu.tr

M. Breuß et al. (eds.), *Perspectives in Shape Analysis*, Mathematics and Visualization, DOI 10.1007/978-3-319-24726-7_5

remains a challenge for unconstrained scenes, but is achievable in many other cases) then shape descriptors offer many advantages to those based on intensity, colour, texture, etc. First, although those latter approaches incorporate more information, offering a richer descriptive power, they are consequently also more sensitive to potentially irrelevant variations in illumination, colouring, etc. For instance, whereas the shape of a typical car is clear cut, cars come in many colours, and so colour (unlike shape) is not helpful to the task of assigning an object to the general class of cars. Second, most shape descriptors can be easily normalised so that they are invariant to many transformations (e.g. translation, rotation, scaling, shearing) without requiring expensive and less reliable methods such as scale-space based image processing. Third, many techniques for shape based analysis provide a compact descriptor, that is not only efficient to store, but is also well suited to efficient matching.

Many shape properties, herein called *shape descriptors*, are known to be very suitable for a numerical evaluation (e.g. shape convexity, ellipticity, elongation, compactness, linearity, sigmoidality, tortuosity, etc.). Methods developed to evaluate a certain shape descriptor will be called *shape measures*. Examples of shape measures already developed are: convexity [26, 32, 42], circularity [8, 16, 25, 45], compactness [18], linearity [11, 36, 40], ellipticity [1, 25, 30, 38, 44], sigmoidality [31], rectilinearity [41], tortuosity [13], and many more. As it can be seen, there are shape descriptors with multiple measures developed for their numerical evaluation. This is because none of the shape measures are ideally suited for all the possible applications.

Apart from the shape measures mentioned, which relate to a certain shape property, there are generic shape measures which are not originally designed to measure a specific shape property/characteristic. Among them are: Fourier descriptors [5, 39], moment invariants [17, 21], shape-illumination invariants [3], and so on. Those measures satisfy some desirable properties (e.g. invariance with respect to some transformations) and their power comes from the fact that, at least in theory, an infinite number of them can be generated and assigned to a given object/shape. A drawback is that their behavior is not well explained and cannot be predicted. This further implies that their suitability for a certain task has to be verified through an intensive experimental study, which is always a time consuming process.

Contrary to the generic shape measures, the measures which do relate to a certain shape property have a well understood and predictable behavior. Their disadvantage is that their number is limited. This further causes a limited discriminative power of the object analysis tools based on such measures, particularly when dealing with huge data sets. In this chapter we consider possibilities of increasing the discriminative power of such tools, with applications in image processing and computer vision tasks. We discuss the following possibilities: (i) An involvement of a tuning parameter; (ii) Allocation of multiple shapes to the objects considered; (iii) A combination of the approaches in (i) and (ii). Our discussion is supported with experimental results.

Throughout this chapter we will assume that all occurring shapes are bounded. In order to avoid discussions on pathological situations, we will say that two shapes are equal if their set differences have area equal to zero. This is obviously not a restriction in practical applications – e.g. a closed ellipse $\{(x,y) \mid x^2 + 3 \cdot y^2 \le 1\}$ and the "open" one $\{(x,y) \mid x^2 + 3 \cdot y^2 < 1\}$ are considered to be the same shape.

The geometric moment $m_{p,q}(S)$ of a given shape S, represented by a planar bounded region, is defined as

$$m_{p,q}(S) = \iint_S x^p\, y^q\, dx\, dy. \tag{5.1}$$

Obviously, $m_{0,0}(S)$ equals the area of S. As a short reminder, the centroid of a given shape S is defined as

$$\left(\frac{\iint_S x\, dx\, dy}{\iint_S dx\, dy}, \frac{\iint_S y\, dx\, dy}{\iint_S dx\, dy} \right) = \left(\frac{m_{1,0}(S)}{m_{0,0}(S)}, \frac{m_{0,1}(S)}{m_{0,0}(S)} \right). \tag{5.2}$$

Since shape does not change under translation, we will assume that all the appearing shapes are positioned such that their centroid coincides with the origin. In other words:

$$m_{1,0}(S) = \iint_S x\, dx\, dy = 0 \quad \text{and} \quad m_{0,1}(S) = \iint_S y\, dx\, dy = 0 \tag{5.3}$$

will be assumed, even if not mentioned, for all the shapes considered.

Finally, $S(\omega)$ will denote the shape S rotated around its centroid by the angle ω.

5.2 Power Increase by Introducing a Tuning Parameter

In this section we discuss a family of circularity measures, introduced as a generalization of the first Hu moment invariant [17], by incorporating one parameter [45]. The role of this introduced parameter is to control the behavior of the circularity measures from the given family. Shape interpretation of the first Hu moment invariant, $I_1(S)$,

$$I_1(S) = \iint_S (x^2 + y^2)\, dx\, dy \tag{5.4}$$

has been analyzed in [45]. It has been shown that the first Hu moment invariant, $I_1(S)$, ranges over the interval $[\frac{1}{2\pi}, \infty)$ and returns the minimum possible value

Fig. 5.1 Fish shapes are ranked with respect to their computed $\mathscr{C}(S)$ circularity values

$\frac{1}{2\pi}$ for circles only. This property has been used to define the new circularity measure, $\mathscr{C}(S)$, for planar shapes:

$$\mathscr{C}(S) = \frac{1}{2\pi} \cdot \frac{m_{0,0}(S)^2}{m_{2,0}(S) + m_{0,2}(S)} = \frac{1}{2 \cdot \pi \cdot I_1(S)}. \tag{5.5}$$

Such a circularity measure $\mathscr{C}(S)$ ranges over the interval $(0, 1]$, produces the value 1 if and only if the considered shape S is a circle, and is invariant with respect to translation, rotation, and scaling transformations. It also might be said that the new circularity measure fits well with our perception of what a circularity measure should be – a quantity which indicates how much a shape given differs from a circle. Shapes with relatively large $\mathscr{C}(S)$ values are nearly circular, while shapes with small $\mathscr{C}(S)$ values have a nearly linear structure. We illustrate this by a small collection of fish shapes and their assigned circularity values, but more examples can be found in [45]. Five fish shapes are listed in Fig. 5.1, in accordance with their computed $\mathscr{C}(S)$ circularity values. The largest circularity 0.9579 is assigned to the shape on the left, which is as expected since this shape is nearly circular. The smallest circularity value 0.1390 is assigned to the shape on the right. Again, such a small circularity comes from the fact that this shape has a nearly linear structure. Our judgment is that we may say that these values, as well as the remaining three values, and the ranking obtained, are in accordance with human perception.

The circularity measure $\mathscr{C}(S)$ is area based, and because of this is robust, i.e. relatively resistant to small shape deformations or to defects caused by noise, for example. Of course, such a (robustness) property is an advantage in many situations but it could be a disadvantage in situations when high precision is required. To avoid such a possible drawback, the measure $\mathscr{C}(S)$ has been modified. A tuning parameter α was introduced [45] to produce a family of circularity measures $\mathscr{C}_\alpha(S)$ as follows:

$$\mathscr{C}_\alpha(S) = \frac{1}{(\alpha + 1) \cdot \pi^\alpha} \cdot \frac{m_{0,0}(S)^{\alpha+1}}{\iint_S (x^2 + y^2)^\alpha dxdy} \tag{5.6}$$

for all $\alpha > 0$[1] and for all bounded compact planar shapes S. Obviously, the measure $\mathscr{C}(S)$ also belongs to the new family of circularity measures defined in (5.6),

[1]For an extension to the circularity measures with $\alpha \in (-1, 0)$, see [45].

since $\mathscr{C}(S) = \mathscr{C}_{\alpha=1}(S)$. All circularity measures from the family $\mathscr{C}_\alpha(S)$ keep the basic desirable properties. They range over $(0, 1]$, with the equality $\mathscr{C}_\alpha(S) = 1$ satisfied for circles only. Measures $\mathscr{C}_\alpha(S)$ are invariant with respect to similarity transformations as well. The main role of the tuning parameter α is to enable control of the sensitivity/robustness properties of $\mathscr{C}_\alpha(S)$. It has been shown that bigger values of α lead to a more sensitive measure $\mathscr{C}_\alpha(S)$. More detailed discussion can be found in [45], but here we give a lemma which supports the previous statement. Indeed, Lemma 5.1 says that for any shape S different from a circle, there is a parameter α such that $\mathscr{C}_\alpha(S)$ is arbitrarily close to 0. In other words, there is a choice of circularity measure $\mathscr{C}_\alpha(S)$ (i.e. the choice of the parameter α) which would penalize, strongly enough, any existing difference between the shape S and a circle.

Lemma 5.1 *For a bounded planar compact shape S, different from a circle, the following equality is true*

$$\lim_{\alpha\to\infty} \mathscr{C}_\alpha(S) = 0. \tag{5.7}$$

Some of the benefits from having the possibility to tune the behavior of circularity measures are illustrated by examples in Fig.5.2. All the four shapes listed can be understood as very similar to a circle. The first shape is a regular 7-gon while the remaining three shapes are obtained from a circle by adding noise. For the second and third shape a different level noise is added to the shape boundary, while salt noise (i.e. holes) is added to the interior of the fourth shape. The circularity $\mathscr{C}(S)$ of all these shapes is very close to 1, and $\mathscr{C}(S)$ can neither distinguish among these shapes nor detect the presence of the obvious irregularities. These irregularities become visible once measures $\mathscr{C}_\alpha(S)$ from the new family are employed. Indeed, looking at the graphs of $\mathscr{C}_\alpha(S)$ (considered as a function in α), displayed in the second row in Fig.5.2, we see that an increase of α leads to a decrease of $\mathscr{C}_\alpha(S)$. After some point, it becomes clearly evident that all the given shapes differ from a

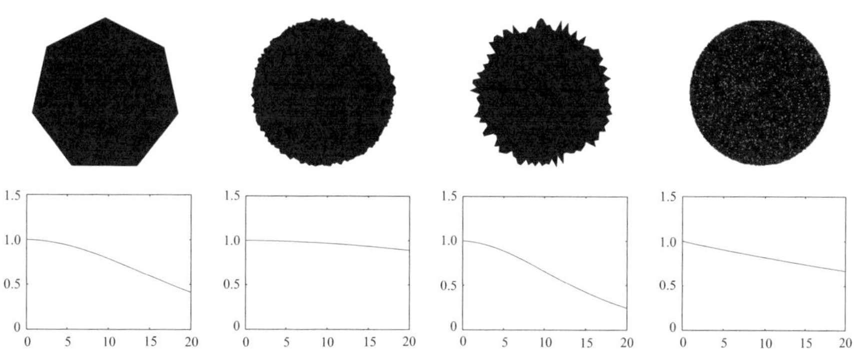

Fig. 5.2 Graphs of the measured circularities $\mathscr{C}_\alpha(S)$, for $\alpha \in [0, 20]$, are given below the corresponding shapes

circle, and also that each of these shapes differs from the others. For example, if we set $\alpha = 20$ then, for all the shapes displayed, the computed $\mathscr{C}_\alpha(S)$ circularities are all mutually different.

Next, we illustrate that some classification accuracies, reached by some of the well known shape measures, can be outperformed by selecting a suitable measure from the family $\mathscr{C}_\alpha(S)$. For this purpose we will use the standard circularity measure, and the circularity measures of Proffitt [25] and Haralick [16]. The standard circularity measure $\mathscr{C}_{st}(S)$ exploits the fact that among all shapes with the same perimeter, the circle has the largest area. It is defined as

$$\mathscr{C}_{st}(S) = \frac{4 \cdot \pi \cdot Area\_of\_S}{(Perimeter\_of\_S)^2}. \tag{5.8}$$

Note that in the following experiments the perimeter of S was calculated for $\mathscr{C}_{st}(S)$ either directly from the pixel boundaries extracted from the images with inter-pixel weights set according to Dorst and Smeulders [9], or alternatively the perimeters were calculated from polygonal approximations of the boundaries [27]. For classification, leave one out testing was performed with a nearest neighbor classifier using Euclidean distances.

For this example, circularity was measured for the set of 54 masses from mammograms, combining images from the MIAS and Screen Test databases [28], see Fig. 5.3. Rangayyan et al. [28] assessed the measures by classifying them as

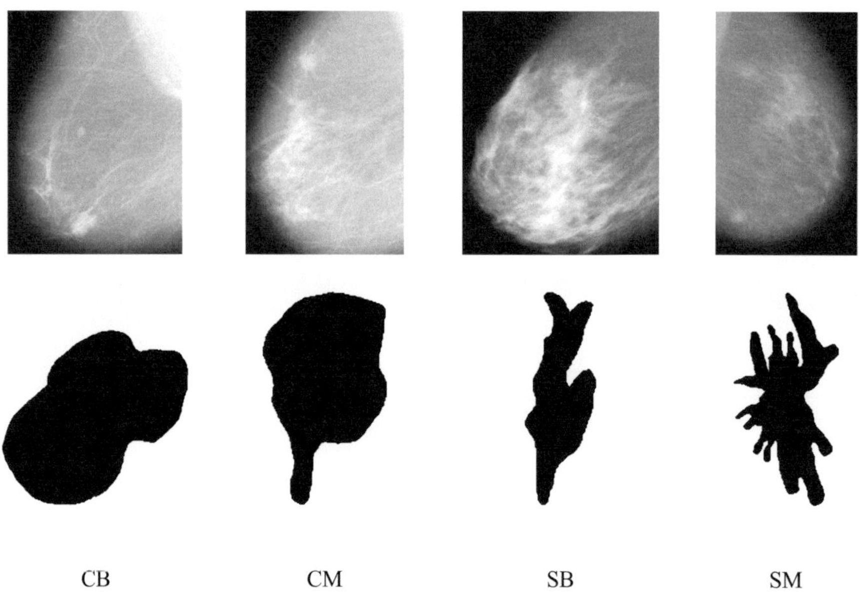

Fig. 5.3 Examples of the four classes of mammographic masses: circumscribed benign (*CB*), circumscribed malignant (*CM*), spiculated benign (*SB*), spiculated malignant (*SM*). The masses were extracted from the mammograms (*top row*), and have been drawn rescaled (*bottom row*)

Table 5.1 Applications of the circularity measures to classification of mammographic masses. The classification accuracies, for three classification tasks, are given for different choices of the circularity measures. The best performing measure was $\mathcal{C}_{\alpha=32}(S)$ (the score highlighted in bold)

Circularity	Mammography		
Measure	circ./spic.	mal./ben.	4 groups
$\mathcal{C}_{\alpha=1/8}(S)$	83.33	66.67	51.85
$\mathcal{C}_{\alpha=1/4}(S)$	85.19	64.81	51.85
$\mathcal{C}_{\alpha=1/2}(S)$	75.93	57.41	42.59
$\mathcal{C}_{\alpha=1}(S)$	68.52	68.52	51.85
$\mathcal{C}_{\alpha=2}(S)$	75.93	68.52	53.70
$\mathcal{C}_{\alpha=4}(S)$	72.22	46.30	33.33
$\mathcal{C}_{\alpha=8}(S)$	79.63	59.26	50.00
$\mathcal{C}_{\alpha=16}(S)$	87.04	57.41	51.85
$\mathcal{C}_{\alpha=32}(S)$	**90.74**	**70.37**	**64.81**
$\mathcal{C}_{st}(S)$ pixel	87.04	59.26	57.41
$\mathcal{C}_{st}(S)$ polygon	85.19	59.26	57.41
Haralick [16]	68.52	46.30	37.04
Proffitt [25]	51.85	42.59	25.93

circumscribed/spiculated, benign/malignant, and CB/CM/SB/SM, in two group and four group classification experiments. Their best shape measure results for the three classification tasks were: (i) circumscribed versus spiculated: 88.9 % achieved by both $\mathcal{C}_{st}(S)$ and a Fourier based shape factor, (ii) benign versus malignant: 75.9 % achieved by the Fourier based shape factor, (iii) four-way discrimination: 64.8 % achieved by both $\mathcal{C}_{st}(S)$ and the Fourier based shape factor.[2] From Table 5.1 we see that the best results from using $\mathcal{C}_{\alpha}(S)$ occurred for $\alpha = 32$ and were respectively better, worse, and equal to Rangayyan et al.'s. The other circularity measures did not perform as well as $\mathcal{C}_{\alpha}(S)$.

5.3 Family of Ellipticity Measures with an Application in an Galaxy Classification Task

Shape ellipticity measures are intensively studied in the literature. An early attempt [38] goes back to 1910. Notice that there are two approaches for how to measure shape ellipticity. The first one assumes that all ellipses are of the same shape, regardless of their axis length ratios, e.g. [1, 30]. Another approach assumes that ellipses whose axis ratios differ also differ in shape, e.g. [2]. It is not possible to say a priori which approach is better. In some applications the first approach would be more appropriate, whilst in some others the second is preferred.

[2] We note that our results for $\mathcal{C}_{st}(S)$ listed in Table 5.1 do not match Rangayyan et al.'s [28] reported accuracies for $\mathcal{C}_{st}(S)$. This can be attributed to several factors: (i) different classifiers were used, and also (ii) different methods for estimating perimeter may have been used.

Ellipticity measures considered in this section, from the family introduced recently [2], assume that ellipses with a different axis length ratio are different in shape. A precise definition follows.

Definition 5.1 Let a bounded planar shape S, whose centroid coincides with the origin, be given. For every $\rho \in (0,1]$ the ellipticity measure $\mathcal{E}_\rho(S)$ of S, is defined as

$$\mathcal{E}_\rho(S) = \frac{1}{2 \cdot \pi} \cdot \frac{m_{0,0}(S)^4}{\min\limits_{\omega \in [0,2\pi]} \iint_{S(\omega)} \left(\frac{x^2}{\rho} + \rho \cdot y^2 \right) dx\, dy}. \tag{5.9}$$

Note 5.1 The formula in (5.9) enables an easy and straightforward numerical computation of $\mathcal{E}_\rho(S)$, with $\rho \in (0,1]$. There is also a closed formula for the computation of $\mathcal{E}_\rho(S)$, derived recently in [46].

All the ellipticity measures $\mathcal{E}_\rho(S)$, from Definition 5.1 have the following properties (for a proof see [2]):

(a) $\mathcal{E}_\rho(S) \in (0,1]$, for any shape S;
(b) $\mathcal{E}_\rho(S) = 1$ if and only if S is an ellipse whose axis length ratio is ρ;
(c) $\mathcal{E}_\rho(S)$ is invariant with respect to similarity transformations.

Theoretical foundations for understanding the behavior of the new ellipticity measures $\mathcal{E}_\rho(S)$ are established in [2]. Here we give a brief discussion. The parameter ρ can be understood as a tuning parameter, because the behavior of the ellipticity measures, from $\{\mathcal{E}_\rho(S) \mid \rho \in (0,1]\}$, depends on the choice of the parameter ρ. For a fixed ρ, the measure $\mathcal{E}_\rho(S)$ indicates how much the considered shape S differs from a perfect ellipse $E(\rho)$ whose axes length ratio is ρ. The highest score, equal to 1, is given only to the $E(\rho)$ ellipses. For all the shapes different from $E(\rho)$, including the ellipses whose axes length ratio differs from ρ, the computed $\mathcal{E}_\rho(S)$ ellipticities are strictly less than 1. Which values of the parameter ρ should be selected depends on the application which is going to be performed. Ellipticity $\mathcal{E}_\rho(S)$, corresponding to one selection of the parameter ρ, can perform well in one application, but also can have a poor performance in another.

In this section, in addition to the use of a tunable ellipticity measures, we consider another possibility to increase the discriminative capacity of shape based object analysis tools. The idea is to assign a number of shapes to an object presented in an image, instead of just a single shape, as is the common approach. Multiple shapes can be assigned in several ways (e.g. as it is done in this section – see Fig. 5.5, and also as it is done in Sect. 5.4 – see Fig. 5.8, or in [29], etc.). In this section we will assign two shapes to each object by using two versions of Otsu's thresholding method [23]: A "global" one (the same threshold level is applied to all pixels) and a "local" one (the original method is applied to blocks of the original image, so that the threshold level applied varies). This means that we allocate two shapes (represented as two binary images) for each object. For each of these two shapes/images we will compute three shape measures, which will comprise the components of the feature vectors used for classification.

This approach will be applied to a galaxy classification task. The elliptical and spiral galaxies, listed in the Nearby Galaxy Catalog [10], are used as the data/shape set. The same data set has been used by many others, and the classification task has been already recognized as a difficult problem [19]. Many approaches have been applied and used to provide an automatic machine galaxy classification, e.g. neural networks approaches [4, 15, 22], fuzzy sets theory [20], geometric shape features [12, 14], shape squareness [34], fractal signatures [19], etc.

The benchmark results, prior to a 100 % classification rate obtained in [2], were 92.3 % and 95.1 %, obtained in [19] by using nearest neighbor and neural network classifiers, respectively.

5.3.1 Ellipticity Measures Used and Classification Results Obtained

Three ellipticity measures were used to perform the galaxy classification task. Two of them are from the family $\mathcal{E}_\rho(S)$:

- $$\mathcal{E}_{\rho=0.7}(S) = \frac{1}{2\cdot\pi}\cdot\frac{m_{0,0}(S)^4}{\min\limits_{\omega\in[0,2\pi]}\iint_{S(\omega)}\left(\frac{x^2}{0.7}+0.7\cdot y^2\right)dx\,dy} \qquad \textbf{(m1)}$$

- $$\mathcal{E}_{\rho=0.9}(S) = \frac{1}{2\cdot\pi}\cdot\frac{m_{0,0}(S)^4}{\min\limits_{\omega\in[0,2\pi]}\iint_{S(\omega)}\left(\frac{x^2}{0.9}+0.9\cdot y^2\right)dx\,dy} \qquad \textbf{(m2)}$$

while the third ellipticity measure used is introduced in [1] and is defined as follows

- $$\mathcal{E}(S) = \frac{1}{2\cdot\pi}\cdot\frac{m_{0,0}(S)^4}{\min\limits_{\omega\in[0,2\pi]}\iint_{S(\omega)}\left(\frac{x^2}{\gamma}+\gamma\cdot y^2\right)dx\,dy} \qquad \textbf{(m3)}$$

  where the parameter γ is defined as

$$\gamma = \frac{\sqrt{\mu_{2,0}(S)+\mu_{0,2}(S)+\sqrt{4\cdot(\mu_{1,1}(S))^2+(\mu_{2,0}(S)-\mu_{0,2}(S))^2}}}{\sqrt{\mu_{2,0}(S)+\mu_{0,2}(S)-\sqrt{4\cdot(\mu_{1,1}(S))^2+(\mu_{2,0}(S)-\mu_{0,2}(S))^2}}}.$$

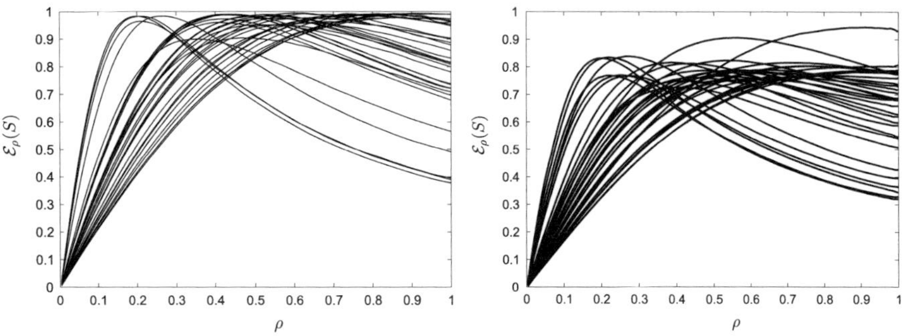

Fig. 5.4 $\mathcal{E}_\rho(S)$, $\rho \in (0, 1]$, graphs for the shapes S obtained by thresholding of 32 randomly selected galaxy images: global thresholding applied (on the *left*) and local thresholding applied (on the *right*)

Ellipticity measures $\mathcal{E}_{\rho=0.7}(S)$ and $\mathcal{E}_{\rho=0.9}(S)$ were selected according to the graphs displayed in Fig. 5.4. Precisely, 32 shapes were selected randomly and then thresholded by both the global and local methods. The graphs of $\mathcal{E}_\rho(S)$, for ρ varying through the interval $(0, 1]$, were computed. The graphs of $\mathcal{E}_\rho(S)$ corresponding to shapes obtained by the global thresholding are on the left in Fig. 5.4, while the graphs $\mathcal{E}_\rho(S)$ for shapes obtained by the local thresholding are on the right in Fig. 5.4. Our hypothesis was: "Since for both $\rho = 0.7$ and $\rho = 0.9$ the values of $\mathcal{E}_\rho(S)$ are "scattered" reasonably well, an efficient discrimination among the galaxy shapes would be enabled by using the functions/measures $\mathcal{E}_{\rho=0.7}(S)$ and $\mathcal{E}_{\rho=0.9}(S)$". Also, the selected parameters are preferred to be reasonably different. It turns out that, at least in this case, the hypothesis was valid.

Thus, each galaxy $\mathbf{g}$ was represented by a 6-dimensional feature vector determined as follows:

$$\left(\mathcal{E}_{\rho=0.7}(S'_{\mathbf{g}}),\ \mathcal{E}_{\rho=0.9}(S'_{\mathbf{g}}),\ \mathcal{E}(S'_{\mathbf{g}}),\ \mathcal{E}_{\rho=0.7}(S''_{\mathbf{g}}),\ \mathcal{E}_{\rho=0.9}(S''_{\mathbf{g}}),\ \mathcal{E}(S''_{\mathbf{g}})\right) \tag{5.10}$$

where $S'_{\mathbf{g}}$ and $S''_{\mathbf{g}}$ are the shapes (i.e. binary images) obtained from the original image (of the galaxy $\mathbf{g}$) thresholded by two selected methods (the global and local one). Some examples are in Fig. 5.5: original images are in the left column, shapes $S'_{\mathbf{g}}$ obtained by the global thresholding are in the middle column, while shapes $S''_{\mathbf{g}}$ obtained by local thresholding are in the right column.

We have used the k-Nearest Neighbour Classifier (k-NN), with $k = 5$. For the training set we have used 4 elliptical and 28 spiral galaxies (e.g. approximately 30 % of the galaxies have been used for the training – the same percentage as in [19]). The classification was performed on the complete data set (galaxies selected for the training were also included). In order to get a reliable indicator about the efficiency of the classification "mechanism" described above, 100 experiments were performed. The experiments were mutually independent – i.e. galaxies for the

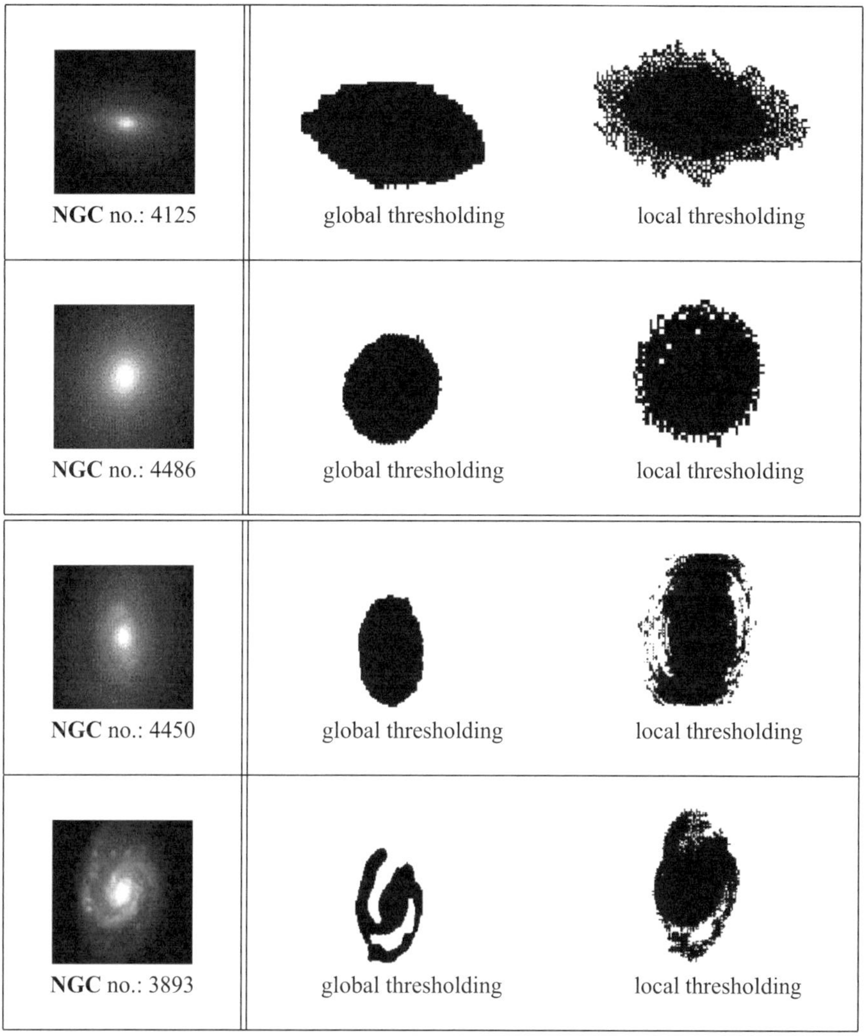

Fig. 5.5 Original images and their **NGC** number are in the *left column*; shapes obtained by the global and local thresholding are in the *middle* and *right column*, respectively

training set (4 elliptical and 28 spiral galaxies) have been selected randomly in every experiment.

The classification results were very good, and outperform the previous accuracies. Among 100 experiments performed, the classification rate of 100 % was achieved 3 times. The average classification rate was 95.6 % – better than both best rates obtained by k-NN and neural network classifiers in [19]. The minimal classification rate of 90.2 % was obtained 4 times. The classification results, for each of the 100 experiments, are displayed in Fig. 5.6. It is worth mentioning

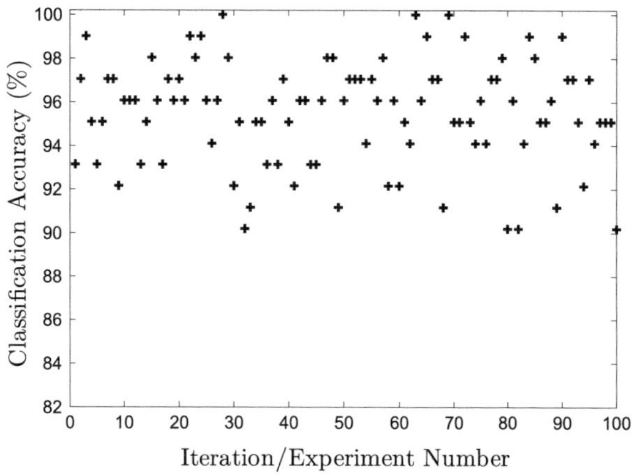

Fig. 5.6 Classification rates obtained for 100 mutually independent galaxy classification experiments

that because the ellipticity measures have predictable behavior, it was expected that good classification results might be expected (galaxy shapes have an elliptical structure). Such a prediction would not be possible if some generic shape measures were used instead. The additional tool which led to the maximum classification is the use of multiple shapes assigned to an object/image. To illustrate the latter statement we provide the classification results in experiments where a single shape is allocated to each galaxy. The same ellipticity measures: $\mathscr{E}_{\rho=0.7}(S)$, $\mathscr{E}_{\rho=0.9}(S)$, and $\mathscr{E}(S)$ were used again. As expected, smaller classification rates were obtained. The classification results, based on 100 mutually independent experiments, are displayed in Fig. 5.7. As it can be seen:

- If the shapes obtained by the global thresholding and the 3-dimensional feature vector

$$\left(\mathscr{E}_{\rho=0.7}(S'_{\mathbf{g}}),\ \mathscr{E}_{\rho=0.9}(S'_{\mathbf{g}}),\ \mathscr{E}(S'_{\mathbf{g}})\right) \tag{5.11}$$

  were used, the following rates were achieved:

  - the average classification rate: 87.5 %;
  - the maximum classification rate: 92.1 %;
  - the minimal classification rate: 82.4 %.

- If single shapes, obtained by the local thresholding method, and the 3-dimensional feature vector

$$\left(\mathscr{E}_{\rho=0.7}(S''_{\mathbf{g}}),\ \mathscr{E}_{\rho=0.9}(S''_{\mathbf{g}}),\ \mathscr{E}(S''_{\mathbf{g}})\right) \tag{5.12}$$

  were used, then the following rates were achieved:

  - the average classification rate: 92.2 %;
  - the maximum classification rate: 96.0 %;
  - the minimal classification rate: 84.3 %.

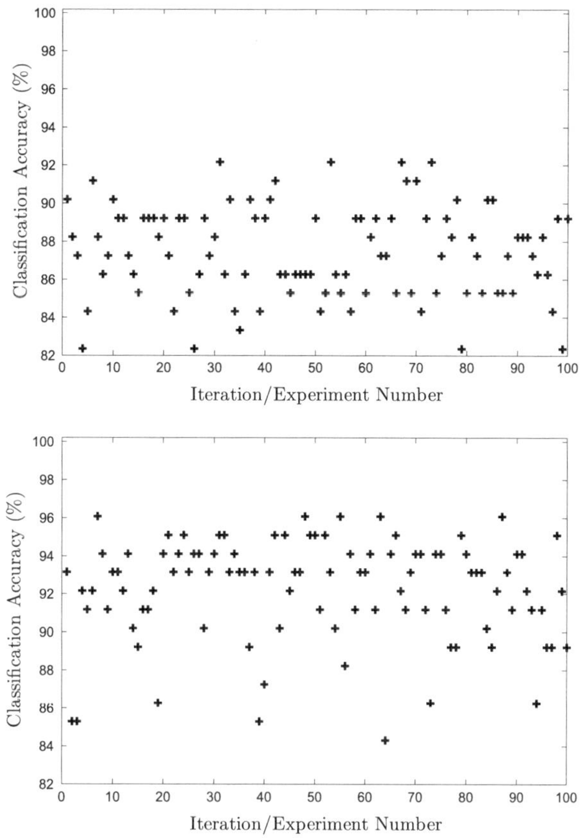

Fig. 5.7 Classification rates obtained for 100 mutually independent, simplified classification experiments. *Top row*: the global thresholding method and the feature vector (5.11) are used. *Bottom row*: the local thresholding method and the feature vector (5.12) are used

5.4 Multiple Shapes Assigned to Boundary Simplification

A final pair of experiments is described in which multiple shapes will be derived for each object. Distinct from the experiments in Sects. 5.2 and 5.3, classification will be performed using boundary based features. Therefore boundary based methods will be employed to generate multiple shapes. This is the most straightforward and appropriate approach if the input data consists of boundaries, and also ensures that the number of components does not change, that open curves remain open, etc.

Our approach to generate multiple shapes from the given data is to perform simplification of the input shapes. This can be applied at different degrees to create an arbitrary number of additional shapes. For the two examples described in this section two approaches are taken: Gaussian blurring and polygonal approximation.

Unlike the previous examples, only boundary information is provided, and there is no additional information such as object intensities. This means that the additional shapes generated will not introduce new information, although there is still a potential benefit to be gained by making different aspects of the data more explicit, whilst suppressing others. Nevertheless, the expected performance gain is likely to be less than in Sect. 5.3.

5.4.1 Closed Curves Example: MPEG-7 CE-1

In [34] a set of five shape based features (namely, a Fourier based triangularity measure [5], roundness based on the ratio of the areas of the shape and its circumscribing circle, rectangularity based on the ratio of the areas of the shape S and its minimum bounding rectangle, ellipticity based on the first affine moment invariant [30], and convexity based on the areas of the shape and its convex hull) along with two squareness measures ($\mathcal{Q}_{\beta=2}(S)$ and $\mathcal{Q}_{fit}(S)$) were combined to achieve a bull's eye test score of 74.74 % when applied to the MPEG-7 CE-1 set of 1400 shapes using a minimum Mahalanobis distance classifier.

A richer feature set can be obtained by expanding the data set to include multiple smoothed versions of the 1400 curves, and using them to compute additional features. For example, Gaussian blurring was applied at scales $\{\sigma = 2, 32, 128\}$, see some examples in Fig. 5.8. When the additional convexity values produced from these scales was included in the classifier then the test score increased to 75.76 %.

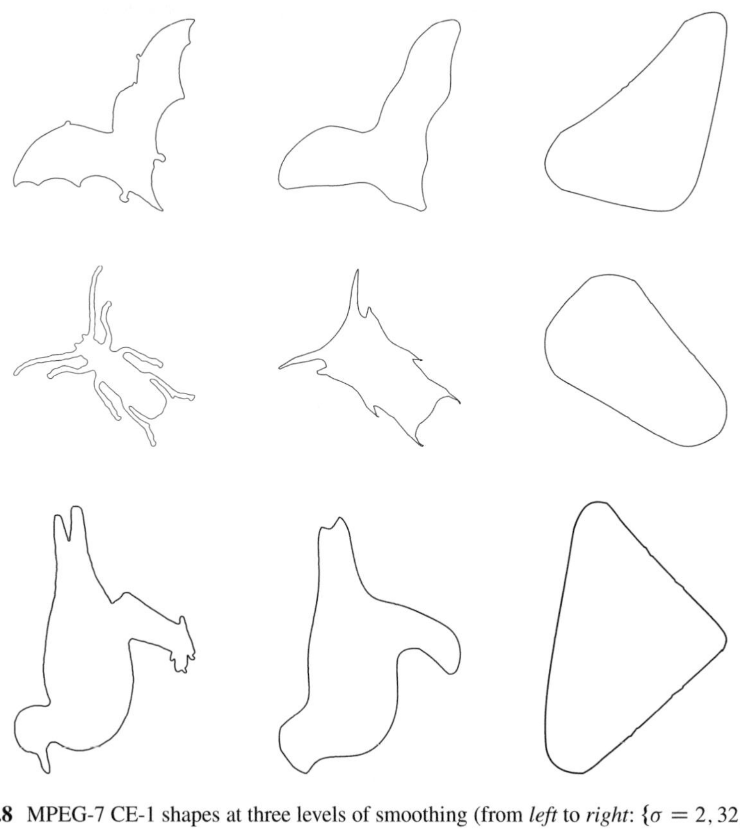

Fig. 5.8 MPEG-7 CE-1 shapes at three levels of smoothing (from *left* to *right*: $\{\sigma = 2, 32, 128\}$)

Fig. 5.9 Characters from the UJI Pen character data set. *Top row*: (**a**) single simple curve; (**b**) two intersecting curves; (**c**) three intersecting curves; (**d**) three non-intersecting curves; (**e**) a mixture of open and closed curves; (**f**) Spanish character; (**g**) non-ASCII character. *Middle row*: examples showing the variability of a single character across different writers. *Lower row*: one character progressively simplified by increasing degrees of polygonal approximation

5.4.2 Open Curves Example: UJI Pen Characters

The next experiment uses the UJI Pen character data set [6], in which handwriting samples were captured with a stylus. Each of the participating 60 writers wrote two samples of 97 characters, that included ASCII, Spanish and other non-ASCII characters, making up 11640 samples in total. Note that some of the characters are multi-stroke, and that of those, their component strokes do not necessarily touch. The top row in Fig. 5.9 illustrates some of the different types of characters in the data set, while the middle row demonstrates the wide variability in handwriting styles for a single character. The creators of the data set have split the characters into disjoint training and test sets created by 40 writers and 20 writers respectively.

We used a Support Vector Machine (SVM) to perform classification of the characters: LIBSVM [7] with a Radial Basis Function (RBF) kernel and default settings. Grid search and five-fold cross validation in the training set were used to obtain the optimized parameters and the model was then applied to the test data.

The challenging nature of the data potentially complicates the processes of feature extraction and/or matching, and in the original paper [6] the experiments were restricted to the ASCII alpha-numeric characters, while more recent work has further restricted the task to 26 classes [37]. In our experiments we will use the full set of 97 character classes. The features need to be chosen such that they can be applied to open or closed boundaries comprised of single or multiple components, ruling out many standard shape measures. In our experiments we have used: anisotropy [35], aspect ratio, convexity [43], linearity [33], line moments (both Hu's first seven rotation, translation and scale moment invariants [17] were used as well as six further moment invariants designed for character recognition [24] which are invariant to change in aspect ratio, but are *not* orientation dependent so that e.g. '6' and '9' can be distinguished), rectilinearity [41] (both the regular version R_1 and a modification in which the measure is *not* maximized over orientation), and the absolute sum of turning angles.

The classification rate obtained was 51.2 % for features extracted from the raw data. Next, the data was simplified using Ramer's polygonal approximation method [27] over a range of scales (distance thresholds of $\{1, 2, 4, 8, 16, 32, 64, 128\}$) – see the bottom row in Fig. 5.9 for some examples. When classification was performed on the data set using features from any single approximation level then no advantage was found, as the classification rate dropped to 36.40 %–50.56 %. However, when the features from several scales were combined – namely the raw data, and Ramer thresholds $\{2, 64\}$ – then an increase of classification to 56.67 % was achieved. This demonstrates the benefit of augmenting the data set by additional alternative versions of the shapes.

Of course, further improvements could be obtained by developing and using additional features, in particular those specific to the stylus and multi-stroke characteristics of the data. Examples are: trajectories (i.e. temporal information), the number of strokes, the distribution of various stroke characteristics within a character, etc.

5.5 Conclusion

In this chapter we have considered some possibilities to increase the discrimination capabilities of shape based tools used in image processing and computer vision tasks. We focused on shape based characteristics/properties with a intuitively clear meaning. Many of these properties, commonly named shape descriptors, are clearly identified (e.g. convexity, linearity, elongation, circularity, sigmodality, etc.) and methods for their computation (i.e. numerical evaluation/estimation) are derived. These methods are called shape measures. It has been noted that a single method for evaluation of a given shape descriptor does not suit all applications. That is why, for several shape descriptors, multiple shape measures have already been developed. Among them, convexity, circularity, and ellipticity are probably the shape descriptors with the largest number of measures developed for their evaluation. Multiple measures, related to the same descriptors, are used (either alternatively or jointly) as components in the feature vectors allocated to the objects/shapes analyzed. More shape measures increase the dimensionality of the space of the feature vectors, and consequently, the potential for greater efficiency in shape based tasks (classification, recognition, matching, etc.) increases as well. But the number of approaches to design a measure to certain shape property is limited. Thus, the question: *"How else can we increase the power of shape descriptor/measure based tools?"* arises. Here, we have discussed some possibilities. In Sects. 5.2 and 5.3, we considered area based shape measures (in which all the shape points are used) and show how incorporating a tuning parameter can lead to an infinite family of circularity and ellipticity measures. In Sects. 5.3 and 5.4, we have illustrated that further improvements can be obtained by assigning multiple shapes to the objects considered. As mentioned, area based measures were used in Sect. 5.3, while in Sect. 5.4 shape boundaries (i.e. operations on them) were used to allocate the multiple shapes to the objects considered, and then boundary based shape measures were employed.

Acknowledgements This work is partially supported by the Ministry of Science of the Republic of Serbia.

References

1. Aktaş, M.A., Žunić, J.: Measuring shape ellipticity. In: Pinz, A., et al. (eds.) Pattern Recognition – Joint 34th DAGM and 36th OAGM Symposium, Graz. Lecture Notes in Computer Science, vol. 7476, pp. 307–316 (2012)
2. Aktaş, M.A., Žunić, J.: A family of shape ellipticity measures for galaxy classification. SIAM J. Imaging Sci. **6**, 765–781 (2013)
3. Arandjelović, O.: Computationally efficient application of the generic shape illumination invariant to face recognition from video. Pattern Recognit. **45**, 92–103 (2012)
4. Bazell, D., Peng, Y.: A comparison of neural network algorithms and preprocessing methods for star-galaxy discrimination. Astrophys. J. Suppl. Ser. **116**, 47–55 (1998)
5. Bowman, E.T., Soga, K., Drummond, T.: Particle shape characterization using Fourier analysis. Geotechnique **51**, 545–554 (2001)
6. Castro-Bleda, M.J., Boquera, S., Gorbe, J., Zamora, F., Llorens, D., Marzal, A., Prat, F., Vilar-Torres, J.M.: Improving a DTW-based recognition engine for on-line handwritten characters by using MLPs. In: Proceedings of the 10th International Conference on Document Analysis and Recognition, Barcelona, pp. 1260–1264 (2009)
7. Chang, C.-C., Lin, C.-J.: LIBSVM: a library for support vector machines. ACM Trans. Intell. Syst. Technol. **2**, 27 (2011)
8. Di Ruberto, C., Dempster, A.: Circularity measures based on mathematical morphology. Electron. Lett. **36**, 1691–1693 (2000)
9. Dorst, L., Smeulders, A.W.M.: Length estimators for digitized contours. Comput. Vis. Graph. Image Process. **40**, 311–333 (1987)
10. Frei, Z., Guhathakurta, P., Gunn, J.E., Tyson, J.A.: A catalog of digital images of 113 nearby galaxies. Astron. J. **111**, 174–181 (1996)
11. Gautama, T., Mandić, D.P., Van Hulle, M.M.: Signal nonlinearity in fMRI: A comparison between BOLD and MION. IEEE Trans. Med. Images **22**, 636–644 (2003)

12. Goderya, S.N., Lolling, S.M.: Morphological classification of galaxies using computer vision and artificial neural networks: a computational scheme. Astrophys. Space Sci. **279**, 377–387 (2002)
13. Grisan, E., Foracchia, M., Ruggeri, A.: A novel method for the automatic grading of retinal vessel tortuosity. IEEE Trans. Med. Imaging **27**, 310–319 (2008)
14. Guo, Q., Guo, F., Shao, J.: Irregular shape symmetry analysis: theory and application to quantitative galaxy classification. IEEE Trans. Pattern Anal. Mach. Intell. **32**, 1730–1743 (2010)
15. Han, M.: The luminosity structure and objective classification of galaxies. Astrophys. J. **442**, 504–522 (1995)
16. Haralick, R.M.: A measure for circularity of digital figures. IEEE Trans. Syst. Man Cybern. **4**, 394–396 (1974)
17. Hu, M.: Visual pattern recognition by moment invariants. IRE Trans. Inf. Theory **8**, 179–187 (1962)
18. Lee, D.R., Sallee, T.: A method of measuring shape. Geogr. Rev. **60**, 555–563 (1970)
19. Lekshmi, S., Revathy, K., Prabhakaran Nayar, S.R.: Galaxy classification using fractal signature. Astron. Astrophys. **405**, 1163–1167 (2003)
20. Mähönen, P., Frantti, T.: Fuzzy classifier for star-galaxy separation. Astrophys. J. **541**, 261–263 (2000)
21. Mei, Y., Androutsos, D.: Robust affine invariant region-based shape descriptors: the ICA Zernike moment shape descriptor and the whitening Zernike moment shape descriptor. IEEE Signal Process. Lett. **16**, 877–880 (2009)
22. Odewahn, S., Stockwell, E., Pennington, R., Humphreys, R., Zumach, W.: Automated star/galaxy discrimination with neural networks. Astron. J. **103**, 318–331 (1992)
23. Otsu, N.: A threshold selection method from gray level histograms. IEEE Trans. Syst. Man Cybern. **9**, 62–66 (1979)
24. Pan, F., Keane, M.: A new set of moment invariants for handwritten numeral recognition. In: Proceedings of the International Conference on Image Processing, Austin, pp. 154–158 (1994)
25. Proffitt, D.: The measurement of circularity and ellipticity on a digital grid. Pattern Recognit. **15**, 383–387 (1982)
26. Rahtu, E., Salo, M., Heikkilä, J.: A new convexity measure based on a probabilistic interpretation of images. IEEE Trans. Pattern Anal. Mach. Intell. **28**, 1501–1512 (2006)
27. Ramer, U.: An iterative procedure for the polygonal approximation of plane curves. Comput. Graph. Image Process. **1**, 244–256 (1972)
28. Rangayyan, R.M, Elfaramawy, N.M., Desautels, J.E.L., Alim, O.A.: Measures of acutance and shape for classification of breast-tumors. IEEE Trans. Med. Imaging **16**, 799–810 (1997)
29. Richardson, E., Werman, M.: Efficient classification using the Euler characteristic. Pattern Recognit. Lett. **49**, 99–106 (2014)
30. Rosin, P.L.: Measuring shape: ellipticity, rectangularity, and triangularity. Mach. Vis. Appl. **14**, 172–184 (2003)
31. Rosin, P.L.: Measuring sigmoidality. Pattern Recognit. **37**, 1735–1744 (2004)
32. Rosin, P.L., Mumford, C.L.: A symmetric convexity measure. Comput. Vis. Image Underst. **103**, 101–111 (2006)
33. Rosin, P.L., Pantović, J., Žunić, J.: Measuring linearity of closed curves and connected compound curves. In: Kyoung Mu Lee et al.: (eds.) 11th Asian Conference on Computer Vision, Daejeon. Lecture Notes in Computer Science, vol. 7726, pp. 310–321 (2012)
34. Rosin, P.L., Žunić, J.: Measuring squareness and orientation of shapes. J. Math. Imaging Vis. **39**, 13–27 (2011)
35. Rosin, P.L., Žunić, J.: Orientation and anisotropy of multi component shapes from boundary information. Pattern Recognit. **44**, 2147–2160 (2011)
36. Stojmenović, M., Nayak, A., Žunić, J.: Measuring linearity of planar point sets. Pattern Recognit. **41**, 2503–2511 (2008)

37. Teja, S.P., Namboodiri, A.M.: A ballistic stroke representation of online handwriting for recognition. In: International Conference on Document Analysis and Recognition, Washington, DC, pp. 857–861 (2013)
38. Tool, A.Q.: A method for measuring ellipticity and the determination of optical constants of metals. Phys. Rev. (Ser. I) **31**, 1–25 (1910)
39. Wang, B.: Shape retrieval using combined Fourier features. Opt. Commun. **284**, 3504–3508 (2011)
40. Žunić, J., Martinez-Ortiz, C.: Linearity measure for curve segments. Appl. Math. Comput. **215**, 3098–3105 (2009)
41. Žunić, J., Rosin, P.L.: Rectilinearity measurements for polygons. IEEE Trans. Pattern Anal. Mach. Intell. **25**, 1193–1200 (2003)
42. Žunić, J., Rosin, P.L.: A new convexity measurement for polygons. IEEE Trans. Pattern Anal. Mach. Intell. **26**, 923–934 (2004)
43. Žunić, J., Rosin, P.L.: Convexity measure for shapes with partially extracted boundaries. Electron. Lett. **43**, 380–382 (2007)
44. Žunić, D., Žunić, J.: Shape ellipticity from Hu moment invariants. Appl. Math. Comput. **226**, 406–414 (2014)
45. Žunić, J., Hirota, K., Rosin, P.L.: A Hu moment invariant as a shape circularity measure. Pattern Recognit. **43**, 47–57 (2010)
46. Žunić, J., Kakarala, R., Aktaş, M.A.: Elliptical shape signature (Submitted)

Chapter 6
Shape Distances for Binary Image Segmentation

Frank R. Schmidt, Lena Gorelick, Ismail Ben Ayed, Yuri Boykov, and Thomas Brox

Abstract Shape distances are an important measure to guide the task of shape classification. In this chapter we show that the right choice of shape similarity is also important for the task of image segmentation, even at the absence of any shape prior. To this end, we will study three different shape distances and explore how well they can be used in a trust region framework. In particular, we explore which distance can be easily incorporated into trust region optimization and how well these distances work for theoretical and practical examples.

6.1 Shape Acquisition and Shape Distances

An important task of shape analysis is the acquisition of shapes that we want to analyze. One classical approach is *binary image segmentation* that can be formulated as an energy minimization approach. In other words, we define an energy function $E : \mathscr{S} \to \mathbb{R}$ that evaluates how well a certain shape $S \in \mathscr{S}$ of a *chosen shape space* $\mathscr{S}$ fits to the observed image and we are interested in the minimizer $S^* := \arg\min_{S\in\mathscr{S}} E(S)$ of the energy E.

The shape space $\mathscr{S}$ is usually equipped with a distance dist: $\mathscr{S} \times \mathscr{S} \to \mathbb{R}_0^+$. The literature is divided on the exact definition of a *distance*. Sometimes, but not always it is equated with a metric. In this chapter we call any positive-definite function

F.R. Schmidt (✉) • T. Brox
Computer Science Department and BIOSS Centre for Biological Signalling Studies, University of Freiburg, Georges-Köhler Allee 52, Freiburg, Germany
e-mail: schmidt@cs.uni-freiburg.de; brox@cs.uni-freiburg.de

L. Gorelick • Y. Boykov
Computer Science Department, University of Western Ontario, London, ON, Canada
e-mail: lenagorelick@gmail.com; yuri@csd.uwo.ca

I.B. Ayed
École de Technologie Supérieure, University of Quebec, Montreal, QC, Canada
e-mail: ismail.benayed@etsmtl.ca

M. Breuß et al. (eds.), *Perspectives in Shape Analysis*, Mathematics and Visualization, DOI 10.1007/978-3-319-24726-7_6

$\mathrm{dist}(\cdot,\cdot)$ a distance. Such functions satisfy

$$\mathrm{dist}(A, B) = 0 \Leftrightarrow A = B.$$

In the literature these functions are also referred to as *pre-metrics*. Any shape distance defines a topology of the shape space. In contrast to finite dimensional metric spaces, these topologies are in general not equivalent to one another. In other words, whether a shape $S \in \mathscr{S}$ is a local minimum of an energy E depends on the chosen shape distance $\mathrm{dist}(\cdot,\cdot)$.

In this chapter, we will explore the influence that a shape distance can have on an image segmentation problem. This influence is only observable if $E(\cdot)$ cannot be minimized globally. Note that in contrast to other image segmentation applications like [7–9, 15], we do not use a shape distance in order to enforce a specific shape prior. The only influence that the shape distance has on our optimization task is the definition of a local minimum of the energy E.

This chapter is organized as follows. In Sect. 6.2, we will revisit binary image segmentation that can be solved globally and its extension to the trust region approach [14]. In Sect. 6.3, we will present different shape distances and explore if they can be used in a trust region framework. In Sect. 6.4, we will show how the chosen shape distance drives the optimization process. Section 6.5 provides a summary of this chapter.

6.2 Binary Image Segmentation

Binary image segmentation is an important task in computer vision. The goal is to distinguish the object from the background within an image. An image is a mapping $I\colon \Omega \to \mathbb{R}^3$ that assigns to every pixel $x \in \Omega$ of the d-dimensional connected image domain $\Omega \subset \mathbb{R}^d$ a color $I(x) \in \mathbb{R}^3$. A binary segmentation can be modelled as a mapping $u\colon \Omega \to \mathbb{B}$ where $\mathbb{B} = \{0, 1\}$ encodes the object ($u(x) = 1$) and the background ($u(x) = 0$) respectively. A segmentation can also be represented as a subset $S \subset \Omega$. The relationship between S and u is described via $x \in S \Leftrightarrow u(x) = 1$. In the following, we call the binary labeling u a *segmentation* and the set S a *shape*.

Given a shape S, one can apply different image filters to object and background in order to emphasize the object (cf. Fig. 6.1). In medical image analysis, S can be used to visualize organs or arteries [23]. Object detection tasks can be addressed better if one works with a segmented object instead of a bounding box [12].

6.2.1 Appearance Models

Classically, the binary image segmentation models the object and the background of an image as a sampling from color distributions $\mathrm{pdf}_{\mathrm{obj}}$ and $\mathrm{pdf}_{\mathrm{bg}}$. Using the notation

Fig. 6.1 If the observed object is easy to distinguish from the background (*left image*), a per-pixel data term works very well in practice. To remove noise from a threshold solution (*central image*), an additional length term is used (*right image*). The resulting energy (6.1) can be easily optimized via a graph cut [1] or a primal-dual approach [5]

$\langle f, g \rangle := \int_{\Omega} f(x) \cdot g(x)\, \mathrm{d}x$, image segmentation can be cast as minimizing the energy

$$E(u) = \langle f, u \rangle \qquad \text{with } f(x) = \log\left(\frac{\mathrm{pdf}_{\mathrm{bg}}(x)}{\mathrm{pdf}_{\mathrm{obj}}(x)}\right).$$

While this energy can be easily optimized via a simple thresholding method, the optimal solution exhibits typically a high amount of noise (cf. Fig. 6.1). Therefore, Mumford and Shah proposed in [18] to add the length of the segmentation's boundary as a regularizing term to the energy, resulting in

$$E(u) = \langle f, u \rangle + \mathrm{len}(\partial S) \qquad \text{with } S = \{x \in \Omega \,|\, u(x) = 1\}. \tag{6.1}$$

A discrete formulation approximates the length term via the *Cauchy-Crofton formula* and minimizes the energy via a graph cut approach [1]. A continuous formulation solves the problem via a primal-dual approach that can be efficiently parallelized on GPUs [5]. In the following, we assume that we work in a computer environment where (6.1) can be easily optimized. Whether the discrete or the continuous formulation is used is not important for the rest of this chapter.

In general, $f : \Omega \rightarrow \mathbb{R}$ can be an arbitrary integrable function that need not to be derived from color distributions. In the past, different attempts have been made to model the appearance of object and background by using more information than just the color information $I(x)$ at a pixel x. Besides using more modalities like depth or infra-red information, it is common to use local features like Fourier features, Gabor features or more general texture features [4, 24]. All these approaches can be seen as an attempt of altering the data term f in (6.1). In practice, these approaches improve the segmentation. Nonetheless, these patch-based approaches become less reliable for pixels close to the object's boundary, since the features will then mix object and background information. In the following, we revisit alternative approaches.

6.2.2 Multiple Models and Holistic Distributions

While the energy (6.1) can be applied very successfully if the appearance of object and background vary considerably, it struggles if certain appearances appear in both, the object and the background regions. It was therefore suggested by Delong et al. [10] not to use one but multiple distributions for object and background (cf. Fig. 6.2). The method computes a sub-labeling $u_0\colon \Omega \to \{1,\dots,K\}$ of the image domain Ω and a binary labeling $u_1\colon \{1,\dots,K\} \to \mathbb{B}$ of these K sub-labels. As a result, the method computes simultaneously a superpixel representation u_0 and its binary segmentation u_1. In conjunction, these two functions induce a binary labeling $u\colon \Omega \to \mathbb{B}$ via $u(x) = u_1 \circ u_0(x)$.

While the resulting minimization problem is the instance of an NP hard problem, the approximation that is obtained via α-expansion [3, 11] proved to be more reliable than the binary segmentation driven by (6.1). Nonetheless, the optimization process can take a long time and is therefore not fit for fast segmentation tasks.

In order to model the appearance of different colors without the need to find an optimal superpixel representation, we advocate the concept of *holistic histograms*. To this end, let us assume that we have pre-detected n appearances in an image. An appearance can be based on color, texture or other features. Further, assume that we can decide for every pixel $x \in \Omega$ whether this appearance is present at x. This results in n appearance detectors $f_i\colon \Omega \to \mathbb{B}$. If we partition an image into disjoint areas of the same color, each f_i would represent the indicator function of one of these areas. Nonetheless, it is also possible that different f_i intersect in certain areas. This is for example the case if we have one detector for "blue pixels" and one feature-based detector for the image class "sky". Given a segmentation u, we can now compute the following histogram

$$h(u) = (\langle f_1, u\rangle, \dots, \langle f_n, u\rangle) \in \mathbb{R}^n. \tag{6.2}$$

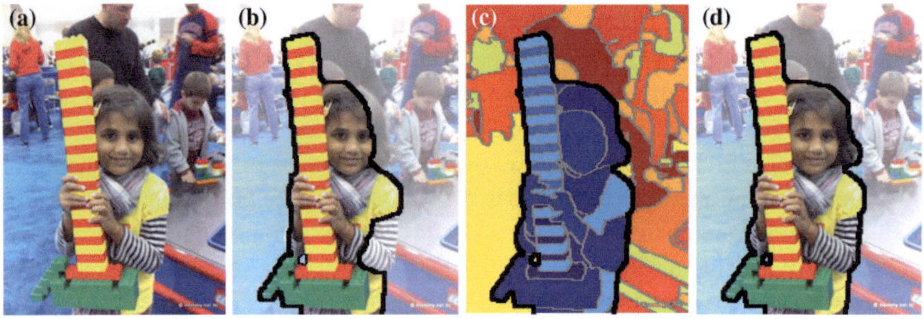

Fig. 6.2 If the object and the background contain similar appearances (**a**), the global optimum of (6.1) does not provide a good segmentation (**b**). Performing a hierarchical segmentation [10] improves the model of the scene (**c**) and provides a more accurate binary segmentation (**d**)

Note that this histogram cannot be pre-computed on the pixel-level. It depends on the whole segmentation u and will change during the optimization process. Since there are also detectors that provide only probabilities about the presence of a certain appearance, we can extend the detectors to $f_i\colon \Omega \to \mathbb{R}_0^+$.

If we want to solve a segmentation task that is scale-invariant, we prefer to work with distributions, i.e., normalized histograms, instead of histograms. Given the appearance detectors f_i as above, we obtain the *holistic distribution*

$$p(u) = \left(\frac{\langle f_1, u\rangle}{\left\langle \sum_{j=1}^{n} f_j, u \right\rangle}, \dots, \frac{\langle f_n, u\rangle}{\left\langle \sum_{j=1}^{n} f_j, u \right\rangle} \right) \in \mathbb{R}^n. \tag{6.3}$$

If a prior distribution $q \in \mathbb{R}^n$ is learned, we would like to use a distribution distance to penalize the deviation of $p(u)$ from the prior q. Combining the Bhattacharyya distance between the distributions with a length term results in the energy

$$E(u) = -\log \left(\sum_{i=1}^{n} \sqrt{\frac{q_i \cdot \langle f_i, u\rangle}{\left\langle \sum_{j=1}^{n} f_j, u \right\rangle}} \right) + \operatorname{len}(u) \tag{6.4}$$

that we want to minimize.

6.2.3 Submodular and Convex Relaxations

Recently, Tang et al. [25] proposed an unsupervised segmentation approach that rewards the L^1-distance between the object's histogram $h(u)$ and the background's histogram $h(1-u)$. Since this results in the minimization of a submodular energy, it can be solved globally and its solution provides for a much better segmentation than the optimization of (6.1). Nonetheless, it cannot be used in order to solve (6.4), which uses distributions instead of histograms.

Nieuwenhuis et al. [19] addressed a problem related to (6.4). Instead of a binary segmentation they addressed a multi-region segmentation, where ratio constraints for each region are encouraged. They addressed this problem by computing the global optimum of an approximation of the original energy with respect to labelings $u_i\colon \Omega \to [0,1]$. Since the *threshold theorem* [6] is not satisfied for the convex function, it cannot be guaranteed that the derived segmentations $\hat{u}_i\colon \Omega \to \mathbb{B}$ is even a local optimum of the approximation.

To guarantee local optimality, Gorelick et al. [14] proposed a method that combines the trust region framework with a class of energies that also includes (6.4). Since we want to explore the relationship between local optimization methods and shape distances, we will focus on the approximation scheme of [14]. After revisiting it in Sect. 6.2.4, in Sect. 6.3, we will study shape distances that define different *trust regions* and thus, compute different local minima.

6.2.4 Trust Region

Trust region methods are used to find a (local) minimum of a function E. Naturally, these methods are only used if it is difficult to find the global optimum of the energy E. The idea is to use an approximation $\tilde{E}$ of E that is exact at a certain feasible solution u_0. If the set of all feasible solutions is equipped with a distance function $\text{dist}(\cdot,\cdot)$, the trust region approach iteratively solves the *trust region sub-problem*

$$\operatorname*{arg\,min}_{\text{dist}(u_0,u)<d} \tilde{E}(u). \tag{6.5}$$

If the solution $\hat{u}$ of this problem reduces the actual energy considerably, i.e.,

$$E(\hat{u}) \leq \alpha E(u_0) \qquad\qquad \text{with } 0 < \alpha < 1,$$

$\hat{u}$ is accepted as a new approximate solution u_0 (cf. Fig. 6.3). Otherwise the region in which we trust the approximation is reduced, i.e., d is multiplied with a factor β, $0 < \beta < 1$. These steps are repeated until the distance d is small enough.

Since we have to minimize the *trust region sub-problem* (6.5) globally, we like to use approximations $\tilde{E}$ that are easy to optimize. If E is differentiable it can be approximated by a linear Taylor approximation. In the case that the space of feasible solutions is a linear space $\mathbb{R}^N$ equipped with the canonical metric $\text{dist}(u_0, u) =$

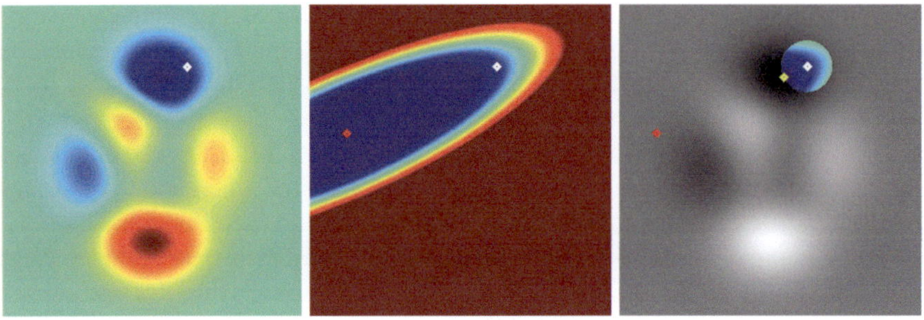

Fig. 6.3 *Left:* For a complex energy E one can use an approximation $\tilde{E}$ that is exact at a point u_0 (*white dot*). *Middle:* The global minimizer u^* (*red dot*) of $\tilde{E}$ will in general not improve the value of the original energy E. *Right:* Trust region approaches trust $\tilde{E}$ in a small vicinity of u_0 (*colored circle*). For a sufficient small vicinity, the optimizer $\hat{u}$ (*yellow dot*) of (6.6) improves the energy E

$\|u_0 - u\|$, a solution of (6.5) is

$$\arg\min_{\|u-u_0\|<d} E(u_0) + \langle E'(u_0), u - u_0 \rangle = u_0 - d \cdot \frac{E'(u_0)}{\|E'(u_0)\|}.$$

Therefore, the trust region approach can be understood as a generalization of the *normalized gradient descent* approach. In practice, second order approximations of E are used [20].

Gorelick et al. [14] used functions E that can be described as the sum of a differentiable function E_1 and a length term. The approximation $\tilde{E}$ only uses a linear approximation for E_1. The length term is not approximated at all, resulting in:

$$E(u) = E_1(u) + \mathrm{len}(u) \qquad \tilde{E}(u) = E_1(u_0) + \langle E_1'(u_0), u - u_0 \rangle + \mathrm{len}(u)$$

To solve the trust region sub-problem (6.5), a Lagrangian formulation[1]

$$\arg\min_u \langle E_1'(u_0), u \rangle + \mathrm{len}(u) + \lambda \,\mathrm{dist}(u_0, u) \tag{6.6}$$

was used and a reciprocal relationship between the Lagrangian factor λ and the distance d was exploited. For more details we refer to [14].

Remark 6.1 Note that in contrast to a gradient descent approach, the length term need not to be approximated, since we can optimize energies of the form (6.1) that also include length terms. If we also approximated the length term, the resulting sub-problem would include a curvature motion as explored in the level set framework [21]. It was shown in [13] that not approximating the length term is beneficial in practice. The resulting method is faster and possesses fewer local minima than the level set approach of [16].

Remark 6.2 The Lagrangian formulation (6.6) uses the current solution u_0 as a prior. If we want to trust $\tilde{E}$ in a smaller vicinity, λ is automatically increased and the prior has a stronger influence. This results in a process where the global optimum of (6.6) is pushed towards u_0 with increasing λ. Note that it is not necessary to *tune the parameter* λ to the application. λ is instead automatically adapted by the trust region framework. This adaptation is driven by the original energy E.

Since the prior in (6.6) depends on the distance $\mathrm{dist}(\cdot, \cdot)$, we explore in the next section different distance functions for shapes. These distances define different sub-problems (6.6) and thus different local minima of E. In order to globally optimize (6.6), we focus on shape distance functions that are affine in u. In these cases, the trust region sub-problem is of the form (6.1), which we can easily optimize.

[1] Since we are only interested in the minimizer, we removed constant terms from the energy.

6.3 Shapes and Shape Distances

In order to avoid shapes $S \subset \Omega$ that can only be created by the set-theoretical *axiom of choice* or sets that are null-sets in the Lebesguean sense, we want to focus on shapes S that are open sets. Since we will also be interested in the boundary ∂S of a shape S, we want to exclude those shapes whose boundaries are empty. With

$$\mathscr{S} := \{S \subset \Omega | S \text{ is open and } \partial S \neq \emptyset\} \tag{6.7}$$

we denote the set of all those shapes. Since the boundary $\partial S := \overline{S} \cap \overline{S^c}$ is the intersection of the closure of S and the closure of its complement S^c, only the empty set and the whole domain Ω are exempted from the shape space $\mathscr{S}$. This is a consequence of Ω being connected.

In order to equip the shape space $\mathscr{S}$ with a distance, we have two choices. We can either define a distance $\text{dist}(S_0, S_1)$ with respect to the whole shapes S_i or with respect to their boundaries ∂S_i. In the first case we speak of *region-based* distances and in the latter case we speak of *contour-based* distances. While contour-based distances proved to be very descriptive [17], it is difficult to incorporate them into image segmentation tasks. The goal of this section is to overcome this limitation of contour-based distances by approximating them in a regional sense.

To study relationships between S and ∂S, we use the following representations.

Definition 6.1 Given a shape $S \in \mathscr{S}$, we denote the *indicator function*, the *signed indicator function*, the *distance function* and the *signed distance function* (cf. Fig. 6.4) as $\mathbb{1}_S, \text{sid}_S, \text{df}_S, \text{sdf}_S \colon \Omega \to \mathbb{R}$ and define them via

$$\mathbb{1}_S(x) := \begin{cases} 1 & , x \in S \\ 0 & , x \notin S \end{cases} \qquad \text{sid}_S(x) := \begin{cases} -1 & , x \in S \\ +1 & , x \notin S \end{cases}$$

$$\text{df}_S(x) := \min_{s \in \partial S} \|x - s\| \qquad \text{sdf}_S(x) := \text{sid}_S(x) \cdot \text{df}_S(x).$$

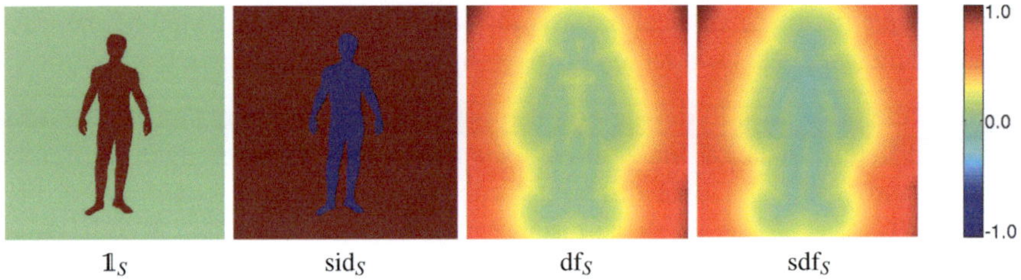

Fig. 6.4 For a shape $S \in \mathscr{S}$, we use different implicit representations, the *indicator function* $\mathbb{1}_S$, the *signed indicator function* sid_S, the *distance function* df_S and the *signed distance function* sdf_S

In Sect. 6.3.1 we will study the Hamming distance $\text{dist}_H(\cdot,\cdot)$ and show its restrictions for the trust region sub-problem (6.6). In Sect. 6.3.2, we will study a contour-based distance $\text{dist}_{L^2}(\cdot,\cdot)$ and explore its regional approximation $\text{dist}_2(\cdot,\cdot)$.

In particular, we will show that both, $\text{dist}_H(u_0, u)$ and $\text{dist}_2(u_0, u)$ are affine in u and can therefore be easily incorporated into (6.6). $\text{dist}_{L^2}(u_0, u)$ on the other hand is not affine in u and cannot be used in the trust region framework. For that reason, we have to approximate it with the distance $\text{dist}_2(u_0, u)$ that is affine in u.

6.3.1 Regional Hamming Distance and Its Restrictions

The Hamming distance of two shapes $A, B \in \mathscr{S}$ is defined as the area of its symmetric difference $A \triangle B := (A \setminus B) \sqcup (B \setminus A)$:

$$\text{dist}_H(A, B) := \text{area}(A \triangle B). \tag{6.8}$$

Using the signed indicator function sid_A, we can rewrite the Hamming distance as

$$\text{dist}_H(A, B) = \int_B \text{sid}_A(x)\,\mathrm{d}x - \int_A \text{sid}_A(x)\,\mathrm{d}x. \tag{6.9}$$

To see that Eqs. (6.8) and (6.9) describe the same function, note that (Fig. 6.5)

$$\begin{aligned}\int_B \text{sid}_A(x)\,\mathrm{d}x - \int_A \text{sid}_A(x)\,\mathrm{d}x &= \int_{\substack{(B\setminus A)\\ \sqcup(B\cap A)}} \text{sid}_A(x)\,\mathrm{d}x - \int_{\substack{(A\setminus B)\\ \sqcup(B\cap A)}} \text{sid}_A(x)\,\mathrm{d}x \\ &= \int_{(B\setminus A)} 1\,\mathrm{d}x - \int_{(A\setminus B)} (-1)\,\mathrm{d}x = \text{area}\,(A \triangle B)\,.\end{aligned}$$

Shapes $A,B \in \mathscr{S}$ Hamming distance sid_A

Fig. 6.5 The Hamming distance $\text{dist}_H(A, B)$ of two shapes $A, B \in \mathscr{S}$ is the area of its symmetric difference. Using the signed indicator function sid_A, this distance becomes affine in B (cf. (6.9))

The advantage of the formulation (6.9) is that it can be integrated into the trust region sub-problem (6.6). Using the notation $\langle f, S\rangle := \langle f, \mathbb{1}_S\rangle = \int_S f(x)\,\mathrm{d}x$, we obtain

$$\mathrm{dist}_H(A,B) = \langle \mathrm{sid}_A, B\rangle + C, \qquad C := -\langle \mathrm{sid}_A, A\rangle\,,$$

which is affine in B. We will show in the following that the Hamming distance is a shape distance that is disadvantageous for shape prior-based image segmentation. For this reason we want to study different shape distances.

Example 6.1 Let us consider two different shapes $A, B \in \mathscr{S}$ and the energy function $E(S) := \langle \mathrm{sid}_A, S\rangle$. Its unique minimizer is $S^* = A$. Adding a weighted shape prior with respect to B leads to the energy

$$\begin{aligned} E_\lambda(S) =& (1-\lambda)\cdot E(S) + \lambda \cdot \mathrm{dist}_H(B,S) \\ =& \langle (1-\lambda)\,\mathrm{sid}_A + \lambda\,\mathrm{sid}_B, S\rangle + \lambda C, \qquad C := -\langle \mathrm{sid}_B, B\rangle\,. \end{aligned}$$

If we denote with S^*_λ a global minimum of $E_\lambda(S)$, we obtain a mapping $m : \lambda \mapsto S^*_\lambda$ that starts at $S^*_0 = A$ and ends at $S^*_1 = B$. One major disadvantage of the used Hamming distance is that m is not a continuous morphing (cf. 1st plot of Fig. 6.6).

Theorem 6.1 *If we define a mapping $m : [0,1] \to \mathscr{S}$ as above, the following holds:*

$$\begin{aligned} m(\lambda) &= A, && \text{if } 0 \le \lambda < \frac{1}{2} \\ m(\lambda) &= B, && \text{if } \frac{1}{2} < \lambda \le 1 \end{aligned}$$

Proof *A minimizer of E_λ is easily found by thresholding $(1-\lambda)\,\mathrm{sid}_A + \lambda\,\mathrm{sid}_B$ at 0. The following observation*

$$(1-\lambda)\,\mathrm{sid}_A(x) + \lambda\,\mathrm{sid}_B(x) = \begin{cases} -1 + 2\lambda & , \text{if } x \in A \setminus B \\ 1 - 2\lambda & , \text{if } x \in B \setminus A \\ -1 & , \text{if } x \in A \cap B \\ 1 & , \text{if } x \notin A \sqcup B \end{cases}$$

proves the theorem. □

Because of this theorem, we cannot use dist_H in (6.6) in order to *push* the segmentation towards a specific shape. As mentioned in Remark 6.2, continuous morphings are essential for a successful trust region computation. With the Hamming distance we can only encode a hard constraint. In order to handle soft constraints, we will explore next a contour-based distance and its region-based approximation.

6.3.2 L^2 Contour Distance and Its Regional Approximation

An L^2 distance between two shapes $A, B \in \mathscr{S}$ can be formulated as

$$\text{dist}_{L^2}(A, B) := \left(\int_{\partial B} \min_{x \in \partial A} \|x - s\|^2 \, \text{d}s \right)^{\frac{1}{2}}. \tag{6.10}$$

This distance only considers the shapes' boundaries. The interior of the shapes is completely ignored. In order to simplify the study of this distance, we will only consider concentric balls B_ρ of radius $\rho > 0$. For these examples, the distance can be computed analytically. Given two concentric balls of radius r and R, we obtain

$$\text{dist}_{L^2}(B_r, B_R)^2 = \int_{\partial B_R} (R - r)^2 \, \text{d}s = 2\pi R \cdot (R - r)^2.$$

The distance $\text{dist}_{L^2}(\cdot, \cdot)$ is not symmetric and thus not a metric. Analogously to Sect. 6.3.1, we want to study the influence that a $\text{dist}^2_{L^2}$-based shape prior has on image segmentation.

Example 6.2 Let us consider the radii $0 < r \leq R$. The unique minimizer of the energy $E(\rho) = \langle \text{sid}_{B_R}, B_\rho \rangle$ is obtained for $\rho^* = R$. Adding B_r as a shape prior, results in the following energy

$$\begin{aligned}
E_\lambda(\rho) =& (1 - \lambda) \langle \text{sid}_{B_R}, B_\rho \rangle + \lambda \, \text{dist}_{L^2}(B_\rho, B_r) \\
=& (1 - \lambda) \langle \text{sid}_{B_R}, B_\rho \rangle + \lambda \int_{\partial B_r} \min_{x \in \partial B_\rho} \|x - s\|^2 \, \text{d}s \\
=& (1 - \lambda) \langle \text{sid}_{B_R}, B_\rho \rangle + \lambda \cdot 2\pi r (\rho - r)^2 \\
=& \begin{cases} \lambda \cdot 2\pi r(\rho - r)^2 - (1 - \lambda)\pi \rho^2 & , \text{if } \rho \leq R \\ \lambda \cdot 2\pi r(\rho - r)^2 + (1 - \lambda)\pi(\rho^2 - 2R^2) & , \text{if } \rho > R \end{cases}
\end{aligned}$$

The global minimum of E_λ is (cf. 2nd plot of Fig. 6.6)

$$\rho^*(\lambda) = \begin{cases} R & , \lambda \leq \frac{1}{2r+1} \\ \min\left(r + \frac{(1-\lambda)r}{2\lambda r - (1-\lambda)}, R\right) & , \lambda > \frac{1}{2r+1}. \end{cases}$$

First of all, this means that ρ^* continuously changes from $\rho^*(0) = R$ to $\rho^*(1) = r$. We are therefore able to continuously push the segmentation to a certain shape prior. Secondly, there is a small range for λ where the shape prior is ignored. This means that a strong data term always overrules the shape prior. Both of these properties are important for the trust region sub-problem (6.6). A major disadvantage of $\text{dist}^2_{L^2}$ over the Hamming distance is the fact that it cannot

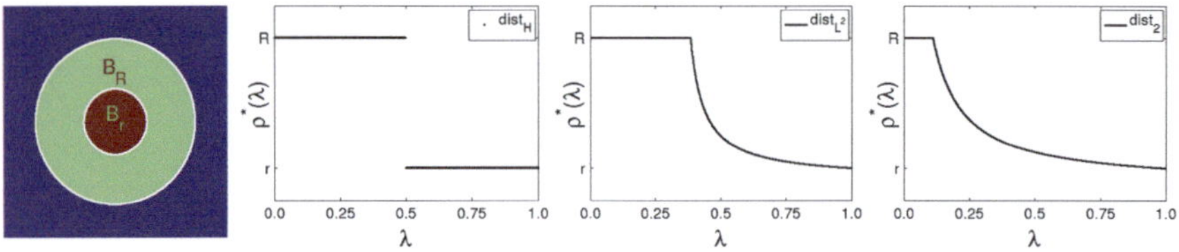

Fig. 6.6 *Image:* As shapes we consider concentric balls B_ρ of radius ρ. Given the radii $0 < r \leq R$, we consider an energy $E_\lambda(\rho) = (1-\lambda)\langle \mathrm{sid}_{B_R}, B_\rho \rangle + \lambda\, \mathrm{dist}(B_r, B_\rho)$. The first term favors $\rho = R$ and the second term favors $\rho = r$. The minimizer ρ^* of E_λ depends on λ. *Plots:* Using dist_H leads to a non-continuous function $\rho^*(\lambda)$. For dist_{L^2} and dist_2, $\rho^*(\lambda)$ becomes continuous

be incorporated into (6.6) in such a way that results in an energy of the form (6.1). This is because $\mathrm{dist}^2_{L^2}(A, B)$ is not affine in B. Therefore, we seek in the following an affine approximation of $\mathrm{dist}^2_{L^2}$.

In order to compute $\mathrm{dist}^2_{L^2}(A, B)$, an explicit matching $\xi\colon \partial B \to \partial A$ between the shapes' boundaries is computed, where $\xi(s) := \arg\min_{x\in\partial A} \|x - s\|$. If we denote the straight line from $\xi(s)$ to s as

$$\ell_s : [0,1] \to \Omega \qquad\qquad \ell_s(t) := (1-t)\cdot \xi(s) + t \cdot s,$$

we observe $\mathrm{df}_A(\ell_s(t)) = t \cdot \|\xi(s) - s\|$. This leads to

$$\begin{aligned}
\mathrm{dist}_{L^2}(A,B)^2 &= \int_{\partial B} \min_{x\in\partial A} \|x-s\|^2 \,\mathrm{d}s = \int_{\partial B} \|\xi(s)-s\|^2 \,\mathrm{d}s \\
&= \int_{\partial B}\int_0^1 2t\,\|\xi(s)-s\|^2 \,\mathrm{d}t\,\mathrm{d}s \\
&= \int_{\partial B}\int_0^1 2\,\mathrm{df}_A(\ell_s(t)) \cdot \left\|\ell_s'(t)\right\| \,\mathrm{d}t\,\mathrm{d}s \\
&= \int_{\partial B}\int_{\ell_s} 2\,\mathrm{df}_A(x)\,\mathrm{d}x\,\mathrm{d}s
\end{aligned}$$

In the last equation, we rewrote the equation in means of the line integral evaluated along the line ℓ_s, which still depends on $\xi(s)$. Since ξ is in general difficult to compute, we want to replace the integration domain $(s,t) \mapsto (1-t)\cdot\xi(s) + t\cdot s$ with a simpler domain. Note that if both A and B are concentric circles, the integration domain is exactly $A \,\triangle\, B$. Therefore, we will approximate $\mathrm{dist}^2_{L^2}$ via

$$\begin{aligned}
\mathrm{dist}_2(A,B) &:= \int_{B\setminus A} 2\,\mathrm{df}_A(x)\,\mathrm{d}x + \int_{A\setminus B} 2\,\mathrm{df}_A(x)\,\mathrm{d}x \\
&= \int_{B\setminus(B\cap A)} 2\,\mathrm{sdf}_A(x)\,\mathrm{d}x - \int_{A\setminus(B\cap A)} 2\,\mathrm{sdf}_A(x)\,\mathrm{d}x
\end{aligned}$$

$$= \int_B 2\,\mathrm{sdf}_A(x)\,\mathrm{d}x - \int_A 2\,\mathrm{sdf}_A(x)\,\mathrm{d}x \qquad (6.11)$$

This distance can be easily integrated into the sub-problem (6.6), because it is affine in B, similar to the Hamming distance formulation (6.9). The main difference between these two distances is that instead of sid_A we use the signed distance function sdf_A.

Note that in general, dist_2 does not approximate $\mathrm{dist}^2_{L^2}$ very well. First of all, the integration domain $(s,t) \mapsto (1-t)\cdot\xi(s) + t\cdot s$ does not always coincide with $A \triangle B$. Even if it does, the explicit parameterization of the integration domain is partly ignored. Only the variation in the direction of ℓ_s is considered correctly. As a result, the distance between two concentric balls becomes

$$\begin{aligned}\mathrm{dist}_2(B_r, B_R) &= \int_{B_R} 2(|x|-r)\,\mathrm{d}x - \int_{B_r} 2(|x|-r)\,\mathrm{d}x = 4\pi\left(\frac{R^3}{3} - \frac{R^2 r}{2} + \frac{r^3}{6}\right)\\ &= 2\pi R(R-r)^2\cdot\left(1 + \frac{r-R}{3R}\right).\end{aligned}$$

The scaling factor $1 + \dfrac{r-R}{3R}$ is the result of the reparametrization and is only negligible if $|R-r| \ll r$. Only in that sense can we speak of dist_2 as an approximation of $\mathrm{dist}^2_{L^2}$. Note that even for balls, dist_2 is only a zeroth order approximation for $\mathrm{dist}^2_{L^2}$.

In order to see whether dist_2 is as useful for shape prior-based image segmentation as $\mathrm{dist}^2_{L^2}$, let us take another look at Example 6.2 of Page 147. If we replace $\mathrm{dist}^2_{L^2}$ with dist_2, the energy E_λ becomes

$$\begin{aligned}E_\lambda(\rho) &= (1-\lambda)\left\langle \mathrm{sid}_{B_R}, B_\rho\right\rangle + \lambda\,\mathrm{dist}_2(B_r, B_\rho)\\ &= (1-\lambda)\int_{B_\rho} \mathrm{sid}_{B_R}(x)\,\mathrm{d}x + \frac{2\lambda\pi}{3}\left(2\rho^3 - 3\rho^2 r + r^3\right)\\ &= \begin{cases} \frac{2\lambda\pi}{3}\left(2\rho^3 - 3\rho^2 r + r^3\right) - (1-\lambda)\pi\rho^2 & ,\text{ if } \rho \le R\\ \frac{2\lambda\pi}{3}\left(2\rho^3 - 3\rho^2 r + r^3\right) + (1-\lambda)\pi(\rho^2 - 2R^2) & ,\text{ if } \rho > R\end{cases}\end{aligned}$$

and its global optimum is realized at (cf. 3rd plot of Fig. 6.6)

$$\rho^*(\lambda) = \begin{cases} R & ,\lambda \le \frac{1}{1+2(R-r)}\\ r + \frac{1-\lambda}{2\lambda} & ,\lambda > \frac{1}{1+2(R-r)}.\end{cases}$$

As for $\mathrm{dist}^2_{L^2}$, ρ^* starts at $\rho^*(0) = R$ and changes continuously to $\rho^*(1) = r$. It also remains at the initial solution $\rho^* = R$ for a certain range of λ. Therefore, we consider dist_2 as a good compromise between dist_H and $\mathrm{dist}^2_{L^2}$ to be used in the trust region framework as proposed in [14].

To use a shape distance that depends on the shapes' signed distance function is not a new concept. Rousson and Paragios [22] used the distance

$$\mathrm{dist}_{\mathrm{sdf}}(A, B) = \left(\int_{\Omega} (\mathrm{sdf}_A(x) - \mathrm{sdf}_B(x))^2 \,\mathrm{d}x \right)^{\frac{1}{2}}$$

to penalize shape dissimilarity. $\mathrm{dist}_{\mathrm{sdf}}$ depends in contrast to dist_2 on the size of the image domain Ω, e.g., $\mathrm{dist}_{\mathrm{sdf}}(B_r, B_R) = (R - r)^2 \,\mathrm{area}(\Omega)$. It is therefore not a general, domain-independent shape measure. In addition, we cannot use $\mathrm{dist}_{\mathrm{sdf}}$ as shape distance for the trust region sub-problem, because it depends on computing the signed distance function of *both shapes*. As a result, $\mathrm{dist}_{\mathrm{sdf}}(A, S)$ is not affine in S. Therefore, the sub-problem (6.6) does not become an energy of the form (6.1).

The distance dist_2 is very different in that respect. If we use $\mathrm{dist}_2(A, \cdot)$ in (6.6), A is known and sdf_A can be pre-computed. This makes dist_2 much easier to handle than $\mathrm{dist}_{\mathrm{sdf}}$. To our knowledge, dist_2 was first applied to a computer vision application by Boykov et al. in [2].

6.4 Experiments

In the following we present two applications of the trust region method. To solve the subproblem (6.6) we use the primal-dual method of [5]. Since only a few iterations are necessary, we will present most of the iterations. By doing so, we substantiate the theoretical results in Sect. 6.3 with practical examples.

6.4.1 Volume Constraint

We consider the energy $E_{\mathrm{Vol}}(u) = (\langle 1, u \rangle - V)^2 + \mathrm{len}(u)$, which penalizes the deviation of the volume $\langle 1, u \rangle$ from the target volume $V > 0$. The additional length term $\mathrm{len}(u)$ guarantees that the global minimum u^* of E_{Vol} describes a circle of radius r that satisfies $2\pi \cdot r^3 - 2Vr + 1 = 0$. For this toy example, we set the target volume V to cover 50 % of the image domain Ω. In Fig. 6.7 we show how the trust region method finds the global optimum in just a few iterations. While the shape from one iteration to the next changes drastically in the beginning, the energy E_{Vol} decreases in each iteration and moves the shape to the global optimum of the energy.

Besides the energy, we also show the derivative E' of the regional energy $E(u) = (\langle 1, u \rangle - V)^2$. If the current solution u_0^k is smaller than the target volume, $E'(u_0^k)$ is constantly negative (blue or cyan in Fig. 6.7) in the image domain and encourages larger segmentations. If on the other hand u_0^k is larger than the target volume, $E'(u_0^k)$ is constantly positive (orange and yellow in Fig. 6.7) and encourages smaller segmentations. Without the distance constraint in (6.6), the approximation $E'(u) + \mathrm{len}(u)$ would either choose $\emptyset$ or Ω as $u_0^{(k+1)}$. Together with the scaled signed distance function that originates from the dist_2 distance, we are able to smoothly change the shape in a way that the overall energy E_{Vol} decreases.

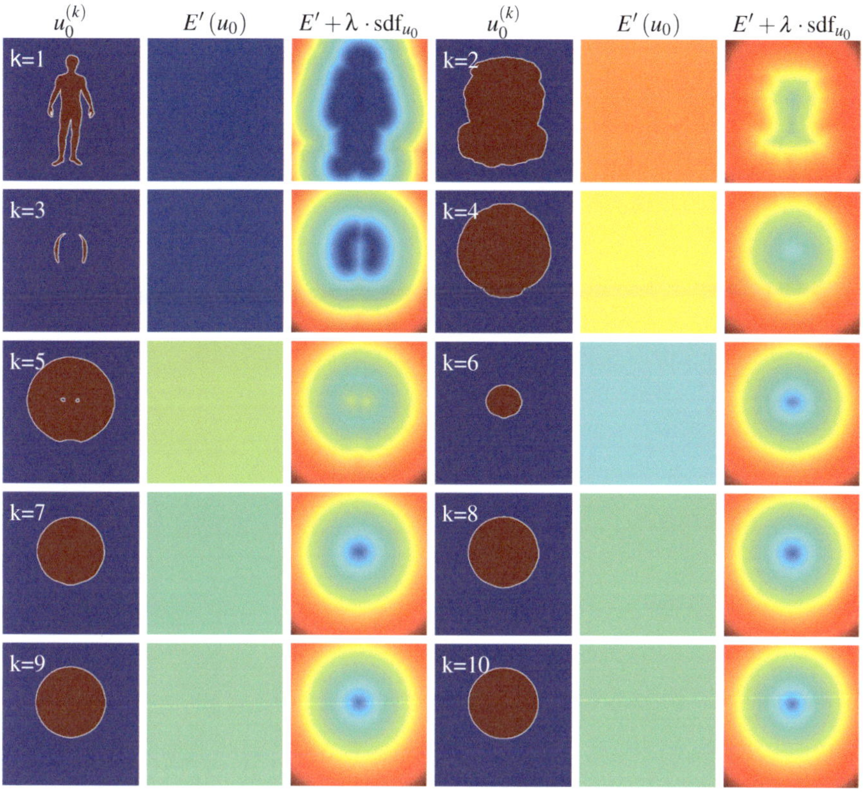

Fig. 6.7 In this toy example we explore the volume constraint $(\langle 1, u\rangle - V)^2$, where $V = \frac{1}{2}|\Omega|$ represents 50 % of the image domain's area. Using a man shaped initialization (1st image of top row), the method computes in a few iterations a circle (last image of last row). Each row represents two iterations of the trust region method. **1st, 4th column:** At each iteration k, we start with a current solution $u_0^{(k)}$. **2nd, 5th column:** The derivative $E'(u_0)$ could define a gradient descent. The global optimum of this energy is a trivial solution (Ω or $\emptyset$). **3rd, 6th column:** $E'(u_0) + \lambda \cdot \mathrm{sdf}_{u_0}$ is the data term for the Lagrangian formulation (6.6) of the trust region approach [14]. λ is chosen automatically by the trust region method. **4th, 1st column:** $\hat{u}$ is the global optimizer of the Lagrangian trust region sub-problem. It becomes u_0 of the next iteration. Note that in the beginning we can experience big jumps with respect to the segmentation. Nonetheless, the energy decreases in each iteration until we reach a local minimum of the original energy, which for this toy example is a global optimum

6.4.2 Distribution Constraint

We consider the energy function (6.4) as introduced in Sect. 6.2.2. For this application, we assume knowledge about the object and describe it with 512 color models (8 per color channel). The results are presented in Fig. 6.8. As in the previous experiment, the data term of the approximation $\tilde{E}$ is in general not very informative

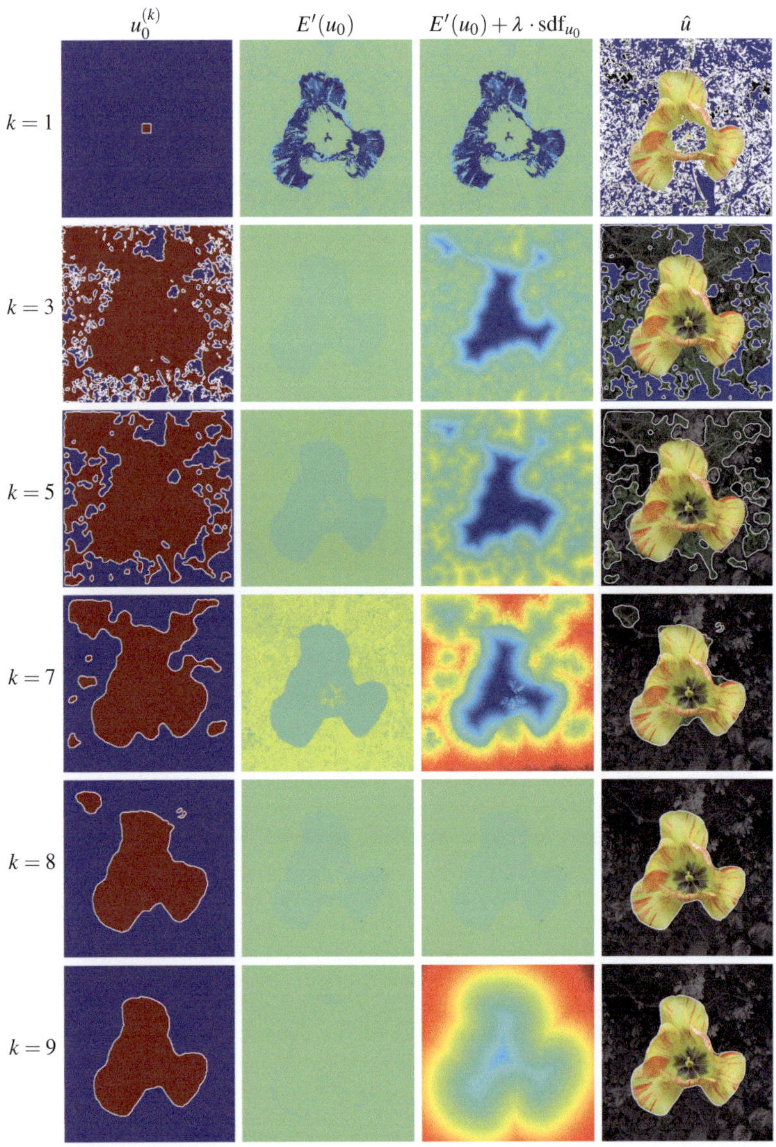

Fig. 6.8 We explore the segmentation task (6.4) that uses a distribution of 512 entries. Using a square as initialization (1st image of top row), the method computes in a few iterations the segmentation of the flower (last image of last row). Each row represents one iteration of the trust region method. **1st column:** At each iteration k we start with a current solution u_0. **2nd column:** The derivative $E'(u_0)$ could define a gradient descent, but it provides only for a weak data term. **3rd column:** $E'(u_0) + \lambda \cdot \mathrm{sdf}_{u_0}$ is the data term for the Lagrangian formulation (6.6) of the trust region approach [14]. λ is chosen automatically by the trust region method. **4th column:** $\hat{u}$ is the global optimizer of the Lagrangian trust region sub-problem. For visualization purposes we set the background in the first two rows to blue and in the remaining rows to the gray-scale of the original image

(cf. 2nd column of Fig. 6.7). Only in combination with the distance dist_2 do we obtain a data term (cf. 3rd column of Fig. 6.8) that helps to improve the segmentation (cf. 4th column of Fig. 6.8).

6.5 Summary

In this chapter we demonstrated that the choice of a shape distance influences the result for image segmentation applications, even at the absence of any shape prior. The importance of the chosen shape distance becomes apparent if we want to deal with local optimization. In particular, we analyzed the behavior with respect to three different distances in the context of the fast trust region image segmentation framework of [14]. In order to obtain good segmentation results, we advocate the use of the distance dist_2.

References

1. Boykov, Y., Kolmogorov, V.: An experimental comparison of min-cut/max-flow algorithms for energy minimization in vision. IEEE Trans. PAMI **26**(9), 1124–1137 (2004)
2. Boykov, Y., Kolmogorov, V., Cremers, D., Delong, A.: An integral solution to surface evolution PDEs via Geo-Cuts. In: Leonardis, A., Bischof, H., Pinz, A. (eds.) European Conference on Computer Vision. LNCS, vol. 3953, pp. 409–422. Springer, Graz (2006)
3. Boykov, Y., Veksler, O., Zabih, R.: Fast approximate energy minimization via graph cuts. IEEE Trans. PAMI **23**(11), 1222–1239 (2001)
4. Brox, T., Rousson, M., Deriche, R., Weickert, J.: Unsupervised segmentation incorporating colour, texture, and motion. In: Petkov, N., Westenberg, M.A. (eds.) Computer Analysis of Images and Patterns. LNCS, vol. 2756, pp. 353–360. Springer, Groningen (2003)
5. Chambolle, A., Pock, T.: A first-order primal-dual algorithm for convex problems with applications to imaging. J. Math. Imaging Vis. **40**(1), 120–145 (2011)
6. Chan, T., Esedoḡlu, S., Nikolova, M.: Algorithms for finding global minimizers of image segmentation and denoising models. SIAM J. Appl. Math. **66**(5), 1632–1648 (2006)
7. Cremers, D., Osher, S.J., Soatto, S.: Kernel density estimation and intrinsic alignment for shape priors in level set segmentation. Int. J. Comput. Vis. **69**(3), 335–351 (2006)
8. Cremers, D., Schmidt, F.R., Barthel, F.: Shape priors in variational image segmentation: convexity, Lipschitz continuity and globally optimal solutions. In: IEEE International Conference on Computer Vision and Pattern Recognition, Anchorage (2008)
9. Cremers, D., Soatto, S.: A pseudo-distance for shape priors in level set segmentation. In: Paragios, N. (ed.) IEEE 2nd International Workshop on Variational, Geometric and Level Set Methods, Nice, pp. 169–176 (2003)
10. Delong, A., Gorelick, L., Schmidt, F.R., Veksler, O., Boykov, Y.: Interactive segmentation with super-labels. In: International Conference on Energy Minimization Methods for Computer Vision and Pattern Recognition. LNCS, vol. 6819, pp. 147–162. Springer, Saint Petersburg (2011)
11. Delong, A., Osokin, A., Isack, H.N., Boykov, Y.: Fast approximate energy minimization with label costs. Int. J. Comput. Vis. **96**(1), 1–27 (2012)

12. Gao, T., Koller, B.P.D.: A segmentation-aware object detection model with occlusion handling. In: IEEE International Conference on Computer Vision and Pattern Recognition, Colorado Springs, pp. 1361–1368 (2011)
13. Gorelick, L., Ayed, I.B., Schmidt, F.R., Boykov, Y.: An experimental comparison of trust region and level sets. arXiv http://arxiv.org/abs/1311.2102 (2013)
14. Gorelick, L., Schmidt, F.R., Boykov, Y.: Fast trust region for segmentation. In: IEEE International Conference on Computer Vision and Pattern Recognition, Portland (2013)
15. Leventon, M., Grimson, W., Faugeras, O.: Statistical shape influence in geodesic active contours. In: IEEE International Conference on Computer Vision and Pattern Recognition, Hilton Head Island, vol. 1, pp. 316–323 (2000)
16. Li, C., Xu, C., Gui, C., Fox, M.D.: Level set evolution without re-initialization: a new variational formulation. In: IEEE International Conference on Computer Vision and Pattern Recognition, pp. 430–436 (2005)
17. Ling, H., Jacobs, D.W.: Shape classification using the innerdistance. IEEE Trans. PAMI **29**(02), 286–299 (2007)
18. Mumford, D., Shah, J.: Optimal approximations by piecewise smooth functions and associated variational problems. Commun. Pure Appl. Math. **42**, 577–685 (1989)
19. Nieuwenhuis, C., Strekalovskiy, E., Cremers, D.: Proportion priors for image sequence segmentation. In: IEEE International Conference on Computer Vision, Sydney, pp. 1769–1776 (2013)
20. Nocedal, J., Wright, S.J.: Numerical Optimization. Springer, Berlin (2006)
21. Osher, S.J., Sethian, J.A.: Fronts propagation with curvature dependent speed: algorithms based on Hamilton–Jacobi formulations. J. Comput. Phys. **79**, 12–49 (1988)
22. Rousson, M., Paragios, N.: Shape priors for level set representations. In: Heyden, A., et al. (eds.) European Conference on Computer Vision, Copenhagen. LNCS, vol. 2351, pp. 78–92. Springer (2002)
23. Schmidt, F.R., Boykov, Y.: Hausdorff distance constraint for multi-surface segmentation. In: European Conference on Computer Vision. LNCS, vol. 7572, pp. 598–611. Springer, Florence (2012)
24. Soares, J.V.B., Cesar, J.J.G.L.R.M., Jelinek, H.F., Cree, M.J.: Retinal vessel segmentation using the 2-d gabor wavelet and supervised classification. IEEE Trans. Med. Imaging **25**(9), 1214–1222 (2006)
25. Tang, M., Gorelick, L., Veksler, O., Boykov, Y.: Grabcut in one cut. In: IEEE International Conference on Computer Vision, Sydney, pp. 1769–1776 (2013)

Chapter 7
Segmentation in Point Clouds from RGB-D Using Spectral Graph Reduction

Margret Keuper and Thomas Brox

Abstract In this chapter, we tackle the problem of segmentation in point clouds from RGB-D data. In contrast to full point clouds, RGB-D data only provides a part of the volumetric information, the depth information of the one view given in the corresponding RGB image. Still, this additional information is valuable for the segmentation task as it helps disambiguating texture gradients from structure gradients. In order to create hierarchical segmentations, we combine a state-of-the-art method for natural RGB image segmentation based on spectral graph analysis with an RGB-D boundary detector. We show that spectral graph reduction can be employed in this case, facilitating the computation of RGB-D segmentations in large datasets.

7.1 Introduction

The decomposition of images into meaningful segments has been in the scope of computer vision research for decades. In many biological and medical applications, segmenting images into meaningful parts is a necessary step for data analysis [16, 17]. In natural images, segmentations help to implement higher-level applications such as scene understanding [21].

Hierarchical segmentations [1] produce image decompositions at different levels of granularity. The finest level segmentations aim at a complete oversegmentation and should have full recall i.e. all object objects are separated. Segments at this level are referred to as superpixels and facilitate further analysis of the data as well as higher-level reasoning [18, 23].

Main applications for the segmentation of RGB-D data so far are *scene classification*, and *support analysis*, i.e. analyzing which object present in the scene is supported by which other part of the scene [12, 29]. Segmentations of image sequences can also be used to improve 3D reconstructions [20, 33].The additional information present in the depth channel helps generating reliable segmentations and disambiguating texture from structure gradients. The main challenges are how

M. Keuper (✉) • T. Brox
University of Freiburg, Freiburg im Breisgau, Germany
e-mail: keuper@cs.uni-freiburg.de; brox@cs.uni-freiburg.de

M. Breuß et al. (eds.), *Perspectives in Shape Analysis*, Mathematics and Visualization, DOI 10.1007/978-3-319-24726-7_7

Fig. 7.1 Hierarchical segmentation of an RGB-D Image. The darker the marked contour, the higher it is in the segmentation hierarchy

to extract the contour cues from the depth channel and how to combine these cues with the RGB derived contour information. Furthermore, the RGB-D data can be large and one potentially wants to apply segmentation not only to single images but to image sequences. Runtime and memory consumption should therefore be kept small.

In image segmentation, the currently leading methods rely on a good boundary probability estimation followed by a spectral analysis of the manifold created from these probabilities. More concrete, the eigenvectors of the graph Laplacian defined by boundary probabilities are used to create a hierarchical partitioning of the image. An obvious way to approach RGB-D segmentation is thus given by combining the top algorithm for boundary detection in RGB-D, the RGB-D version of [7], with a spectral method generating hierarchical segmentations [1, 2] (Fig. 7.1). A similar approach has been implemented in [12]. However, spectral methods can easily become expensive in runtime and memory consumption, making an application to large RGB-D datasets challenging or impossible. We show how this issue can be handled using the spectral graph reduction technique from [11].

7.2 Overview on Existing Work

The first step of successful image segmentation algorithms is to find reliable contour cues. This can be achieved by computing local image gradients [4]. More recent methods make use of oriented contour cues [22] computed on image patches such that also texture gradients can be represented. Structured random forests [7] are a state-of-the-art method for boundary probability computation. They make use of the correlation between local image information computed on image patches and the boundary inside these patches. Using a random forest learning framework, they produce a structured output, i.e. a boundary patch, around every pixel from which a boundary probability map can be computed for the whole image. Finally, deep learning based methods [3] make use of convolutional neural networks to produce highly reliable boundary probabilities that can be used to generate image segmentations.

For the segmentation of natural RGB images, methods relying on a spectral analysis step [28] and a subsequent hierarchical segmentation produce state-of-the-art results [1, 2, 14] but are expensive in terms of runtime and memory consumption. To reduce runtime and memory consumption of such methods, several algorithms for approximate eigenvector computation have been proposed. In Fowlkes et al. [10], the Nyström method is employed, producing a sampled solution that can be extrapolated to the original problem. Similarly, Chen and Cai [5] use landmark points to represent the original problem. Generally, an issue with sampling points from the original problem is that points formerly linked via transitivity might get disconnected. In Arbeláez et al. [2], this probem is addressed by computing the eigenvectors of a subsampled, squared affinity matrix, i.e. the affinities are propagated to neighboring points before sampling. Still, the solution must be extrapolated from the sampled points. A different approach has been presented in Galasso et al. [11], Taylor [30]. Both works use superpixels to agglomerate points in the graph instead of sampling. In Galasso et al. [11] it is shown how to adjust the weights in the reduced graph of agglomerated points such that the original normalized cut problem is not altered. In the proposed RGB-D segmentation framework, we are using this spectral graph reduction technique in the spectral analysis step.

Some segmentation methods presented for RGB-D data [29] omit the spectral analysis step and put more emphasis on extracting the additional information present in the depth channel. In Silberman et al. [29], surface normals are used to align the scene and extract 3D planes. The boundary strength is then predicted from RGB-D, position features, and features from the plane information using a boosted decision tree. The methods presented in Kim et al. [19], Zheng et al. [34] attempt to solve a voxel-wise 3D segmentation of the scene from RGB-D data. Both use heuristics to estimate the volumetric information from RGB-D. In Zheng et al. [34] the actual segmentation is done using cues from physics i.e. optimizing the energetic status of a scene. The method of Kim et al. [19] is based on a conditional random field that jointly infers geometric and semantic structures.

In Gupta et al. [12], a depth-aware contour detection is performed using geometric contour cues like depth gradients and concave and convex normal gradients. Additionally, the depth information as well as a spectral gradient are input to an additive kernel support vector machine (SVM). The probabilistic output of the SVM defines the contour strength. The remaining pipeline including the spectral analysis is kept as in Arbeláez et al. [1]. The work presented in Gupta et al. [13] builds upon Gupta et al. [12] for RGB-D segmentation and reuses all the above mentioned depth features defined in Gupta et al. [12]. Additionally, Gupta et al. [13] include the RGB boundary detection results of Dollár and Zitnick [7] for better boundary localization and employ the learning framework of Dollár and Zitnick [7].

The method from Karpathy et al. [15] directly works on sparse 3D point clouds generated from the kinect camera. On these large datasets, the fast and greedy graph-based segmentation method of Felzenszwalb and Huttenlocher [9] is employed on normal gradients.

7.3 Image Segmentation Using Spectral Methods

Pixels of an image are represented by a set of points $P = \{p_1, \ldots, p_n\}$ and a graph $\mathcal{G} = (\mathcal{V}, \mathcal{E})$ with vertex set $\mathcal{V}$ and undirected edges $e \in \mathcal{E}$, such that each point $p_i \in P$ is represented by a vertex $v_i \in \mathcal{V}$. Each edge $e_{i,j}$ between vertices v_i and v_j is weighted by the pairwise affinity w_{ij} of these vertices computed from the underlying image data. Spectral methods seek to partition according to the minimal ratio cut (RCut) of $\mathcal{V}$ into sets A and B with $A \cup B = \mathcal{V}$ and $A \cap B = \emptyset$

$$\mathrm{RCut}(A, B) = \frac{\mathrm{cut}(A, B)}{|A|} + \frac{\mathrm{cut}(A, B)}{|B|}, \tag{7.1}$$

where $|.|$ denotes the cardinality of a set, or they aim at partitioning according to the normalized cut (NCut)

$$\mathrm{NCut}(A, B) = \frac{\mathrm{cut}(A, B)}{\mathrm{vol}(A)} + \frac{\mathrm{cut}(A, B)}{\mathrm{vol}(B)}, \tag{7.2}$$

with $\mathrm{cut}(A, B) = \sum_{i \in A, j \in B} w_{ij}$ and $\mathrm{vol}(A) = \sum_{i \in A, j \in \mathcal{V}} w_{ij}$. These objectives are optimized using spectral clustering [6, 24, 31]. In image segmentation, NCut is the more desired objective [27, 28, 32, 32]. As shown in von Luxburg [31], the normalized cut problem can be formulated equivalently as

$$\begin{aligned} &\min f'Lf \quad \text{subject to} \quad Df \perp \mathbf{1}, \quad f'Df = \mathrm{vol}(\mathcal{V}), \\ &\text{and} f_i = \begin{cases} \sqrt{\frac{\mathrm{vol}(B)}{\mathrm{vol}(A)}} & \text{if } v_i \in A \\ -\sqrt{\frac{\mathrm{vol}(A)}{\mathrm{vol}(B)}} & \text{if } v_i \in B \end{cases}, \end{aligned} \tag{7.3}$$

where $L = D - W$ is the graph Laplacian with W being the matrix containing the pairwise affinities w_{ij} between vertices v_i and v_j and D being the degree matrix with $d_{ij} = 0 \; \forall i \neq j$ and $d_{ii} = \sum_{j \in \mathcal{V}} w_{ij}$. With $\mathbf{1}$ we denote the constant one vector having the same length as f and $\perp$ denotes orthogonality.

Finding the exact solution to the NCut problem is NP-hard. However, by relaxing the problem and allowing f to take arbitrary real values and substituting $g := D^{1/2}f$, the problem reduces to a generalized eigenvalue problem, i.e. to finding the generalized second eigenvector of the normalized graph Laplacian $L_{\text{sym}} = I - D^{-\frac{1}{2}}WD^{-\frac{1}{2}}$.

$$\lambda_2 = \min g' L_{\text{sym}} g = \min \sum_{i,j=1} w_{ij} \left(\frac{g_i}{\sqrt{d_{ii}}} - \frac{g_j}{\sqrt{d_{jj}}} \right)^2$$
$$\text{subject to} \quad g \perp D^{1/2}\mathbf{1},\ \|g\|^2 = \text{vol}(\mathscr{V}) \tag{7.4}$$

This formulation allows for the application of the Rayleigh-Ritz theorem and thus for the numerical computation of the generalized eigenvalues and eigenvectors, i.e. the relaxated partition function f. The complexity of this computation is in general $O(n^3)$ [5].

The eigenvectors related to the NCut play a crucial role in state-of-the-art segmentation algorithms [1, 2, 14]. In the following, we will have a closer look at Arbeláez et al. [1] since the spectral analysis commonly used in all these methods was originally defined here.

The workflow of Arbeláez et al. [1] basically consists of four steps: a boundary detection step (*mPb*), the *globalization* of the boundary cues *gPb*, the computation of an oriented watershed transform (*OWT*), and finally the generation of the region hierarchy called ultrametric contour map (*UCM*). The newer algorithms presented in Arbeláez et al. [2], Isola et al. [14] and Gupta et al. [12] are similar in the last three steps but differ in how the input to the globalization step is generated.

The globalization itself consists of spectral clustering and represents the computational bottleneck for these algorithms, limiting their application to small images (e.g. 640×480 px) or images patches that need to be stitched together after segmentation.

For a better understanding, let us briefly sketch the original *gPb* computation of Arbeláez et al. [1]. In a first step, multiscale boundary probabilities *mPb* are generated from color, brightness, and texture gradients at different scales and orientations. The second step consists of a *spectral analysis* of the feature space induced by the *mPb* and results in so-called *global* boundary probabilities (*gPb*). Instead of the *mPb* any distance measure inducing boundary probabilities between at least directly neighboring pixels can be used in theory. The method of Arbeláez et al. [2] uses a combination of *mPb* and the boundary probabilities from the RGB version of Dollár and Zitnick [7] while in Isola et al. [14] boundary probabilities are derived from pointwise mutual information. Both improve over Arbeláez et al. [1] on standard benchmarks [1].

A pixel adjacency graph $\mathscr{G} = (\mathscr{V}, \mathscr{E})$ is built such that every pixel, represented by a vertex $v_i \in \mathscr{V}$, is connected to its neighbors with maximal distance r by an undirected edge $e_{ij} \in \mathscr{E}$ weighted by the affinity between v_i and v_j. Hereby, the sparse affinity matrix W is built based on the boundary information in *mPb*. For every pair of pixels within a spherical neighborhood with radius r, w_{ij} is computed from the maximal value of *mPb* on the line $\bar{ij}$ connecting the two pixels i and j (the intervening contour cue) as

$$w_{ij} = \exp\left(-\max_{p\in\bar{ij}} \max_{\theta} \frac{mPb(p,\theta)}{\rho}\right), \tag{7.5}$$

where ρ is a constant. Thus, W represents the pixel affinity graph, where every pixel i is represented by a node v_i and every edge between two nodes v_i and v_j is indicated by a positive matrix entry w_{ij}. For the graph Laplacian $L = D - W$ with D being the *degree matrix*, the generalized eigenvalue problem is solved. Instead of directly clustering the image pixels according to these eigenvectors as suggested in the literature [31], the spectral information is used in a soft way. On every resulting eigenvector, the spatial gradients are computed, yielding filter responses that contain the spectral contour information *sPb*. The linear combination with learned weighting of the local, original boundary cues and the more global *sPb* finally forms the global probability of boundaries *gPb*.

The method proposed in Arbeláez et al. [2] improves over Arbeláez et al. [1] not only by using richer boundary cues. Also, the computation of the *gPb*, *OWT*, and *UCM* is repeated for three different scale segmentations. The computation of the eigenvectors is approximated using a fast sampling method. The final *UCM* is computed from the different scales such that the boundary localization of the finest resolution is combined with the hierarchical information from coarser scales (compare Fig. 7.2).

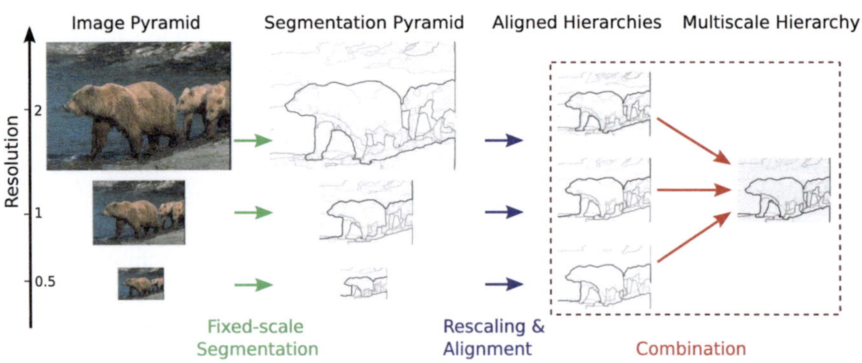

Fig. 7.2 Workflow from Arbeláez et al. [2] for the generation of multiscale segmentation hierarchies. The boundary probability computation and spectral analysis steps are part of the first step in the depicted workflow: *Fixed-Scale Segmentation*. The visualization is following Arbeláez et al. [2]

7.4 Adaptation for RGB-D Data

The affinity matrix computation in Arbeláez et al. [2] is based on boundary probabilities computed using structured random forests [7]. Structured random forests [7] can predict a local boundary patch from an image patch. To generate boundary probabilities for an entire image, boundary patches are predicted from several decision trees for every second pixel and averaged. The quality of these boundary probabilities can be improved using a multiscale detection [7], i.e. by applying the boundary detection step at multiple scales, resizing and averaging the resulting edge maps. To further improve the boundary probability maps, Dollár and Zitnick [8] propose an edge sharpening procedure, basically warping the boundary prediction to the original image patch. This sharpening step helps improve the exact boundary localization. The extension of this boundary prediction framework to RGB-D data proposed by Dollár and Zitnick [7] is straightforward: The depth information is treated as a fourth channel in the same learning framework. Some results are displayed in Fig. 7.3. For the affinity matrix computation according to (7.5), we set ρ to 0.12.

Fig. 7.3 RGB-D boundary probabilities produced by Dollár and Zitnick [8] on examples from NYU Depth (v2)

7.5 Spectral Graph Reduction

For larger problems, the runtime and memory consumption of the eigenvector computation (7.3) can be prohibitive. In Galasso et al. [11] it is shown how to adjust the weights in the reduced graph of agglomerated points such that the original NCut problem is not altered. We will briefly summarize this work in the following.

We assume we are given a user or image-driven point pre-grouping (for example superpixels) $G = \{I_1, \dots, I_m\}$ with $I_1 \cup \dots \cup I_m = P$ and $I_i \cap I_j = \emptyset \ \forall 1 \leq i, j \leq m$. Constraining the partitioning to this given grouping, we reduce $\mathscr{G}$ to a new graph $\bar{\mathscr{G}} = (\bar{\mathscr{V}}, \bar{\mathscr{E}})$, such that the optimal partitioning of $\bar{\mathscr{V}} = \{v_{I_1}, \dots, v_{I_m}\}$ according to the NCut is as similar as possible to the partitioning in the original, unreduced graph. Compare Fig. 7.4. Assuming we can find the optimal solution in both cases, the problem can be formulated as

$$\begin{aligned} \forall \bar{A} = \{v_{I_\ell}, \dots, v_{I_k}\} \equiv \{p_i, \dots, p_j\} \equiv \{v_i, \dots, v_j\} = A \\ \text{and} \quad \bar{B} \equiv B \quad \text{respectively} \\ \text{with} \quad \bar{A} \cup \bar{B} = \bar{\mathscr{V}} \quad \text{and} \quad \bar{A} \cap \bar{B} = \emptyset : \\ \mathrm{NCut}(\bar{A}, \bar{B}) \stackrel{!}{=} \mathrm{NCut}(A, B), \end{aligned}$$

i.e. the value of the NCut must not change for any possible partitioning. According to Galasso et al. [11] and Rangapuram and Hein [25], this can be achieved by setting

$$w_{IJ} = \sum_{i \in I} \sum_{j \in J} w_{ij} \tag{7.6}$$

for all v_I and v_J in $\bar{\mathscr{V}}$. This is equivalent to imposing must-link constraints to all the grouped points in the original graph [25].

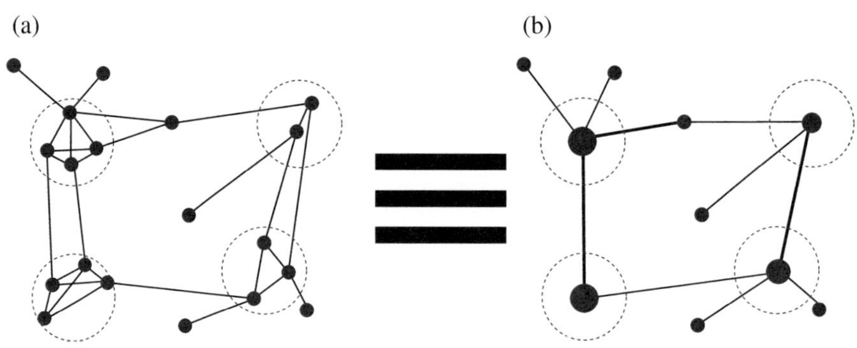

Fig. 7.4 In the original graph, close-by nodes are agglomerated and linked by must-link constraints. Given the correct handling of edge weights, these agglomerated nodes can be treats as one single node, yielding a reduced graph with exactly the same normalized cut cost. (**a**) Original graph. (**b**) Equivalent graph (Visualization according to [11])

In Galasso et al. [11], this method was applied to image segmentation in the framework of Arbeláez et al. [1]. To reduce the complexity of the spectral analysis step in Arbeláez et al. [1], the pixels with low gradients are pre-grouped by the watershed regions of the *mPb* to build the reduced graph $\mathcal{G}^Q = (\mathcal{V}^Q, \mathcal{E}^Q)$. For $\mathcal{G}^Q$ affinities w^Q_{IJ} are computed preserving the NCut (7.6). The remaining original workflow from Arbeláez et al. [1] was pursued on the reduced graph. By this approach, runtime and memory consumption were reduced by factor two with practically no loss in precision or recall.

Similarly, we want to replace the spectral analysis step of Arbeláez et al. [2] with a spectral analysis on a reduced graph. An overview of the workflow in Arbeláez et al. [2] was given is Fig. 7.2. The spectral analysis is part of the *fixed-scale segmentation* step and is only approximated in Arbeláez et al. [2] using sampling. As in Galasso et al. [11] we agglomerate the pixels based on the watershed regions. These watershed regions are computed on the RGB-D boundary probabilities from Dollár and Zitnick [7], reducing the number of nodes from 238,000 to 3,500 on average, depending of the structures in the image. On the reduced graph, the spectral analysis can be computed correctly. The workflow of the *fixed-scale segmentation* step without and with spectral graph reduction is depicted in Fig. 7.5. The remaining steps are kept as in Arbeláez et al. [2].

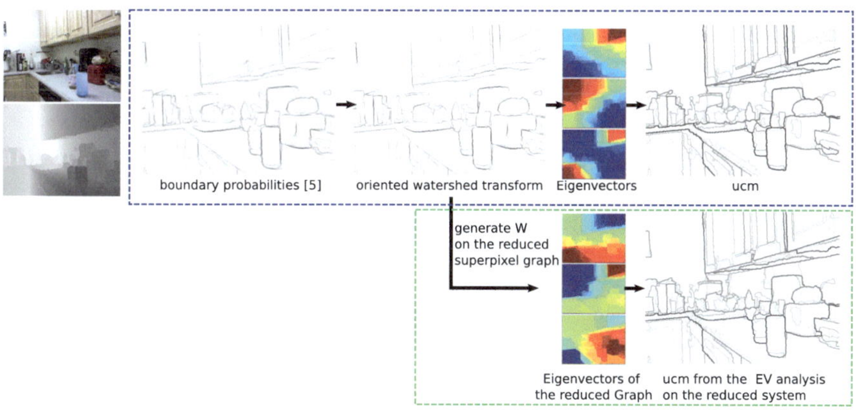

Fig. 7.5 Workflow of the *fixed-scale segmentation* step with the adaptation to RGB-D data. The *upper column* shows the original workflow. A weighted, oriented watershed transform is computed from boundary probabilities. From this *OWT*, an affinity matrix is computed using the intervening contour cue. Eigenvectors of the corresponding graph Laplacian are computed to generate the *UCM* segmentation. We propose to compute a smaller graph $\mathcal{G}^Q = (\mathcal{V}^Q, \mathcal{E}^Q)$ from the *OWT* with one vertex for every watershed region, preserving the original NCut problem. The eigensystem is solved in the reduced graph

7.6 Experiments and Results

We evaluate the effect of the spectral graph reduction on the NYU Depth dataset (v2) [29] using the boundary metrics proposed in Arbeláez et al. [1]. The dataset contains 1,449 pairs of RGB and depths images. For every image, two annotations are provided: a semantic labeling and an instance labeling. In [12], the dataset has been adapted for evaluation with the metrics from Arbeláez et al. [1]. Therefore, the image boundaries were cut off and the ground truth was cleaned from double contours. The dataset is split into 795 training and 654 test images. The images in the resulting dataset have a size of 425 × 560 pixels. In Ren and Bo [26], a second adaptation of the dataset is proposed for the evaluation. The data is further downsampled and a different split into training and test images is proposed. For our evaluation, we stick to the version of Gupta et al. [12] which contains the larger data. An example of the original data and annotations is given in Fig. 7.6. In order to measure boundary precision and recall, a distance threshold has to be set to determine which boundary candidates are accepted as correct detections. In the evaluation metrics provided in Arbeláez et al. [1] his threshold is given in terms of pixel distance as a fraction of the image diagonal. In Gupta et al. [12], this value is set to 0.011. We compare the effect of the spectral graph reduction on the multiscale method for *UCM* generation of Arbeláez et al. [2] (M-UCM) against the

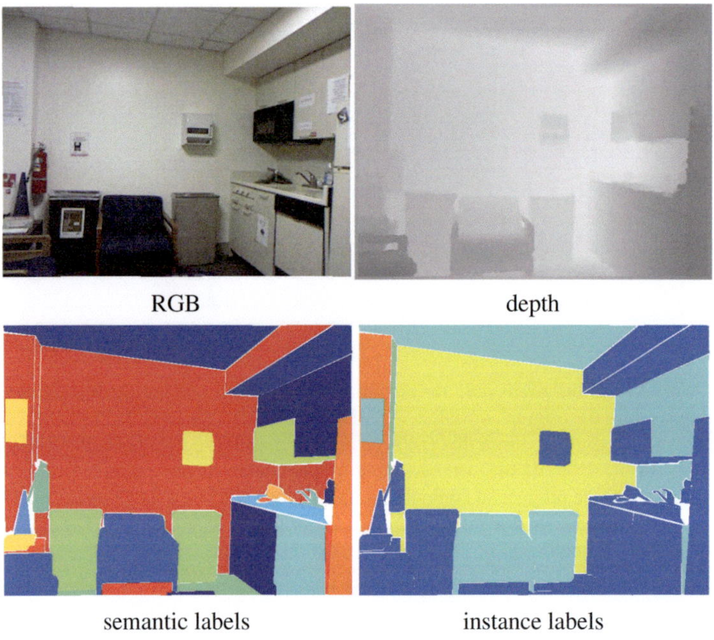

Fig. 7.6 An image with RGB and Depth channel from NYU Depth dataset (v2) [29] with the original label and instance annotations

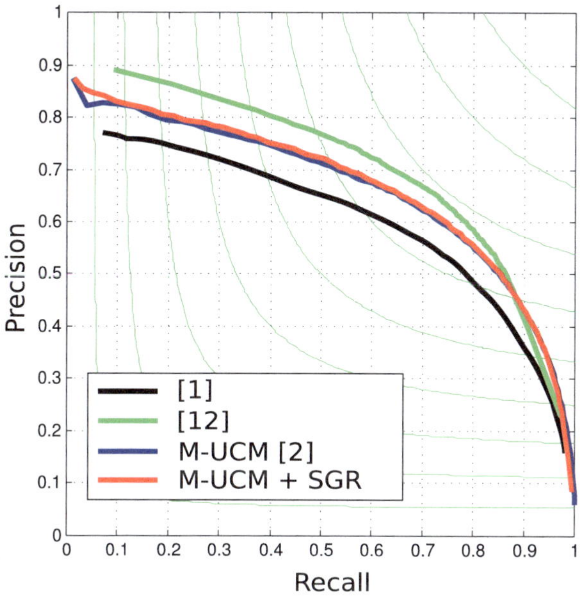

Fig. 7.7 Boundary precision and recall on the NYU Depth dataset (v2). In *blue*, we show the performance of Arbeláez et al. [2] (M-UCM) using boundary probabilities from Dollár and Zitnick [7]. In *red*, we show our result, applying spectral graph reduction (M−UCM+SGR). SGR improves over the RGB-D adapted version of Arbeláez et al. [2]

original method. Instead of the graph reduction, Arbeláez et al. [2] reduce runtime by using an approximate eigenvector computation. In both cases we use the RGB-D boundary probabilities from Dollár and Zitnick [7]. The boundary precision/recall curve is plotted in Fig. 7.7. Basically, the performance is very similar. The spectral graph reduction improves only a little over [2] (M-UCM). The average precision is 0.65 for M-UCM, 0.67 for the proposed setup. With the same threshold for all segmentations, i.e. the optimal dataset score (ODS), the resulting f-measure is 0.66 for both. If we allow different thresholds per dataset (optimal individual score (OIS)), the f-measure is 0.70. This improves over the baseline method [1] by 0.03 for the ODS and 0.12 in average precision (numbers for Arbeláez et al. [1] taken from Gupta et al. [12]). Still, Gupta et al. [12] who build upon specially designed depth gradients and solve the full eigensystem correctly for the spectral analysis perform better at equal error rate, with an f-measure at ODS of 0.69. In the high recall range, their curve drops below ours (compare Fig. 7.7). For results on some examples compare Fig. 7.8.

For an image at the original scale (425 × 560 pixels), our method takes about 2.87 s on a 3.30 GHz CPU, Arbeláez et al. [2] need slightly more (3.02 s). Solving the full eigensystem amounts at 12.97 s of computation time per image. In Gupta et al. [12], computing the RGB and depth features already takes about 135 s.

Fig. 7.8 Results of the segmentation with spectral graph reduction on some expamples from NYU Depth (v2). We show the overlay of the segmentation at the optimal dataset scale (ODS) and the full *UCM*

7.7 Conclusion

We have presented a fast method for hierarchical segmentation in RGB-D data. In order to produce reliable segmentations, we combine a state-of-the-art method for image segmentation [2] with an RGB-D boundary detection [7]. We could show that by applying spectral graph reduction in that framework, we would not only reduce runtime but also improve on the average precision.

Acknowledgements We acknowledge funding by the ERC Starting Grant VideoLearn.

References

1. Arbeláez, P., Maire, M., Fowlkes, C.C., Malik, J.: Contour detection and hierarchical image segmentation. IEEE TPAMI **33**(5), 898–916 (2011)
2. Arbeláez, P., Pont-Tuset, J., Barron, J., Marques, F., Malik, J.: Multiscale combinatorial grouping. In: CVPR, Columbus (2014)
3. Bertasius, G., Shi, J., Torresani, L.: Deepedge: a multi-scale bifurcated deep network for top-down contour detection. CoRR (2014). abs/1412.1123
4. Canny, J.: A computational approach to edge detection. IEEE TPAMI **8**, 679–698 (1986)
5. Chen, X., Cai, D.: Large scale spectral clustering with landmark-based representation. In: AAAI, San Francisco (2011)
6. Chung, F.R.K.: Spectral graph theory, vol. 92. CBMS, Providence (1997)
7. Dollár, P., Zitnick, C.L.: Structured forests for fast edge detection. In: ICCV, Sydney (2013)
8. Dollár, P., Zitnick, C.L.: Fast edge detection using structured forests. CoRR (2014). abs/1406.5549
9. Felzenszwalb, P.F., Huttenlocher, D.P.: Efficient graph-based image segmentation. IJCV **59**(2), 167–181 (2004)
10. Fowlkes, C., Belongie, S., Chung, F., Malik, J.: Spectral grouping using the nyström method. IEEE TPAMI 26(2), 214–225 (2004)
11. Galasso, F., Keuper, M., Brox, T., Schiele, B.: Spectral graph reduction for efficient image and streaming video segmentation. In: CVPR, Columbus (2014)
12. Gupta, S., Arbelaez, P., Malik, J.: Perceptual organization and recognition of indoor scenes from RGB-D images. In: CVPR, Portland (2013)
13. Gupta, S., Girshick, R., Arbelaez, P., Malik, J.: Learning rich features from RGB-D images for object detection and segmentation. In: ECCV, Zurich (2014)
14. Isola, P., Zoran, D., Krishnan, D., Adelson, E.H.: Crisp boundary detection using pointwise mutual information. In: ECCV, Zurich (2014)
15. Karpathy, A., Miller, S., Fei-Fei, L.: Object discovery in 3d scenes via shape analysis. In: ICRA, Karlsruhe (2013)
16. Keuper, M., Bensch, R., Voigt, K., Dovzhenko, A., Palme, K., Burkhardt, H., Ronneberger, O.: Semi-supervised learning of edge filters for volumetric image segmentation. In: 32nd DAGM Symposium, Darmstadt. LNCS, pp. 462–471. Springer (2010)
17. Keuper, M., Padeken, J., Heun, P., Burkhardt, H., Ronneberger, O.: Mean shift gradient vector flow: a robust external force field for 3d active surfaces. In: ICPR, Istanbul (2010)
18. Keuper, M., Schmidt, T., Rodriguez-Franco, M., Schamel, W., Brox, T., Burkhardt, H., Ronneberger, O.: Hierarchical markov random fields for mast cell segmentation in electron microscopic recordings. In: ISBI, Chicago (2011)

19. Kim, B., Kohli, P., Savarese, S.: 3d scene understanding by voxel-CRF. In: ICCV, Sydney (2013)
20. Kundu, A., Li, Y., Dellaert, F., Li, F., Rehg, J.M.: Joint semantic segmentation and 3d reconstruction from monocular video. In: ECCV, Zurich (2014)
21. Li, L., Socher, R., Fei-Fei, L.: Towards total scene understanding: classification, annotation and segmentation in an automatic framework. In: CVPR, Miami (2009)
22. Martin, D.R., Fowlkes, C.C., Malik, J.: Learning to detect natural image boundaries using local brightness, color and texture cues. IEEE TPAMI **26**(5), 530–549 (2004)
23. Morath, V., Keuper, M., Rodriguez-Franco, M., Deswal, S., Fiala, G., Blumenthal, B., Kaschek, D., Timmer, J., Neuhaus, G., Ehl, S., Ronneberger, O., Schamel, W.: Semi-automatic determination of cell surface areas used in systems biology. Front. Biosci. Elite **5**, 533–545 (2013)
24. Ng, A.Y., Jordan, M.I., Weiss, Y.: On spectral clustering: analysis and an algorithm. In: NIPS, Vancouver (2001)
25. Rangapuram, S.S., Hein, M.: Constrained 1-spectral clustering. In: AISTATS, La Palma (2012)
26. Ren, X., Bo, L.: Discriminatively trained sparse code gradients for contour detection. In: NIPS, Lake Tahoe (2012)
27. Shi, J., Malik, J.: Normalized cuts and image segmentation. In: CVPR, San Juan (1997)
28. Shi, J., Malik, J.: Normalized cuts and image segmentation. IEEE TPAMI **22**(8), 888–905 (2000)
29. Silberman, N., Hoiem, D., Kohli, P., Fergus, R.: Indoor segmentation and support inference from rgbd images. In: ECCV, Florence (2012)
30. Taylor, C.J.: Towards fast and accurate segmentation. In: CVPR, Portland (2013)
31. von Luxburg, U.: A tutorial on spectral clustering. Stat. Comput. **17**(4), 395-416 (2007)
32. Weiss, Y.: Segmentation using eigenvectors: a unifying view. In: ICCV, Corfu (1999)
33. Xiao, J., Owens, A., Torralba, A.: Sun3d: a database of big spaces reconstructed using sfm and object labels. In: ICCV, Sydney (2013)
34. Zheng, B., Zhao, Y., Yu, J.C., Ikeuchi, K., Zhu, S.: Beyond point clouds: scene understanding by reasoning geometry and physics. In: CVPR, Portland (2013)

Part II
Sparse Data Representation and Machine Learning for Shape Analysis

Chapter 8
Shape Compaction

Honghua Li and Hao Zhang

Abstract We cover and discuss techniques that are designed for compaction of shape representations or shape configurations. The goal of compaction is to reduce storage space, a fundamental problem in many application domains. We consider compaction both at the representation level (i.e., digital storage) and in physical domains (i.e., physical storage). Shape representation compaction focuses on reducing the memory space allocated for storing the shape geometry data, whilst shape compaction techniques in the physical domain reduce the physical space occupied by shape configuration. We use the term *shape configuration* to refer to how a shape, real or conceptual, is physically modeled (e.g., design and composition of its parts) and spatially arranged (e.g., shape parts positioning and possibly in relation to other shapes). In this paper we briefly cover the representation compaction techniques whilst placing our focus on the less explored realm of shape compaction approaches on physical configurations.

8.1 Introduction

Memory space is valuable in digital environment. Digital models of 3D shapes are widely used in a vast number of industrial and scientific applications. Typically the same shape admits multiple mathematical representations which may vary significantly in storage cost. Among them the most compact ones in terms of storage cost are usually more preferable since they can reduce the cost of storage, transmission, computation and visualization, as well as facilitate shape understanding and intelligent shape processing.

Physical space is also costly and thus the demand for compact products is strong in practice. Objects that can change the arrangement of their parts or their spatial relation with other shapes (the so-called *shape configuration*) to save space when storing or transporting them, are of great value for survival (e.g., fire fighter equipment, army weapons and tools), camping in the wild (e.g., tent, pocket knife), living (e.g., IKEA furniture) and leisure (e.g., LEGO assembling toys).

H. Li (✉) • H. Zhang
Simon Fraser University, Burnaby, BC, Canada
e-mail: howard.hhli@gmail.com;haoz@cs.sfu.ca

M. Breuß et al. (eds.), *Perspectives in Shape Analysis*, Mathematics and Visualization, DOI 10.1007/978-3-319-24726-7_8

In this paper, we use the term *shape compaction* to refer to techniques that can either assist human beings to reduce the storage size of shapes on both representation level and configuration domain, or automatically accomplish this goal. While numerous algorithms have been proposed for compaction of shape representations in literature, including simplification, abstraction, compression, etc., the compaction of shape configuration is still a realm that remains unexplored.

8.1.1 Compaction of Shape Representation

Given shape representation R_0 of a 3D object S, *representation compaction* is to (1) reorganize the data of R_0 to reduce its storage space or (2) find a new representation R of S which occupies less storage space subject to some criteria.

Shape representations are mathematical models conveying the geometry of 3D objects, and their size is measured as the amount of memory required to store such models. There are two key factors that influence the data size of a shape representation: the number of low-level primitives, and the statistical redundancy in geometric data. Shape representation compaction approaches addressing the former factor fall into the category of shape simplification and abstraction, while methods addressing the latter are usually regarded as shape compression techniques.

Shape simplification and abstraction The basic idea is to find a proxy with fewer primitives to represent the original object that consists of many finer primitives. Shape simplification aims to preserve the geometric fidelity within a prescribed error tolerance, while shape compaction has more freedom to modify the topology or geometry as long as the new generated representations are perceptually equivalent to the original shapes.

Compression Data compression techniques either exploit statistical redundancy in the underlying data to represent data more concisely (lossless), or modify the data in a subtle manner such that the statistical redundancy is enhanced (lossy). Mesh compression is the application of data compression on polygonal meshes. Typical mesh compression algorithms encode the connectivity and geometry data separately. Both natural and man-made objects present huge amount of regular and repeated substructures, which are usually captured by symmetries within the shape. Traditional mesh compression approaches do not explicitly utilize this statistical redundancy on the structure level. In a recent trend of research, several hierarchical representation techniques have been proposed to compactly represent complex shapes with rich symmetries in their structures.

8.1.2 *Compaction of Shape Configuration*

Size reduction of physical storage space is significantly different from that of memory space occupied by shape representations. The redundancy in digital models can be efficiently encoded to reduce the total storage space, which however isn't useful for physical storage reduction at all. For example, a shape with reflective symmetry can be compactly represented by half of its geometry and the associated reflection plane. In contrast, the two identical halves (in terms of reflection) both need to physically exist and thus occupy the same amount of space.

An intriguing problem about compact shapes is: what makes some objects more amenable to saving space than others? In an excellent introduction to space-saving designs, [37] discussed twelve collapsible principles. Collapsible objects are able to adjust in size by switching between two opposite configurations: one unfolded and *functional*, the other folded for *storage*. The existence of functional and storage configurations only makes it possible for an object to be collapsible. To be *practically* collapsible, the transformation between these two configurations must be feasible and easy to conduct.

Shape configuration is the arrangement of shape parts and/or the spacial relationships between shapes.

Collapsible objects can save space either individually involving the organization of parts within a shape, which is called *intra-shape configuration*, or cooperatively involving spacial relationships among multiple objects, which is called *inter shape configuration*. The chairs in Fig. 8.1a, b demonstrate two examples of intra-shape collapsing strategies: folding and decompose-and-pack. The stackable chair in Fig. 8.1c has a set of identical chairs involved while storing them. The outdoor tea table set in Fig. 8.1d consists of one tea table and four seats, which as a group can be packed compactly when not in use.

Given a 3D object S, *configuration compaction* is to find a new 3D object T such that (1) T is close to S, (2) T is able to change configuration to save space.

The problem has a trivial solution if sufficiently large perturbation from the source object S is allowed (e.g., let T be a cube). The "closeness" between two objects needs to be formalized such that it preserves the essence, i.e., structure and functionality of the original shape S. We classify compaction techniques into two categories based on the type of shape configuration they attempt to tackle.

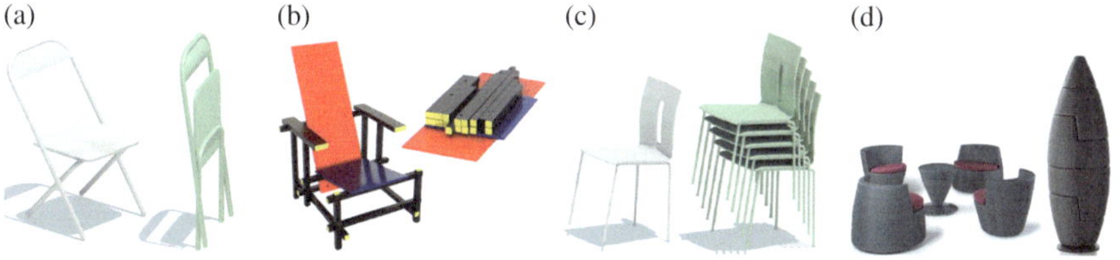

Fig. 8.1 Four collapsible mechanisms with the functional(*left* in each cell) and storage(*right* in each cell) configurations. (**a**) Folding. (**b**) Decompose-pack. (**c**) Stacking. (**d**) Group packing

Intra-shape configuration is the arrangement of shape parts within a shape. Compaction approaches in this category produce new shapes that preserve the essences of the original shapes in terms of either geometric appearance or functionality. Shape parts of the output can be transferred into a storing configuration which takes much less physical storage space than the original shape.

Inter-shape configuration indicates the internal relations between multiple shapes. Given a set of shapes, finding out the optimal configuration itself is a very challenging problem. Moreover there are algorithms that can modify the original shapes subtly such that the final packing result can be more space-saving.

8.2 Simplification and Abstraction

Given a representation, simplification and abstraction approaches output another representation option for the underlying shape which consists of fewer primitives than the original one. Dozens of simplification algorithms have been proposed by researchers in computer graphics. A detailed review of simplification techniques in literature is beyond the scope of this paper. Interested readers should refer to [14, 28, 29] for a broader survey on simplification approaches.

Representing complex objects with low bit budget goes beyond the capability of a error-metric-driven simplification method and the answer often lies in the area of human perception and cognition. Given a shape S, the goal of shape abstraction is to produce a proxy $\mathscr{S}$ such that perceptually S and $\mathscr{S}$ are comparable, but representationally $|\mathscr{S}| \ll |S|$. Note that S and $\mathscr{S}$ are likely to be quite different from a purely geometric point of view. These compact representations are visually more appealing than the detailed original models, which might appear visually cluttered. Therefore they are widely used for prototyping and concept communication.

The boundary between shape simplification and abstraction sometimes is blurry. Simplification with extremely low bit budget can be considered as abstraction, and abstraction at a very fine level may produce comparable results to simplification. The key characteristic of abstraction is that it directly extracts the shape defining features of objects which usually are inspired from human perception and cognition (Fig. 8.2).

Curve networks Sparse characteristic feature curves are typically sufficient for humans to identify a shape. Despite the fact that CG lines (image intensity edges, geometric ridges and valleys, suggestive contours, and apparent ridges) seem likely to succeed in conveying shapes [5, 6], they are usually not well organized and might be view-dependent. De Goes et al. [8] proposed the so-called *exoskeleton* to convey both the perceptual and the geometric structure of a 3D model. They first segment the input shape into parts, and further divide the shape surface into patches. The boundaries of these resulted patches form the exoskeleton.

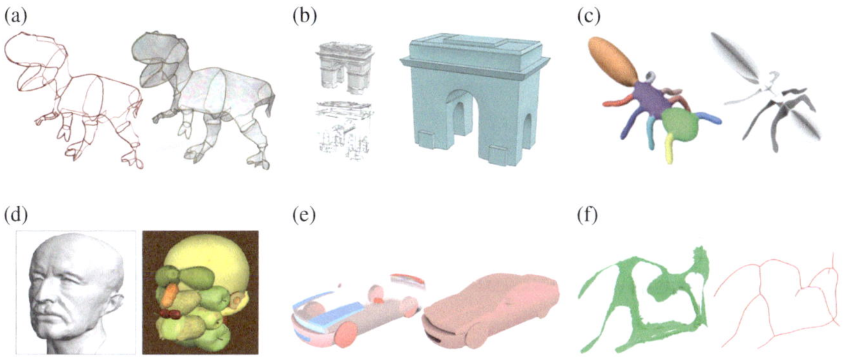

Fig. 8.2 Various shape abstractions. (**a**) Exoskeletion. (**b**) Surface model. (**c**) Slices. (**d**) Collage. (**e**) Subvolume. (**f**) Skeletons

Surface models Unlike simplification approaches which operate at low-level primitives and usually do not preserve shape defining features under extreme simplification, [33] extracts a sparse network of space curves that capture the essential characteristic features of a given man-made object, from which a abstract surface model can be reconstructed. Their method operates in two steps. First, a closed manifold envelope surface that wraps the entire input model is extracted from the voxelization of the input object. Second, they extract a network of curves or vectors from the computed envelope.

Planar sections Inspired by section planes in medical and engineering visualization which illustrate the interior details of complex shapes, [32] proposed an approach for generating shape proxies consisting of planar sections. In their method, planes are progressively selected to maximally capture shape features weighted by their importance, which is learned from the user study trying to discover how humans define planar section representations for various 3D shapes.

Collage Collage is an abstract and expressive visual style that build a new whole by assembling given primitives in a database. In a collage, both the parts and the whole can be easily recognized. Gal et al. [13] created 3D collages that express the target shape using a database of objects as primitive building block. In a parallel thread of work, [47] generated animation collage from mesh animation. In a recent work, [19] developed an algorithm for creating a collage which represents a given image with multiple Internet images. Note that the primitives used for collage are usually more complex than simple geometry primitives, therefore the collage techniques are considered as shape abstraction approaches solely because the little number of primitives.

Subvolumes Yumer and Kara [53] proposed an abstraction method that is built on *subvolumes*. The most abstract form is generated first and more details that are represented by volume chunks can be added or subtracted to the current abstraction. The main contribution of this work is that they can generate a spectrum of abstractions for each shape, and rely on the co-analysis on the associated shape collections to determine the "right" abstraction.

Skeletons The most well-known skeletal shape representation is probably the medial axis transform (MAT) [3]. In computer graphics, curve skeletons [7] are more broadly utilized due to their compactness and ease of manipulation. We refer the interested readers to recent advances on curve skeleton extraction [2, 18, 45, 46] for more details.

8.3 Compression

Data compression means to encode information using fewer bits than the original representation. Compression can be either lossless or lossy. Lossless compression is conducted by eliminating the statistical redundancy in the data. Some information lost is acceptable in lossy compression. By modifying in a subtle way, the data could be more amenable to coding, thus higher compression rate can be achieved.

The output of shape compression has to be decoded to be used, which is never a free lunch. However, shape compression has the advantage of using a given budget of storage space to represent more detailed shapes. Moreover, compression techniques can be used together with shape simplification and abstraction to obtain more compact shape representations. We refer interested readers to [1, 40] for a deeper and broader review of mesh compression techniques.

With the recent advance on shape structure analysis [36], compression techniques have been proposed to address data redundancy at structure level. Repeating substructures in digital models can be explicitly encoded to reduce its space complexity [39]. Due to the nested nature of symmetries, the simple strategy may encode the same symmetry multiple times. A hierarchical encoding, however, can reflect the nested structure and produce a more compact representation of the entire shape.

As a recent advance, there has been three pieces of work that develop hierarchical representation of single objects or complex scenes to address this type of structural redundancy.

Folding mesh Simari et al. [43] used a *folding tree* data structure to encode the reflective symmetries within a mesh by recursively applying a symmetry detection algorithm. The data structure encodes the non-redundant regions of the original geometry as well as the reflection planes. The folding tree can eventually be unfolded to recover the original shape approximately, see Fig. 8.3 (top right).

Symmetry and instancing Martinet [30] proposed the hierarchical assembly graph (HAG) to represent the structural information in scenes. A HAG is a directed graph, in which each node denotes an *object* and a arc denotes the sub-part relation between two objects. An object is defined as a *closed frequent pattern*, which is a part of the scene that does not have subpart having higher frequency than itself, see Fig. 8.3 (left).

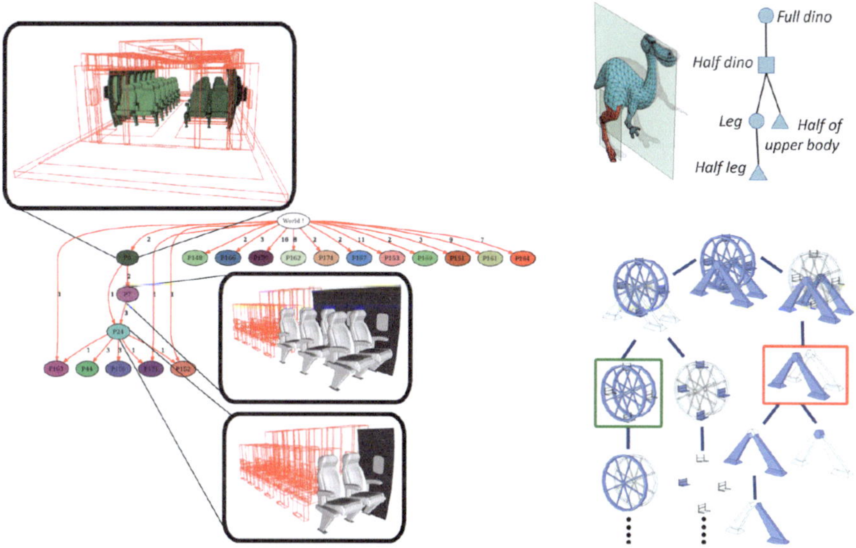

Fig. 8.3 Structural shape compression techniques: hierarchical assembly graph (*left*), folding mesh (*top right*), and symmetry hierarchy (*bottom right*)

Symmetry hierarchy Wang et al. [50] described an analogous method to construct the symmetry hierarchical (SYMH) organization of object parts by using perceptual grouping criteria. The input mesh is initially segmented into parts which are refined by symmetries detected in the original shape. An initial graph is built to encode inter-part symmetry and connectivity relations among the resulting segments, as well as self-symmetry for individual segments. The symmetry hierarchy is then constructed from the initial graph via graph contraction, which either groups parts by symmetry, or assembles connected sets of parts. The order of graph contraction is determined by a set of rules designed to respect human perceptions and the principle of compactness. See Fig. 8.3 (bottom right) for an example.

The HAG proposed by [30] is a directed graph which is different from the tree structures described by the other two. The advantage of a graph structure is that different part of the shape can share the same set of geometry primitives stored in leaf nodes. Primitives geometry represented by leaf nodes are building blocks when establishing hierarchical representations. Although different algorithms have been explored, finding the "best" primitive geometries still remains an open problem.

8.4 Compaction of Intra-shape Configuration

Shapes can reduce size individually by changing their own configuration. Given a 3D object, compaction of intra-shape configurations is to find another object that is close to the input in terms of geometrical appearance, structural form, or functional essence, but also is able to adjust its configuration to meet the requirement of space-saving. This can be achieved by either modifying the original shape or creating a new shape via approximation. In this section we discuss two mechanisms – folding and decomposing – that are frequently utilized for shape compaction.

8.4.1 Folding

Folding via hinges is a popular collapsible principle that impacts many tools in our daily lives. Generally speaking a hinge is a movable joint that connects two objects and typically allows rotation between them. The most popular form of folding is probably paper folding [9, 20], with origami [31] being the best known instance.

Pop-up design Popups are paper arts that can be closed down to a flat surface and opened up again without tearing the paper or introducing new creases other than those in the design. A popup is collapsible since it has both functional and storing configurations, one of which can be easily transformed into the other without extra forces other than holding and turning two support pages.

Origamic architectures, also called paper architectures (PA), are paper buildings created by cutting and folding from a single piece of paper. The simple mechanisms of parallel PA enabled development of automated algorithms to construct a PA from an input 3D model [25, 35] as well as interactive tools [34], see Fig. 8.4a. Li et al.

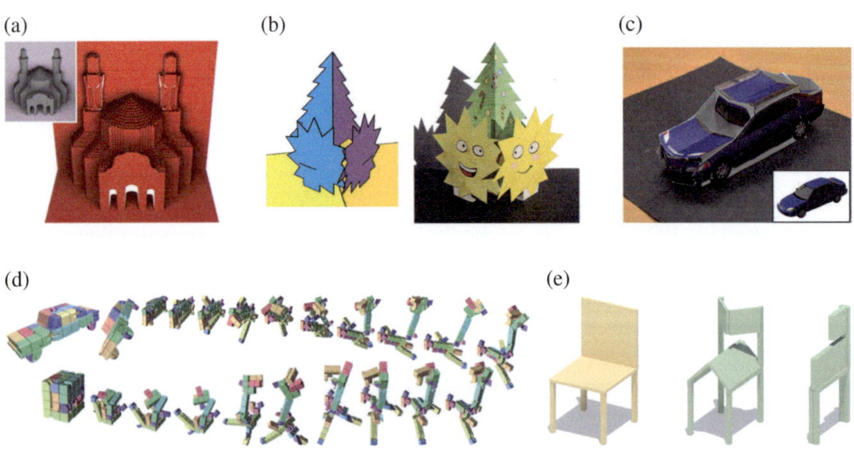

Fig. 8.4 Shape compaction techniques utilizing the folding mechanisms. (**a**) Popup. (**b**) V-style popup. (**c**) Multi-style popup. (**d**) Boxelization. (**e**) Foldabilization

[24] studied the general v-style popups, which contain two more parallel groups of planes with multiple pieces of paper, see Fig. 8.4b. Ruiz et al. [41] extend the pop-up design to multi-style by fitting volumetric primitives and mapping to selected mechanisms, see Fig. 8.4c.

Foldable puzzle design In a recent work, [54] approximates the input shape using a voxel-tree that can fold from the input shape into a cube. The goal of boxelization is to find a physically achievable solution for transforming a shape into a cube. Compactness is one of the objective terms in the optimization procedure and shape compaction is therefore achieved as a by-product. Their algorithm involves three major steps: finding a good voxelization, finding the tree structure that can form the input and target shapes' configurations, and finding a non-intersecting folding sequence.

Foldable furniture design Space-saving furniture designs are ubiquitous in our lives and folding is perhaps the most popular mechanism observed and practiced [37]. However the design process of foldable furniture has to follow the trial-and-error iteration, which is usually both tedious and time consuming. Here we pose an open *foldabilization* problem: given a 3D furniture, how to apply a minimum amount of modification to the input to allow it to be folded? Figure 8.4e provides an example solution: by introducing hinges on the seat and back and shrinking the back, the modified chair is able to fold into a flat configuration.

8.4.2 Decomposing

The functional configuration of an object usually leaves large amount of free space among its parts, which increases the cost for fabrication or storing. Decomposing provides an option to reorganize shape parts to reduce this unused space.

Decompose-and-print Layered printing has been widely used in 3D printers. Usually support structure has to be printed together with the object itself to allow complex shapes to be fabricated, which however causes material cost and takes longer time to print. The amount of support material depends on the free space within the projection volume of an object. Inspired by pyramidal shapes which always have solid projection volume with respect to the given base, [17] proposed an algorithm to decompose the input 3D model into approximately pyramidal parts, see Fig. 8.5a. By printing each pyramidal parts individually and gluing, the original object can be fabricated. The pyramidal composition is more compact than the original object in terms of projection volume.

Decompose-and-pack Decompose-and-pack is a time-honored collapsible principle. A number of separate parts are assembled into a whole to perform functions, and then later are dismantled again into its parts for storage. An excellent example is flat pack furniture which supports almost the entire business of IKEA.

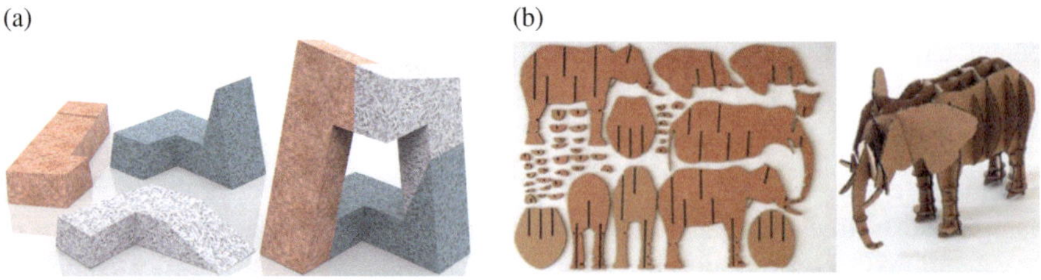

Fig. 8.5 Shape compaction techniques using the decomposing mechanism. (**a**) Approximate pyramidal decomposition. (**b**) Cardboard sculpture

The cardboard sculpture is another example where cardboard pieces have prefabricated slits along which they can be slid to assemble the whole shape. Obviously these cardboard pieces can be stored much more compact than as a whole. Hildebrand et al. [15] proposed an algorithm to automatically generate cardboard sculptures with guaranteed constructibility, see Fig. 8.5b.

Given an arbitrary 3D model, searching for the decomposition and packing strategy that leads to the most compact packing remains an open problem.

8.5 Inter-shape Configuration Compaction

Shapes can work cooperatively to save space. This group strategy involves changing the spatial relations with other shapes. A set of objects can be packed more compactly under rigid transformations as long as the unused space within one shape's bonding volume can be used by another shape, see Fig. 8.1c, d.

Without modifying input shapes in any way, the problems we are discussing here degenerate into the classic *nesting problems*. As a specific type of *cutting and packing (C&P) problems*, nesting problems consider packing irregular shapes in order to optimize the packing volume. The problem is NP-hard and as a result solution methodologies usually utilize heuristics. The term "compaction" was also used by [26] to refer to a simultaneous motion of the components that generates a more densely packed layout.

A dense nesting is possible only if the irregular shapes can fit into each other very well. An extreme case is tiling [49], where each tile can exactly fit into its neighbours such that all tiles together can cover the entire plane. However arbitrary shapes usually do not have such nice properties. In many cases the input geometries do not have to keep unchanged but their essences, e.g. main features and functionality. In fact, allowing subtle changes to the input shapes can greatly improve nesting results [21, 23].

In this section, we first briefly overview the challenges and state-of-the-art solutions of nesting problems, then follow up with techniques that modify and optimize the geometry of input shapes for more compact packing results.

8.5.1 Without Shape Modification

The topic of cutting and packing covers a variety of problems of a common logical structure which is usually classified under the heading of packing, packaging, layout, configuration, container stuffing, pallet loading or spatial arrangement in the literature. Dyckhoff and Wäscher et al. [10, 51] introduced a useful typology of C&P problems, where C&P problems can be classified into regular packing and irregular packing, the latter is also called *nesting problems*.

The nesting problem is usually abstracted as an optimization problem where an assignment of the positions and orientations of components that minimizes an objective is sought. Comparing to regular packing [27], irregular components increase the complexity of the solution space. The problem is a NP-hard combinatorial problem [38] such that meta-heuristics are typically used to generate acceptable solutions. Hopper and Turton [16] reviewed these meta-heuristic algorithms, in particular genetic algorithms, for both 2D regular and irregular packing problems. As research progressed, new breakthroughs have been achieved in recent years. Timmerman [48] compared different optimization methods using benchmarks and concludes that extended local search [22] is the best method currently available.

3D nesting problem shares most characteristics with its two-dimensional counterpart, but the geometric complexity of 3D irregular components makes it a more challenging problem. Cagan et al. [4] reviewed a spectrum of approaches ranging from deterministic algorithms to stochastic algorithms proposed for solving 3D layout problems. The geometric representation and interference detection approaches of 3D components are also discussed in that survey. Most algorithms are originally designed for 2D nesting problems and have the potential to be extended to 3D [11, 44]. In contrast, the extended pattern search algorithm [52] was particularly designed for 3D nesting problems.

8.5.2 With Shape Modification

Modifying the input shapes is not necessary, but when applied it has the potential to improve the nesting density. Shape modification is not always possible or allowed, since traditionally nesting is an independent post procedure after the design of a product has been fixed. If nesting quality is not considered during the product design, a nesting algorithm solely is doomed to fail on finding very dense packing layout. In fact products that are successful in space saving are originally designed to be so. Instead of barely relying on the designer's experience and letting the designers improve their design in a trial-and-error iteration, computational algorithms can be designed to either assist designers to speed up the iterations or automatically modify the design in a subtle manner to achieve more compact packing layout.

Escherization Tiling is a special case of 2D nesting problems because each component (tile) can exactly fit into its neighbors such that the entire plane can be

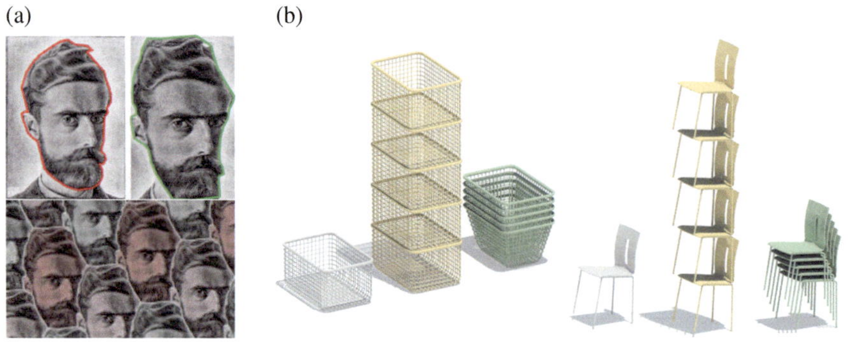

Fig. 8.6 Shape compaction techniques with modification. (**a**) Escherization. (**b**) Stackabilization

covered seamlessly. The Dutch artist M.C. Escher spent his career on producing a notebook with more than a hundred of ingenious and playful designs of tiling [42]. Inspired by Escher's work, [21] presented a solution to the "Escherization" problem: given a closed figure, find a new closed figure that is similar to the input and tiles the entire plane, see Fig. 8.6a. Their approach utilizes a simulated annealer to optimize over a parameterization of the *isohedral* tilings, which is flexible enough to encompass nearly all of Escher's own tilings.

Stackabilization Stacking objects on top of each other is a common strategy performed by humans to save space. The nesting layout of a stack along a stacking direction can be achieved by repeated application of a translation and a possible rotation on object copies until two adjacent objects are just touching each other without overlap or gaps. One of the most celebrated examples of stackable objects are chairs [12].

Li et al. [23] first introduced the geometric problem of stackabilization: how to geometrically modify a 3D object so that it is more amenable to stacking? They consider the class of stackings that involves only translation in the stacking configuration. The main challenge in stackabilization lies in the desire to modify the input geometry only subtly so that the intended functionality and aesthetic appearance of the original object are not significantly affected.

8.6 Conclusion

This is the first general introduction on shape compaction techniques, at both the digital representation level and the physical configuration domain. These two compaction categories share the same goal of finding economy solutions for storing and transporting objects, which is beneficial in a large range of applications. They also share the spirit of utilizing shape modification to facilitate the compaction results. In particular, simplification and abstraction of shape representation would have strong connection to shape compaction of intra-shape configuration, e.g. popup and cardboard sculpture.

Due to the strong practical demands on compact digital and physical objects, more effort from researchers is expected to commit in this realm. To conclude this paper we list a few possible future directions along this thread of research.

Because of the conceptual nature, shape abstraction is worth more creative investigation. Structure analysis has attracted tremendous attention recently, which provides opportunities for finding better structural compression approaches.

By now 3D nesting problems have not drawn comparable amount of attention from researchers as that in 2D. The needs arising in the product layout, rapid prototyping, and efficient use of resources (e.g., 3D printing material) justify the development of efficient nesting approaches for 3D components with complex geometry.

Most intra-shape configuration compaction approaches, e.g. popup designs, and boxelization, can only approximate the appearance of the input in a very rough manner. The reason is that the feasibility of particular collapsibility usually serves as hard constraint, while sacrificing the appearance and even the essential of the given 3D model. An open problem is how to develop generic computational approaches for generating collapsible objects that can preserve the functionality or at least the structure of the input.

Generally speaking, compaction of shape configuration is a relatively unexplored area with numerous open problems waiting to be studied. Solving these problems will benefit a huge amount of practical applications which are sensitive to physical storage space.

References

1. Alliez, P., Gotsman, C.: Recent advances in compression of 3d meshes. In: Advances in Multiresolution for Geometric Modelling, pp. 3–26. Springer, Berlin/London (2005)
2. Au, O.K.-C., Tai, C.-L., Chu, H.-K., Cohen-Or, D., Lee, T.-Y.: Skeleton extraction by mesh contraction. ACM Trans. Graph. **27**(3), 44 (2008)
3. Blum, H.: A transformation for extracting new descriptors of shape. In: Wathen-Dunn, W. (ed.) Models for the Perception of Speech and Visual Form, pp. 362–380. M.I.T. Press, Cambridge (1967)
4. Cagan, J., Shimada, K., Yin, S.: A survey of computational approaches to three-dimensional layout problems. Comput. Aided Des. **34**, 597–611 (2002)
5. Cole, F., Golovinskiy, A., Limpaecher, A., Barros, H.S., Finkelstein, A., Funkhouser, T., Rusinkiewicz, S.: Where do people draw lines? ACM Trans. Graph. (Proc. SIGGRAPH) **27**(3), 88 (2008)
6. Cole, F., Sanik, K., DeCarlo, D., Finkelstein, A., Funkhouser, T., Rusinkiewicz, S., Singh, M.: How well do line drawings depict shape? ACM Trans. Graph. **28**(3), 28 (2009). Proc. SIGGRAPH
7. Cornea, N.D., Silver, D., Min, P.: Curve-skeleton properties, applications, and algorithms. IEEE Trans. Vis. Comput. Graph. **13**(3), 530–548 (2007)
8. De Goes, F., Goldenstein, S., Desbrun, M., Velho, L.: Technical section: exoskeleton: curve network abstraction for 3d shapes. Comput. Graph. **35**(1), 112–121 (2011)

9. Demaine, E.D., O'Rourke, J.: Geometric Folding Algorithms: Linkages, Origami, Polyhedra. Cambridge University Press, Cambridge/New York (2007)
10. Dyckhoff, H.: A typology of cutting and packing problems. Eur. J. Oper. Res. **44**(2), 145–159 (1990)
11. Egeblad, J., Nielsen, B.K., Odgaard, A.: Fast neighborhood search for two- and three-dimensional nesting problems. Eur. J. Oper. Res. **183**(3), 1249–1266 (2007)
12. Fiell, C., Fiell, P.: 1000 Chairs. Taschen, New York (2000)
13. Gal, R., Sorkine, O., Popa, T., Sheffer, A., Cohen-Or, D.: 3D collage: expressive non-realistic modeling. In: NPAR: Proceedings of the 5th International Symposium on Non-Photorealistic Animation and Rendering, San Diego, pp. 7–14 (2007)
14. Heckbert, P.S., Garland, M.: Survey of polygonal surface simplification algorithms. In: Multiresolution Surface Modeling Course SIGGRAPH'97, Los Angeles (1997)
15. Hildebrand, K., Bickel, B., Alexa, M.: crdbrd: shape fabrication by sliding planar slices. Comput. Graph. Forum **31**, 1583–592 (2012)
16. Hopper, E, Turton, B.C.H.: A review of the application ofmeta-heuristic algorithms to 2d strip packing problems. Artif. Intell. Rev. **16**(4), 257–300 (2001)
17. Hu, R., Li, H., Zhang, H., Cohen-Or, D.: Approximate pyramidal shape decomposition. In: Proceedings of SIGGRAPH, Vancouver (2014)
18. Huang, H., Wu, S., Cohen-Or, D., Gong, M., Zhang, H., Li, G., Chen, B.: L1-medial skeleton of point cloud. ACM Trans. Graph. **32**, 65:1–65:8 (2013)
19. Huang, H., Zhang, L., Zhang, H.-C.: Arcimboldo-like collage using internet images. ACM Trans. Graph. **30**(155), 1–8 (2011)
20. Jackson, P.: Folding Techniques for Designers: From Sheet to Form. Laurence King Publishing, London (2011)
21. Kaplan, C.S, Salesin, D.H.: Escherization. In: Proceedings of the 27th annual conference on Computer graphics and interactive techniques, SIGGRAPH'00, New Orleans, pp. 499–510. ACM Press/Addison-Wesley Publishing Co. (2000)
22. Leung, S.C., Lin, Y., Zhang, D.: Extended local search algorithm based on nonlinear programming for two-dimensional irregular strip packing problem. Comput. Oper. Res. **39**(3), 678–686 (2012)
23. Li, H., Alhashim, I., Zhang, H., Shamir, A., Cohen-Or, D.: Stackabilization. ACM Trans. Graph. **31**(6), 158:1–158:9 (2012)
24. Li, X.-Y., Ju, T., Gu, Y., Hu, S.-M.: A geometric study of v-style pop-ups: theories and algorithms. ACM Trans. Graph. **30**(4), 98:1–10 (2011)
25. Li, X.-Y., Shen, C.-H., Huang, S.-S., Ju, T., Hu, S.-M.: Popup: automatic paper architectures from 3d models. ACM Trans. Graph. **29**(4), 111:1–9 (2010)
26. Li, Z.: Compaction algorithms for non-convex polygons and their applications. Ph.D. thesis, Harvard University (1994)
27. Lodi, A., Martello, S., Monaci, M.: Two-dimensional packing problems: a survey. Eur. J. Oper. Res. **141**(2), 241–252 (2002)
28. Luebke, D., Watson, B., Cohen, J.D., Reddy, M., Varshney, A.: Level of Detail for 3D Graphics. Elsevier Science Inc., New York (2002)
29. Luebke, D.P.: A developer's survey of polygonal simplification algorithms. IEEE Comput. Graph. Appl. **21**(3), 24–35 (2001)
30. Martinet, A.: Structuring 3D geometry based on symmetry and instancing information. Ph.D. thesis, INP Grenoble (2007)
31. McArthur, M., Lang, R.J.: Folding Paper: The Infinite Possibilities of Origami. Turtle Publishing, Tokyo (2013)
32. McCrae, J., Singh, K., Mitra, N.J.: Slices: a shape-proxy based on planar sections. ACM Trans. Graph. **30**(6), 168:1–168:12 (2011)
33. Mehra, R., Zhou, Q., Long, J., Sheffer, A., Gooch, A., Mitra, N.J.: Abstraction of man-made shapes. ACM Trans. Graph. **28**(5), 137:1–137:10 (2009)
34. Mitani, J., Suzuki, H.: Computer aided design for origamic architecture models with polygonal representation. In: Proceedings of Computer Graphics International, Crete, pp. 93–99 (2004)

35. Mitani, J., Suzuki, H., Uno, H.: Computer aided design for origamic architecture models with voxel data structure. Trans. Inf. Process. Soc. Jpn. **44**(5), 1372–1379 (2003)
36. Mitra, N.J., Wand, M., Zhang, H., Cohen-Or, D., Bokeloh, M.: Structure-aware shape processing. In: EUROGRAPHICS State-of-the-art Report, Girona (2013)
37. Mollerup, P.: Collapsible: The Genius of Space-Saving Design. Chronicle, San Francisco (2001)
38. Nielsen, B.K., Odgaard, A.: Fast neighborhood search for the nesting problem. Technical Report 03/03, Department of Computer Science, University of Copenhagen, Universitetsparken 1, DK-2100 Copenhagen Ø (2003)
39. Pauly, M., Mitra, N.J., Wallner, J., Pottmann, H., Guibas, L.: Discovering structural regularity in 3D geometry. ACM Trans. Graph. **27**(3), 43:1–11 (2008)
40. Peng, J., Kim, C.-S., Jay Kuo, C.C.: Technologies for 3d mesh compression: a survey. J. Vis. Commun. Image Represent. **16**(6), 688–733 (2005)
41. Ruiz Jr., C.R., Le, S.N., Yu, J., Low, K.-L.: Multi-style paper pop-up designs from 3d models. Comput. Graph. Forum (Special Issue of Eurographics) **33**(2), 487-496 (2014)
42. Schattschneider, D., Escher, M.C.: Visions of Symmetry. W.H. Freeman, New York (1990)
43. Simari, P., Kalogerakis, E., Singh, K.: Folding meshes: hierarchical mesh segmentation based on planar symmetry. In: Symposium on Geometry Processing, Cagliari, pp. 111–119 (2006)
44. Stoyan, Y., Romanova, T.: Mathematical models of placement optimisation: two- and three-dimensional problems and applications. In: Fasano, G., Pintér, J.D. (eds.) Modeling and Optimization in Space Engineering. Springer, New York (2013)
45. Tagliasacchi, A., Alhashim, I., Olson, M., Zhang, H.: Mean curvature skeletons. Comput. Graph. Forum **31**(5), 1735–1744 (2012)
46. Tagliasacchi, A., Zhang, H., Cohen-Or, D.: Curve skeleton extraction from incomplete point cloud. ACM Trans. Graph. **28**(3): 71, 9 (2009)
47. Theobalt, C., Rössl, C., de Aguiar, E., Seidel, H.-P.: Animation collage. In: Symposium on Computer Animation, San Diego, pp. 271–280. Eurographics (2007)
48. Timmerman, M.: Optimization methods for nesting problems. Master's thesis, University West (2013)
49. van Lemmen, H.: Tiles: 1000 Years of Architectural Decoration. Harry N. Abrams, Inc., New York (1993)
50. Wang, Y., Xu, K., Li, J., Zhang, H., Shamir, A., Liu, L., Cheng, Z., Xiong, Y.: Symmetry hierarchy of man-made objects. Comput. Graph. Forum (Special Issue of Eurographics) **30**(2), 287–296 (2011)
51. Wäscher, G., Haußner, H., Schumann, H.: An improved typology of cutting and packing problems. Eur. J. Operat. Res. **183**(3), 1109–1130 (2007)
52. Yin, S, Cagan, J.: An extended pattern search algorithm for three-dimensional component layout. ASME J. Mech. Des. **122**(1), 102–108 (2000)
53. Yumer, M.E., Kara, L.B.: Co-abstraction of shape collections. ACM Trans. Graph. **31**, 158:1–158:11 (2012). Proceedings of SIGGRAPH Asia 2012
54. Zhou, Y., Sueda, S., Matusik, W., Shamir, A.: Boxelization: folding 3d objects into boxes. ACM Trans. Graph. **33**(4), 71:1–71:8 (2014)

Chapter 9
Homological Shape Analysis Through Discrete Morse Theory

Leila De Floriani, Ulderico Fugacci, and Federico Iuricich

Abstract Homology and persistent homology are fundamental tools for shape analysis and understanding that recently gathered a lot of interest, in particular for analyzing multidimensional data. In this context, discrete Morse theory, a combinatorial counterpart of smooth Morse theory, provides an excellent basis for reducing computational complexity in homology detection. A discrete Morse complex, computed over a given complex discretizing a shape, drastically reduces the number of cells of the latter while maintaining the same homology. Here, we consider the problem of shape analysis through discrete Morse theory, and we review and analyze algorithms for computing homology and persistent homology based on such theory.

9.1 Introduction

Recently, in shape analysis, the computation of topological features, which provide global information about a shape, has drawn particular attention, specifically in analyzing medium- and high-dimensional data sets, where pure geometric tools are usually not sufficient. Morse theory [44] and its discrete counterpart [28] have been recognized as important tools for analyzing shapes in several application domains, including physics, chemistry, medicine and geography, thus studying the

L. De Floriani (✉)
Department of Computer Science, Bioengineering, Robotics, and Systems Engineering, University of Genova, Genova, Italy
e-mail: leila.defloriani@unige.it

U. Fugacci
Department of Computer Science and UMIACS, University of Maryland, College Park, MD, USA
e-mail: fugacci@umd.edu

F. Iuricich
Department of Geographical Sciences and UMIACS, University of Maryland, College Park, MD, USA
e-mail: iurif@umd.edu

M. Breuß et al. (eds.), *Perspectives in Shape Analysis*, Mathematics and Visualization, DOI 10.1007/978-3-319-24726-7_9

relationships between the topology of a manifold and the critical points of a real-valued function defined on it (scalar fields).

Discrete Morse theory [28] offers a way to compute and represent Morse complexes in an efficient way. When working with cell complexes, a scalar function is not naturally defined, though Morse complexes can still be computed. The description obtained in such cases is related to the notion of homology, a topological invariant counting the number of independent non-bounding k-cycles characterizing the shape.

Independently on whether the scalar function is defined or not, the problem of computing Morse complexes reduces to the definition of a valid gradient vector field (Forman gradient) on the shape. Here, we review the main algorithms in the literature for computing a Forman gradient and its homology. We distinguish between two classes of algorithms, that we call *constrained* and *unconstrained* algorithms. *Constrained algorithms* compute a Forman gradient conforming with the scalar function defined on the shape. *Unconstrained algorithms* do not have limitations regarding the scalar function and they have to agree only with the topology of the shape itself.

Discrete Morse theory is not the only tool to obtain efficient methods for homology and persistent homology computation. Other techniques to efficiently retrieve homology can be roughly classified into approaches based on simplification operators [20, 46], distributed approaches [7, 41], which are based on a decomposition of the shape, and approaches based on hierarchical models [15]. Here, we focus on unconstrained algorithms, discussing how discrete Morse theory can be combined with homology computation based on reductions and coreductions. We also believe that discrete Morse theory could be combined with distributed and hierarchical methods in order to further improve complexity of such algorithms.

The remainder of this paper is organized as follows. In Sect. 9.2, we present some background notions on homology, persistent homology and discrete Morse theory. In Sect. 9.3, we describe a compact encoding for a Forman gradient. In Sect. 9.4, we review and classify classical methods for computing a Forman gradient on a cell complex. In Sect. 9.5, we discuss how to extract Morse and Morse-Smale complexes from a Forman gradient, and we present a combinatorial representation of such complexes. Finally, in Sect. 9.6, we discuss how discrete Morse theory can be used to efficiently compute homology, homological generators and persistent homology of a cell complex, showing some application domains where homological information plays a crucial role.

9.2 Background Notions

This section is devoted to the presentation of some background notions that we will use in the rest of the paper, such as cell and simplicial complexes and regular grids, homology, persistent homology and discrete Morse theory.

9.2.1 Cell Complexes, Simplicial Complexes and Regular Grids

Cell complexes [38, 42] are used as a discretization and modeling tool in a wide range of application domains.

Intuitively, a cell complex defines a decomposition of a shape into simple subsets, called *cells*, glued together along their boundaries. We define the k-disk as $D^k = \{x \in \mathbb{R}^k : |x| \leq 1\}$ and the $(k-1)$-sphere as $\mathbb{S}^{k-1} = \{x \in \mathbb{R}^k : |x| = 1\}$. A *$k$-cell* (or cell of dimension k) is a homeomorphic image of the open k-disk $int(D^k) = D^k \setminus \mathbb{S}^{k-1}$. A space $\Gamma \subseteq \mathbb{R}^n$ is called a *cell complex* if Γ is a finite disjoint union of cells such that, for each k-cell σ of Γ, there exists a map $\Phi_\sigma : D^k \to \Gamma$ restricting to a homeomorphism $\Phi_\sigma|_{int(D^k)} : int(D^k) \to \sigma$ and taking the $(k-1)$-sphere $\mathbb{S}^{k-1}$ into Γ^{k-1}, where, for each i, Γ^i (called the *i-skeleton* of Γ) is the union of the cells of Γ with dimension less than or equal to i.

We define the *dimension* of a cell complex Γ, denoted as $dim(\Gamma)$, to be the largest dimension of a cell of Γ.

Let Γ be a cell complex, and let σ and τ be two cells of Γ. σ is called a *(proper) face* of τ if σ is contained in the boundary of the cell τ; τ is called a *(proper) coface* of σ. Any cell of Γ, which is not a face of any cell of Γ, is called a *top* cell. The *star* of a cell $\sigma \in \Gamma$ is the set of cells $\tau \in \Gamma$ which are cofaces of σ. A k-cell σ is said to be *adjacent* to a k-cell σ' if σ and σ' share a $(k-1)$-face (see Fig. 9.1b). The *link* of a cell $\sigma \in \Gamma$, denoted as $Lk(\sigma)$, is the set of cells $\tau \in \Gamma$ such that τ is a face of a coface of σ, and is not incident in σ (see Fig. 9.1c). A cell complex is said to be *regular*, if, for each k-cell σ, map $\Phi_\sigma : D^k \to \Gamma$ is a homeomorphism. In other words, the boundary of each cell has no identification. Because of their importance in the applications, in the following we will just consider regular cell complexes and write cell complex in place of regular cell complex.

In many applications, however, simplicial complexes and regular grids are extensively used to discretize a shape or the domain of a scalar field. Cell complexes encompass both of them.

Simplicial complexes can be viewed as a special case of cell complexes, in which the cells are simplices. A *simplex* of dimension k, or a *k-simplex*, is the convex hull of $k+1$ affinely independent points in $\mathbb{R}^n$ which are called the *vertices* of the simplex.

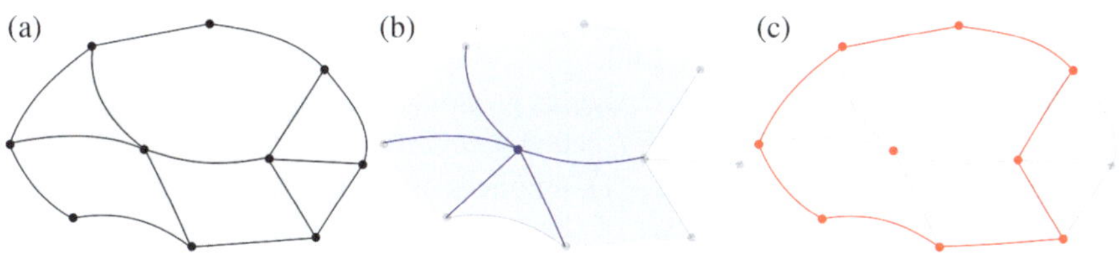

Fig. 9.1 (**a**) Cell complex composed of 2-cells bounded by 1-cells (*bold lines*) and 0-cells (*full dots*). (**b**) The star of the blue vertex is composed of five 1-cells (*blue lines*) and five 2-cells (*blue surfaces*) incident in it. (**c**) The link of the central red vertex is composed of the eight 1-cells (*red lines*) in the neighborhood of the vertex as well as the eight 0-cells (*red dots*) incident in them

Given a k-simplex τ, any simplex σ, which is the convex hull of a non-empty subset of the points generating τ, is called a *face* of τ. A *simplicial complex* Σ is a finite set of simplices, such that each face of a simplex in Σ belongs to Σ, and the non-empty intersection of any two simplices in Σ is a face of both. Examples of simplicial complexes are triangle or tetrahedral meshes.

We consider an *axis-parallel k-dimensional hyper-cube* ρ in $\mathbb{R}^n$ is the Cartesian product of n closed intervals, where exactly k of them are non-degenerate with equal length, i.e., $\rho = \{p = (x_1, \ldots, x_n) \in \mathbb{R}^n \mid x_i \in [a_i, b_i]\}$, where $\#\{i \mid a_i < b_i\} = k$ and, for such i, $b_i - a_i$ is constant. We say that hyper-cube ρ is generated by intervals $[a_i, b_i]$. Usually, intervals have integer endpoints and unit length, i.e., $a_i \in \mathbb{Z}$, and $b_i = a_i$ or $b_i = a_i + 1$. Given a hyper-cube ρ, generated by intervals $[a_i, b_i]$, $i = 1, \ldots, n$, any hyper-cube ρ' generated by intervals $[a'_i, b'_i]$, with either $a'_i = a_i$ and $b'_i = b_i$, or $a'_i = b'_i = a_i$, or $a'_i = b'_i = b_i$, is called a *face* of ρ. Hyper-cube ρ' is a *proper face* of ρ if $\rho' \neq \rho$.

A *regular (hyper-cubic) grid* in $\mathbb{R}^n$ is a finite collection H of hyper-cubes of different dimensions, such that:

- for any hyper-cube $\rho \in H$, all hyper-cubes that are proper faces of ρ are in H;
- for any pair of hyper-cubes $\rho_1, \rho_2 \in H$, either $\rho_1 \cap \rho_2 = \emptyset$, or $\rho_1 \cap \rho_2$ is a hyper-cube of H;

and the domain of H is a hyper-cube in $\mathbb{R}^n$. A 2D regular grid is also called a *square grid*, and a 3D regular grid a *cubic grid*.

9.2.2 Simplicial and Cellular Homology

For the sake of brevity, we will present only the notion of *simplicial homology* which can be suitably extended to cell complexes obtaining *cellular homology* [42]. Both simplicial and cellular homology theories are special cases of the *singular homology theory* defined for topological spaces [38].

A *chain complex* $C_* = (C_k, d_k)_{k\in\mathbb{N}}$ is a collection of Abelian groups C_k and a collection of group homomorphisms $d_k : C_k \to C_{k-1}$ such that $d_k d_{k+1} = 0$ (or, equivalently, $\operatorname{Im} d_{k+1} \subseteq \ker d_k$). Given a simplicial complex Σ, it is possible to define the notion of *simplicial homology* of Σ by associating a chain complex $C_*(\Sigma) = (C_k(\Sigma), \partial_k)_{k\in\mathbb{N}}$ with Σ [48].

Chain groups $C_k(\Sigma)$ are the free Abelian groups generated by the k-simplices of Σ and maps $\partial_k : C_k(\Sigma) \to C_{k-1}(\Sigma)$, called *boundary maps*, encode the boundary relations between the k- and the $(k-1)$-simplices of Σ and are defined as follows. Having chosen an order on the set of the vertices of Σ, we can uniquely write each k-simplex σ of Σ as $[v_0, v_1, \ldots, v_k]$, where σ is generated by $v_0, \ldots, v_k$ and $v_0 < v_1 < \cdots < v_k$. We can define $\partial_k : C_k(\Sigma) \to C_{k-1}(\Sigma)$ by setting, for each k-simplex $\sigma = [v_0, \ldots, v_k]$, $\partial_k(\sigma) = \sum_{i=0}^{k} (-1)^i [v_0, \ldots, \widehat{v_i}, \ldots, v_k]$, where $\widehat{v_i}$ means that the vertex v_i is not present. We denote as $Z_k(\Sigma) = \ker \partial_k$ the group of the k-cycles of Σ and as $B_k(\Sigma) = \operatorname{Im} \partial_{k+1}$ the group of the k-boundaries of Σ. Since $\partial_k \partial_{k+1} = 0$, we can define the kth *homology group of Σ with coefficients in $\mathbb{Z}$* as the kth homology group of the chain complex $C_*(\Sigma)$, i.e.,

$$H_k(\Sigma) = H_k(C_*(\Sigma)) = \frac{Z_k(\Sigma)}{B_k(\Sigma)}.$$

Intuitively, homology detects the presence of holes in a shape. Specifically, a non-null element in a homology group is a cycle not representing the boundary of any collection of simplices of Σ.

As a consequence of the theorem of structure for finitely generated Abelian groups (see [3], Chapter 12), the homology groups of a simplicial complex Σ can be expressed as $H_k(\Sigma) \cong \mathbb{Z}^{\beta_k}\langle c_1, \ldots, c_{\beta_k}\rangle \oplus \mathbb{Z}_{\lambda_1}\langle c'_1\rangle \oplus \cdots \oplus \mathbb{Z}_{\lambda_{p_k}}\langle c'_{p_k}\rangle$, with $\lambda_{i+1} \mid \lambda_i$ and with λ_i non-invertible. We call β_k the kth *Betti number* of Σ, $\bigoplus_{i=1}^{p_k} \mathbb{Z}_{\lambda_i}$ the *torsion part* of $H_k(\Sigma)$ and $c_1, \ldots, c_{\beta_k}, c'_1, \ldots, c'_{p_k}$ the *generators* of $H_k(\Sigma)$.

Simplicial homology is a topological invariant which provides global quantitative and qualitative information about a shape. For each k, the kth Betti number β_k measures the number of independent non-bounding k-cycles in Σ. In dimension 0, β_0 counts the number of connected components of the complex, in dimension 1, β_1 counts the number of its tunnels and its holes, in dimension 2, β_2 counts the number of voids or cavities, and so on.

Given an arbitrary Abelian group A, we can define the kth homology group with coefficients in A of Σ as $H_k(\Sigma; A) = H_k(C_*(\Sigma) \otimes_{\mathbb{Z}} A)$, where $\otimes_{\mathbb{Z}}$ denotes the tensor product of Abelian groups. If we consider $A = \mathbb{Z}_2$, $C_*(\Sigma) \otimes_{\mathbb{Z}} \mathbb{Z}_2 = (C_k(\Sigma) \otimes_{\mathbb{Z}} \mathbb{Z}_2, \partial_k \otimes_{\mathbb{Z}} \mathbb{Z}_2)_{k \in \mathbb{Z}}$ is the chain complex whose groups $C_k(\Sigma) \otimes_{\mathbb{Z}} \mathbb{Z}_2$ are the $\mathbb{Z}_2$-vector spaces generated by the k-simplices of Σ and homomorphisms $\partial_k \otimes_{\mathbb{Z}} \mathbb{Z}_2$ are the boundary maps ∂_k of Σ considered modulo 2. It can be proven (see [2], Chapter X) that, for simplicial complexes embeddable in $\mathbb{R}^3$, each homology group is free, and, thus, its torsion part is trivial. For this reason, the $\mathbb{Z}$-homology groups of a simplicial complex Σ embeddable in $\mathbb{R}^3$ can be retrieved by just computing the homology of Σ with $\mathbb{Z}_2$ coefficients.

Independently of the coefficient group chosen, homology computation is computationally expensive. *Smith Normal Form (SNF) reduction* [1, 48], an algorithm similar to Gauss elimination, represents the classical tool to compute homology, allowing to retrieve the whole homological information by reducing the matrices representing the boundary maps ∂_k of the input simplicial complex. The complexity of such algorithm is super-cubical in the number of simplices of Σ and, thus, is impractical when working with large datasets.

9.2.3 Persistent Homology

Persistent homology [24, 30, 61] is an important tool in topological shape analysis, which aims at overcoming intrinsic limitations of classical homology by allowing a multi-scale approach to shape description. The possibility to retrieve essential topological features of a shape has led to an increasing development of persistent homology in various applications, such as biology and chemistry [13, 18, 57], automatic classification of images [4, 10, 12], and coverage of sensor networks [56]. Similarly to classical homology, persistent homology can be defined for chain and cell complexes. In spite of this, it is typically introduced for simplicial complexes or regular grids.

Let Σ be a simplicial complex. A *filtration* of Σ is a finite sequence of subcomplexes $\{\Sigma_m \mid 0 \leq m \leq M\}$ of Σ such that $\emptyset = \Sigma_0 \subseteq \Sigma_1 \subseteq \cdots \subseteq \Sigma_M = \Sigma$. The *associated chain filtration* is defined as the following sequence of chain complexes

$$C_*(\Sigma_1) \xrightarrow{i_1} C_*(\Sigma_2) \xrightarrow{i_2} \dots \xrightarrow{i_{M-1}} C_*(\Sigma_M)$$

where maps i_m arise from inclusion of groups.

For $p \in \mathbb{N}$, we denote as $i_{m,p} : C_*(\Sigma_m) \to C_*(\Sigma_{m+p})$ the composition $i_{m+p-1} \cdot \ldots \cdot i_m$ when it makes sense. The *p-persistent* kth *homology group of* Σ_m is defined to be

$$H_k^p(\Sigma_m) = \frac{i_{m,p}(Z_k(\Sigma_m))}{i_{m,p}(Z_k(\Sigma_m)) \cap B_k(\Sigma_{m+p})}.$$

Informally, $H_k^p(\Sigma_m)$ consists of the k-cycles included from $C_k(\Sigma_m)$ into $C_k(\Sigma_{m+p})$ modulo boundaries. Persistent homology is a more powerful tool than classical homology, since it allows capturing the changes in homology of a filtered shape by retrieving the cycles that are non-boundary elements in a certain step of the filtration and that will turn into boundaries in some subsequent step. The persistence of a cycle during the filtration gives quantitative information allowing to distinguish significant and irrelevant cycles of a shape.

Persistent homology of a filtered complex can be computed by utilizing an *SNF* reduction algorithm but, focusing on persistent homology with coefficients in a field (such as $\mathbb{Z}_2$), it can be retrieved more efficiently [61]. In this case, the proposed algorithm is incremental and it partitions the simplices into creators and destroyers of homology classes pairing simplices associated with the same homology class.

9.2.4 Discrete Morse Theory

Forman theory [28] is a discrete counterpart of Morse theory, generalizing the results of the smooth theory from the context of manifolds to cell complexes. This goal is achieved by considering a function (also called a *Forman function*) defined over all the cells of a cell complex. A discrete function F, defined on all the cells of Γ, is called a *Forman function* if, for any k-cell σ, all the $(k-1)$-faces of σ have a lower F value than σ, and all the $(k+1)$-faces have a higher F value than σ, with at most one exception. A k-cell is called a *critical with index k* (or a *k-saddle*) if there is no exception. Specifically, a 0-saddle is called a *minimum* and, if $d = dim(\Gamma)$, a d-saddle a *maximum*.

Figure 9.2a shows a Forman function F defined on a simplicial 2-complex. Each simplex is labeled with the corresponding value of function F. Vertex 0 is critical (minimum), since F has higher value on all edges incident to it. Triangle 9 is critical (maximum), since F has lower value on all edges incident to it. Edge 5 is critical (saddle), since F has higher value on the incident triangles, and lower values on its extreme vertices.

The unique exception to the above rule which holds for a non-critical cell σ permits to pair σ with either one of its faces, or one of its cofaces. A *discrete vector field* V on a cell complex Γ is a collection of pairs (σ, τ) such that each cell of Γ is in at most one pair of V. A Forman function F induces a discrete vector field V_F called the *gradient vector field* (or *Forman gradient*) of F on Γ consisting of the collection of pairs (σ, τ), where σ is a $(k-1)$-cell and τ is a k-cell, coface of σ such that $F(\sigma) \geq F(\tau)$. Such pair can be depicted as an arrow going from σ (tail) to τ (head).

Each cell is a head or a tail of at most one arrow, and critical cells are those cells that are neither the head nor the tail of any arrow.

A *V-path* (or *gradient path*) is a sequence $\sigma_1, \tau_1, \sigma_2, \tau_2, \ldots, \sigma_r, \tau_r$ of $(k-1)$-cells σ_i and k-cells τ_i, $i = 1, \ldots, r$ with $r \geq 1$, such that $(\sigma_i, \tau_i) \in V$, σ_{i+1} is a face of τ_i, and $\sigma_i \neq \sigma_{i+1}$. A V-path with $r > 1$ is *closed* if σ_1 is a face of τ_r different from σ_{r-1}.

There is a correspondence between Forman functions and discrete vector fields without closed V-paths [27]. Namely, a discrete vector field V is the gradient vector field of a discrete Morse function F if and only if V has no closed paths. Figure 9.2b shows the gradient vector field V_F corresponding to the Forman function F in Fig. 9.2a.

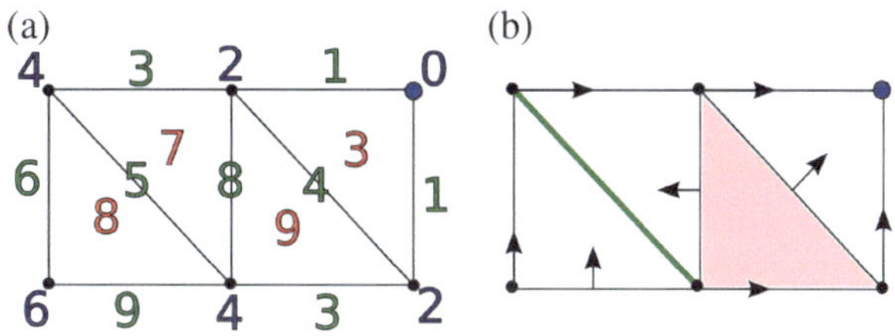

Fig. 9.2 (**a**) A Forman function defined on a simplicial complex and (**b**) the corresponding Forman gradient

Given a cell complex Γ endowed with a gradient vector field V, we can obtain a compact homology-equivalent model for Γ, called the *discrete Morse complex* and denoted as $\mathscr{M}_* = (\mathscr{M}_k, \tilde{\partial}_k)_{k\in\mathbb{N}}$. Its chain groups $\mathscr{M}_k$ are generated by the critical cells of Γ and the boundary maps $\tilde{\partial}_k$ can be retrieved by following the V-paths of the gradient vector field.

As proven by Thm. 8.2 in [28], the discrete Morse complex of a gradient vector field V on Γ and the cell complex Γ are homologically equivalent. In the proof of this result given in [36], the equivalence is demonstrated by providing a chain equivalence between the two complexes which allows recovering the homology generators of Γ through the knowledge of the homology generators of the discrete Morse complex $\mathscr{M}_*$.

As shown in [45], discrete Morse theory can be used to efficiently compute persistent homology. Similarly to what happens for homology groups, we can establish an equivalence between persistent homology of a cell complex Γ and of a discrete Morse complex associated with it. In this context, if we want persistent homology information to be preserved, a compatibility condition between the chosen filtration and the Forman gradient must to be satisfied. More precisely, given a cell complex Γ and its filtration $F = \{\Gamma_m \,|\, 0 \leq m \leq M\}$, a gradient vector field V of Γ is said a *filtered gradient vector field of* F if, for each pair $(\sigma, \tau) \in V$ there exists $m \in \{1, \ldots, M\}$ such that $\sigma, \tau \in \Gamma_m$ and $\sigma, \tau \notin \Gamma_{m-1}$. If such condition holds, the persistent homology groups of Γ and of the discrete Morse complex induced by V are equivalent.

9.3 Encoding the Forman Gradient Vector Field

Defining a compact representation for a Forman gradient V corresponds to compactly encoding all the cells paired in V based on the representation used for the complex on which V is defined. In this section, we describe the most common representation used for cell complexes, the *Incidence Graph (IG)*, and the encoding of the Forman gradient V on it. Then, we describe how the *IG* and the corresponding Forman gradient can be efficiently encoded on a regular grid using bit-vectors. Finally, we describe how to encode the Forman gradient on a simplicial complex described through a compact data structure.

An *Incidence Graph* (*IG*) [23] is a topological graph-based data structure describing the Hasse diagram of a cell complex Γ, which is the partial order set of the cells of Γ and their boundary relations. Thus, the *IG* is a graph, whose nodes correspond to the cells of Γ and such that an arc connects two nodes of consecutive dimension, if the corresponding cells γ and γ' are mutually incident, i.e., if γ is a face or a coface of γ'.

The arcs of the *IG* encode all the possible pairings that can be defined on Γ by considering two cells of consecutive dimension. As discussed in Sect. 9.2.4, a Forman gradient V can be described as a pairing between the cells of Γ such that each cell is involved in at most one pair (see Fig. 9.3c). Thus, a Forman gradient V

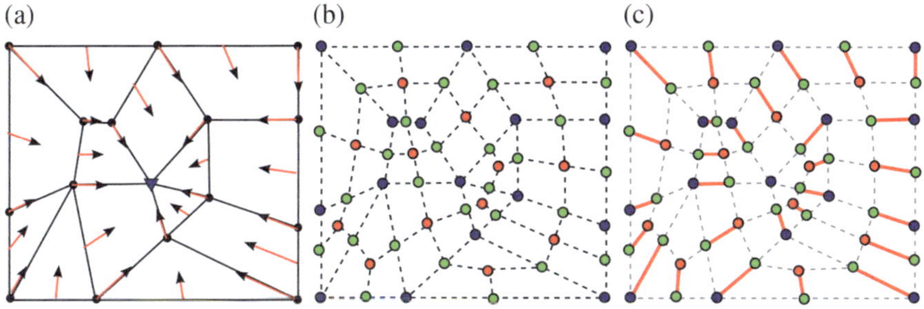

Fig. 9.3 (**a**) Example of a cell complex and a Forman gradient V. *Red arrows* indicate cell pairings and the *blue triangle* indicate the minimum. (**b**) *IG* corresponding to the cell complex in (**a**) with nodes (*colored points*) and arcs (*dotted lines*). (**c**) Arcs depicted in *red* are the subset of arcs in the *IG*, involved in a pairing in V. Notice that all the nodes are connected to at most one *red* arc except for the *blue* node corresponding to the minimum

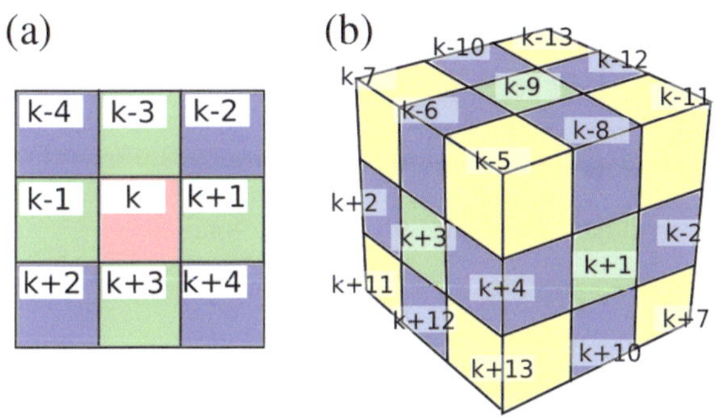

Fig. 9.4 *8-connectivity* (**a**) and *26-connectivity* (**b**) for a *square* and *cubic grid*, respectively

is a subset of the pairings in the set of arcs in the *IG*, and it can be encoded on the *IG* by adding 1 bit flag for each arc indicating whether such pairing is also a valid pair in V.

In a regular grid, a d-cell is connected through a $(d-1)$-face to the $2d$ neighboring elements lying in the directions of the Cartesian axes (known as the *4-connectivity* model for 2D grids), or to the $3^d - 1$ elements lying in the axis-parallel and diagonal directions (*8-connectivity* model for 2D grids) [53]. Figure 9.4 illustrates the *8-connectivity* (a) and *26-connectivity* (b) for a square and cubic grid, respectively. Considering the central hyper-cube ρ of dimension d, hyper-cubes sharing a $(d-1)$-face with ρ are depicted in green, hyper-cubes sharing a $(d-2)$-face with ρ are depicted in purple and hyper-cubes sharing a $(d-3)$-face are depicted in yellow. Indices shown in Fig. 9.4 indicate the position of the hyper-cube with respect to the index k of ρ.

Having fixed a connectivity model, all the cells of a regular grid are indexed. Thus, given a cell ρ all the faces/cofaces/adjacents cells of ρ are retrieved through arithmetic operations on the index of ρ. A Forman gradient V defined on the regular grid is, thus, compactly defined as a bit-vector on the same indexing schema [31].

When dealing with a simplicial complex, a data structure encoding all simplices and their incidence relations, like the *IG*, is definitively too verbose. On the other hand, there exist data structures for simplicial complexes which are much more compact and scale well with the dimension, by encoding only its vertices and top simplices [9, 17]. The use of such data structures makes the computation of the Forman gradient and of the Morse and Morse-Smale complexes on simplicial complexes of large size feasible [26, 29, 59]. On the other hand, an encoding for the Forman gradient is required which associates the gradient pairs only with the top simplices. We call such an encoding *compact gradient* [59].

In a d-dimensional simplicial complex Σ, a d-simplex σ has $\binom{d+1}{k+1}$ faces of dimension k, and each face has in turn $(k+1)$ faces of dimension $(k-1)$. Since each k-simplex can be paired with any of the simplices on its boundary or coboundary, there are $\sum_{k=1}^{d-1} \binom{d+1}{k+1} \cdot (k+1)$ possible pairs in the restriction of the Forman gradient V to σ. Adding the $d+1$ additional pairs from a $(d-1)$-simplex on the boundary of σ to an adjacent d-simplex provides a total of $\sum_{k=1}^{d-1} \binom{d+1}{k+1} \cdot (k+1) + d + 1$ possible pairs. In Fig. 9.5, these pairs are shown with arrows for a 2-simplex (triangle) and 3-simplex (tetrahedron). Red arrows indicate pairings between the d-simplex and its faces, green arrows are between $(d-1)$-simplices and their faces and blue arrows between 1-simplices (edges) and vertices. We refer to such set of pairs as *local frame*.

Because in a discrete vector field each simplex can be involved in at most one pair, there are significantly fewer valid local frame configurations than the possibilities provided by the bit flag representation. Thus, a local frame can be *compressed* by representing only the valid configurations. Since each such pair within a local frame encodes a single bit of information (i.e., the presence or absence of that pair), each local frame can be encoded using a bit-vector per d-simplex.

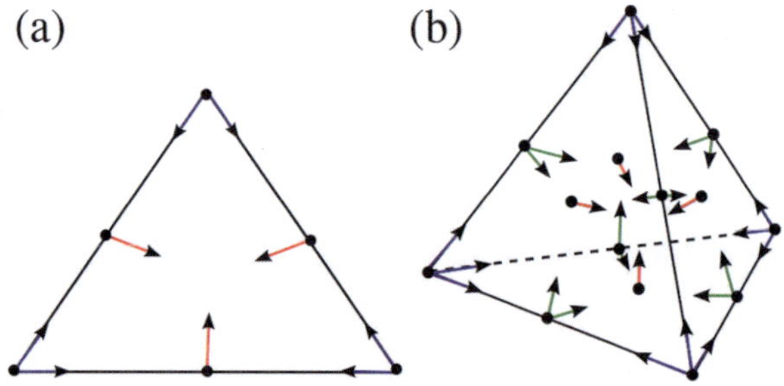

Fig. 9.5 Set of pairings that can be defined inside a triangle (**a**) or tetrahedron (**b**)

In a simplicial 2-complex, the encoding associates with a triangle (2-simplex) $\sigma \in \Sigma$ a subset of the pairs involving its faces. A triangle has $\sum_{i=1}^{2} \binom{3}{k+1} \cdot (k+1) + 3 = 3 \cdot 2 + 1 \cdot 3 + 3 = 12$ possible pairs for a total of $2^{12} = 4{,}096$ cases. However, for a Forman gradient, there are only 97 valid cases for a triangle. Thus, all possible configurations can be encoded by using only 1 byte per triangle. Similarly, in 3D, there are 32 arrows for a total of $2^{32} = 4{,}294{,}967{,}296$ possible configurations, and the valid ones are only 51,030, thus they can be represented with 2 bytes per tetrahedron.

9.4 Computing a Forman Gradient Vector Field

The algorithms proposed in the literature for computing a Forman gradient V have been applied in two large areas: scalar field analysis via topological features and homology and persistent homology computation. The two areas have led to algorithms with different peculiarities. Algorithms developed for scalar field analysis [11, 31, 32, 34, 39, 52, 54, 55] compute a combinatorial gradient simulating the gradient of the function defined at the vertices of the scalar field. We call them *constrained algorithms* (see Sect. 9.4.1).

Algorithms for homology computation, instead, are based on reduction operators used to reduce the dataset in an homologically consistent way [5, 21, 29, 36, 40, 46, 47]. The simplification of the datasets is performed by removing pair of cells but, working with discrete Morse theory, this can be seen as the construction of pairings for a Forman gradient V. Specifically, the gradient is obtained as a set of paired cells (arrows of the gradient) and set of unpaired cells (critical cells). Since the rules for applying the reduction operators are purely combinatorial, we call these algorithms *unconstrained.*

9.4.1 Constrained Approaches

Computing a Forman gradient, i.e., simulating the gradient of a function defined on a scalar field dataset, is a challenging task recently addressed in the literature [11, 31, 32, 34, 39, 52, 54, 55]. Datasets produced in this area are generally characterized by a huge number of points regularly distributed on a 2D or 3D domain. Since topological feature analysis is based on the critical points (simplices) identified on the function under consideration, such algorithms must be designed so as to minimize the number of spurious critical points. This means computing a combinatorial gradient as much similar as possible to the one of the smooth function. Moreover, because of the size of the available datasets, some of them [52, 54, 55, 59] have been defined to be easily parallelizable or have been specifically developed for distributed computation.

In [11], the first constrained algorithm is proposed adapting the unconstrained algorithm described in [40] when a scalar function f (i.e., the discrete Connolly's function [16]) is defined on the vertices of a triangle mesh Σ, and then extended to its edges and triangles. Roughly speaking, the Connolly function can be considered as an analog of the mean curvature within a fixed size neighborhood of each point. Let the primal H and dual graphs H_D be two graphs having as arcs the edges of Σ. The nodes of H are in one-to-one correspondence with the vertices of Σ while the nodes of H_D correspond to the triangles of Σ. Then, a spanning tree on H_D is built for each maximum of f processing the edges by increasing function value and thus obtaining a spanning forest T_D. Dually, a spanning tree is created, for each local minimum, on H obtaining the spanning forest T. The Forman gradient V of Σ is computed by considering T and T_D. Roots of $T(/T_D)$ are the minima(/maxima) of V and edges that do not belong to either T and T_D are the saddles. Starting from each root of T, paths to the leaves are visited in a depth-first manner. The visit on T induces a pairing between each node and the arc to its father, or equivalently a paring on V of the type (vertex,edge). Dually, for T_D the pairings created are of the type (edge,triangle). The algorithm can be extended to d-dimensional complexes but only by restricting to the computation of the pairings between 0-simplices and 1-simplices (forming V-paths connecting minima to 1-saddles) and between $(d-1)$-simplices and d-simplices (forming V-paths connecting maxima to $(d-1)$-saddles).

King et al. [39] propose one of the first constrained algorithms addressing the problem of minimizing the number of critical points. Let the *lower link* $Lk^{-}(\sigma)$ of a simplex the subset of the link of σ (see Sect. 9.2.1) containing only simplices with a lower function value than σ. The algorithm builds the Forman gradient on a tetrahedral mesh Σ working locally in the lower link $Lk^{-}(v)$ of each vertex v. The pairing is extended to the cone $(v; Lk^{-}(v))$, which is the simplex generated by the union of the vertices of v and $Lk^{-}(v)$. The number of critical points introduced by this method is arbitrary large and, thus, a simplification step for reducing the number of critical cells, locally to each lower link, is expected [28].The algorithm proposed by Gyulassy et al. in [33] is one of the first algorithms defined in a dimension independent way and implemented for regular grids. Function f defined on the input vertices is extended on all cells of Γ in such a way that, for each cell τ and each face σ, the function value of τ is slightly larger than the value of σ. The Forman gradient V is then computed sweeping over the cells of Γ according to increasing dimension and function values. A cell, that is not yet critical or paired, is inserted in V as critical if it has no unpaired cofaces. Otherwise, it is paired with the coface of lowest function value. Since the different k-cells in Γ may have the same function value, the resulting process is not deterministic and some unnecessary critical cells may be produced by the algorithm. This problem has been addressed in [54] and [55] where the algorithms, defined for 2D and 3D regular grids respectively, produce pairs independently of the order in which the cells are considered. Such approaches are based on the definition of a new function called *weighted discrete function* and they provide a basis for a parallelization of the algorithm also. In [32], a similar algorithm is proposed which focuses on improving the poor geometric approximation of the gradient caused by the local assignment of the gradient arrows. This is especially useful in scalar field analysis, but not for homology computation.

Robins et al. proposed in [52] a dimension-independent algorithm for cell complexes Γ with scalar values given at the vertices of the complex. In [52], an implementation is provided for regular grids, while in [26, 59] the same algorithm has been implemented for simplicial complexes in 2D and in 3D in combination with a compact representation for the underlying complex and for the Forman gradient. Let the *lower star* of a cell γ be the subset of the star of γ (see Sect. 9.2.1) containing only cells with a lower function value than γ. The lower star of each vertex v in γ is processed independently, thus leading to a straight-forward parallel implementation. Each cell inside the star is processed in ascending order of function values and of dimension. Similarly to [33], each cell is always considered after its faces but here, pairings between cells are defined based on homotopy expansion. Two cells, k-cell σ and $(k+1)$-cell τ, are paired via homotopy expansion when: σ have no unpaired boundary cells and τ has only one unpaired boundary cell (i.e., σ). As shown in [52], the critical cells identified by the algorithm in the 3D case are in one-to-one correspondence with the topological changes in the lower level sets of the scalar function. This behavior is the one to be expected in a smooth Morse setting. This makes this algorithm one of the best topologically correct algorithms for computing a Forman gradient.

Table 9.1 summarizes the algorithms discussed in this section. Algorithms [32, 54, 54] are not indicated in the table since they are improvements of the idea presented in [33].

The only dimension-specific algorithms are the one in [11], specifically defined for 2D simplicial complexes, and the one in [39]. The gradient computation in the algorithm by Kings et al. [39] could be extended to higher dimensions but the simplification step could be problematic in higher dimensions (as described in Sect. 9.6). All of them are implemented for specific complexes (regular grids [33, 52] or simplicial complexes [11, 39]). Algorithms implemented for regular grids are typically used for the analysis of gridded volume datasets.

However, since they all rely on discrete Morse theory, they can be easily adapted to cell complexes.

Table 9.1 Summary of the reviewed algorithms. For each of them the expected input and the worst time complexity are indicated. Note that $|X|$ denotes the cardinality of set X, and X_0 is the set of the vertices of X

Algorithm	Input	Time complexity
Cazals et al. [11]	2D simplicial complex Σ	$O(\vert\Sigma\vert(log\vert\Sigma\vert + \alpha(\vert\Sigma\vert)))$
King et al. [39]	3D simplicial complex Σ	$O(\vert\Sigma_0\vert s)$
Gyulassy et al. [33]	nD cell complex Γ	$O(\vert\Gamma\vert log\vert\Gamma\vert)$
Robins et al. [52]	nD cell complex Γ	$O(\vert\Gamma_0\vert c)$

We can classify the algorithms described above into two groups based on their time complexity. The algorithms in [39, 52] are based on an implicit subdivision of the cells of the complex into independent sets (based on the vertices). Both algorithms consider each cell, in the independent set, exactly once. Since all the operations are performed in constant time, the complexity depends only on the number of simplices s [cells c] in each independent set. In most of applications, s and c are considered negligible with respect to the number of vertices v and thus the complexity of the entire process is considered linear. The algorithms in [11, 32, 33], instead, require as initial step a sorting of the simplices. In [11], all the edges are sorted (with $O(|\Sigma|log|\Sigma|)$ complexity) and a further step, for the forest creation, is performed in $O(|\Sigma|\alpha(|\Sigma|))$ with $\alpha(\cdot)$ the inverse of Ackerman's function. Also algorithms in [32, 33] sort the cells of the cell complex Γ based on the Forman function.

9.4.2 *Unconstrained Approaches*

Several algorithms have been proposed for computing a Forman gradient on a cell complex without any constraint, such as scalar values at the vertices of the complex.

The algorithm by Lewiner et al. [40] is the first algorithm of this kind proposed in the literature with the aim of providing a combinatorial descriptor for 3D shapes. It has been defined on triangle meshes and then extended to general 2-complexes Γ. The algorithm is similar to the one described in [11] (see Sect. 9.4.1), but here the spanning forests are built by considering a spanning tree for each connected component of the shape, without ordering the edges of complex Γ based on a function value.

With the exception of this latter algorithm, most of the unconstrained algorithms are based on two simplification operators, called *reduction* and *coreduction*. Those operators are homology-preserving operators which delete a pair of cells from a cell complex Γ while preserving the homology groups of Γ. In the context of discrete Morse theory, the removal of a pair of incident cells can be seen as a pairing and, thus, it can be used for building a Forman gradient V on Γ. Here, we describe these algorithms presenting a dual strategy for computing the Forman gradient [5, 37].

Let Γ be a cell complex and let σ be a k-cell of Γ. We call *(immediate) coboundary* of σ with respect to Γ the set $cbd_\Gamma\sigma$ of the $(k+1)$-dimensional cofaces of σ. Moreover, we call *(immediate) boundary* of σ with respect to Γ the set $bd_\Gamma\sigma$ of the $(k-1)$-dimensional faces of σ.

A *reduction* corresponds to a deformation retraction of a cell, which is a face of only one other cell onto the complex. A pair (σ, τ) of cells of Γ is a reduction pair if $cbd_\Gamma\sigma = \{\tau\}$. The pair of cells can be removed from Γ without affecting the homology groups of the cell complex.

The algorithm proposed in [5], builds a Forman gradient on a cell complex Γ by randomly applying reduction operators to Γ. The algorithm starts setting the working dimension to d, where d is the maximum dimension among the cells of Γ.

As long as there are available reductions between a d-cell σ and one of it faces τ, the algorithm removes the two cells from Γ and adds the pair (σ, τ) to V. When no more reduction is feasible, a d-cell is excised, which becomes a critical cell in V. When both the set of available reductions and the set of top cells is empty the working dimension is decreased by one.

The algorithm proposed in [36, 37] is based on a homology-preserving operator dual to *reduction*, called a *coreduction*. In a similar fashion to the algorithm proposed in [5], *coreductions* are used in [36] for the construction of a Forman gradient, on a cell complex Γ. Let a free cell be a cell with an empty immediate boundary. The algorithm starts by setting the working dimension to 0. All the available coreduction pairs between a 0-cell and a 1-cell are excised from the complex and the corresponding pairing is added to V. When no more coreduction is possible, a free 0-cell is excised from the cell complex and added as critical to V. When no more 0-cells are available, the working dimension is increased by one.

The two approaches can be considered dual to each other. In particular, it has been proven in [29] that the two methods can produce the same Forman gradient. More formally, any Forman gradient obtained through a sequence of reductions and removals of top cells can also be obtained through a sequence of coreductions and removals of free cells and vice versa. The two operators can also be combined to represent a powerful preprocessing tool to efficiently compute the homology of a cell complex, as described in [21, 46, 47]. In [29], an algorithm has been proposed to build a gradient vector field by executing reduction and coreduction pairs in an interleaved way. It has been shown that any interleaved approach still produces a Forman gradient and that such a gradient can be obtained through a reduction or a coreduction pairing.

In [29], two implementations of such methods based on different data structures have been compared and the trade off between using a verbose data structure encoding all the boundary/coboundary relations (the Incidence Graph (*IG*)) and a compact data structure, encoding only the vertices and top simplices of a simplicial complex (*IA* data structure*) [9], has been shown. The first implementation considered is *Perseus* [49], an *IG*-based software for persistent homology computation. In *Perseus*, an algorithm involving both reductions and coreductions has been developed. Since in the *IG* all the incidence relations for each simplex are explicitly encoded, both operators are computationally efficient and their usage can be interleaved. In practical applications however, only a subset of the boundary/coboundary relations are explicitly stored in order to decrease the storage cost. Then, in order to avoid inefficiency, only one simplification operator has to be chosen. We will choose reductions or corrections depending on whether we can retrieve the boundary or coboundary relations faster, respectively.

For this reason, the IA^*-based implementation proposed in [29] performs correction operators only. In both implementations, the homology of each complex has been retrieved by computing the Forman gradient and extracting the homology generators. To evaluate performances, for each dataset, the maximum amount of memory required by the two algorithms and the timings for computing homology has been computed.

Table 9.2 Comparison between timings (in seconds) for the homology computation algorithms based on the *IA** and *IG* data structures. For dataset Elephant, the *IG* implementation runs out of memory. *d* indicates the maximum dimension for the simplices in the dataset

Dataset	*Buddha*	*Elephant*	*Fertility*	*Skull*	*Neghip*	*Klein*	*Sphere*
d	2			3		7	9
*IA**	300	120	62	10	32	149	138
IG	304	–	110	12	47	9	12

The storage cost required by the two implementations (*IG*/*IA**), considering the maximum amount of memory used at runtime, has been compared. Working with triangle meshes, the *IG*-based implementation occupies 3 times more memory than the *IA**, it increases to 4 with tetrahedral meshes and it occupies 17 and 24 times more when working with 7- and 9- complexes, respectively. As expected, the storage cost for the *IA** is dependent of the number of top simplices in the simplicial complex only. The *IG* representation, instead, limits the maximum size of the input complex that can be handled when working in higher dimensions. However, the timings provided by the two structures are still comparable for 2- and 3-complexes while the *IA** becomes slower in higher dimensions as showing in Table 9.2.

9.5 Computing the Morce Incidence Graph

The *Morse Incidence Graph* (*MIG*) [8, 14, 25, 35] is an efficient graph-based representation for the boundary maps represented by a Forman gradient V. The *MIG* associated with V is a graph $G = (N, A, \mu)$ such that each k-node in N is in one-to-one correspondence with a critical k-cell in V and there is an arc joining a k-node σ with a $(k+1)$-node τ if and only if there is a V-path connecting k-saddle σ to $(k+1)$-saddle τ. Each arc connecting a k-node σ to a $(k+1)$-node τ is labeled with the number of V-paths connecting σ to τ. The label, denoted as $\mu((\sigma, \tau))$, is also called the *multiplicity* of arc (σ, τ).

The Morse Incidence Graph has been originally defined for representing the incidence relations between the cells of the Morse and Morse-Smale complexes. We refer to [59] for a complete description of the relations among a Forman gradient V, the Morse and Morse-Smale complexes that V implicitly represent and the corresponding Morse Incidence Graph.

The *MIG* is computed by traversing the V-paths of the *compact gradient* V defined on the given complex Γ. A node is created for each critical cell, and an arc between two nodes is created if there is a V-path in V connecting them. Since only the connection of critical cells is needed, ad-hoc strategies can be used to reduce the number of cells traversed. In 2D, the set of V-paths between saddles and minima are visited by starting from each critical 1-cell and following the gradient paths until a minimum is reached. Such paths never branch and, thus, a limited number of 1-cells are visited in practice during their traversal. The set of V-paths between saddles and maxima are obtained in a similar way by considering the 1-cells and the 2-cells of Γ The two subgraphs of the *MIG* connecting minima and 1- saddles and maximal and $(d-1)$-saddles, called *extrema graphs*, is performed in a dimension-independent way, leading to the same reduction in complexity. In dimension three or higher, a new step is introduced to compute the saddle connectors, i.e., the arcs of the *MIG* between k-saddles σ and $(k+1)$-saddles τ, with $k \neq 0, d-1$. All the gradient paths starting from τ are considered, and all the traversed $(k+1)$-cells are marked as visited. Then, starting from σ, the same process is performed visiting the gradient paths in reverse order and considering only the $(k+1)$-cells previously marked as visited.

In three and higher dimensions, gradient paths can branch and merge potentially resulting in many-to-many adjacency relationships between critical k-cells and critical $(k+1)$-cells. Let us consider a simplicial 3-complex Σ with v vertices, whose Forman function contains $O(v)$ critical 1-cells, each of which connects to $O(v)$ critical 2-cells. This produces a discrete Morse complex containing $O(v^2)$ gradient paths between critical 1- and 2-simplices. Since the number of critical 1- and 2-cells is bounded by v, the number of traversals for any cell during the breadth-first search is also bounded by v and so the complexity of the whole extraction process is $O(v^3)$.

A simple solution proposed in [31, 55, 59] aims at reducing the time complexity of the above algorithm by slightly increasing the space complexity. This is achieved by storing the k-cells visited during a gradient path traversal. In this way, no cell is ever visited twice and the time complexity drops to $O(v^2)$. This method works well when the computation of the saddle connectors is sequential, but there is a high memory increase for a parallel implementation. The algorithm proposed in [55] is based on a priority queue which allows counting the number of times a cell is visited, i.e., each cell is inserted in the queue only a constant number of times and the complexity of the resulting algorithm has been proven to be $O(v^2 \log v)$. This algorithm is especially well suited for parallel implementations.

9.6 Homology and Persistent Homology Computation

As mentioned before, a discrete Morse complex associated with a cell complex Γ provides a compact homology-equivalent model of Γ. This equivalence allows us to compute the homology of Γ by applying the *SNF* reduction algorithm [1, 48] on the discrete Morse complex $\mathscr{M}_*$. Since the number of critical cells generating $\mathscr{M}_*$ is usually negligible with respect to the number of cells of Γ, this method considerably improves the efficiency of homology computation. This procedure can be applied to compute homology of a cell complex Γ with coefficients different from $\mathbb{Z}$. The homological equivalence still holds for homology with coefficients in any arbitrary Abelian group, but typically, only coefficients in $\mathbb{Z}_2$ are considered in the applications. The chain complex to be considered in this case, and on which the *SNF* reduction algorithm is applied, is $\mathcal{M}_*(\Sigma) \otimes_{\mathbb{Z}} \mathbb{Z}_2$. This chain complex is generated by the same critical cells as $\mathcal{M}_*$ and its boundary maps $\tilde{\partial} \otimes_{\mathbb{Z}} \mathbb{Z}_2$ can be obtained, as described in Sect. 9.5, by considering the $\mathbb{Z}$-coefficient boundary maps $\tilde{\partial}$ modulo 2.

9.6.1 Computing Homology Generators

The homological equivalence between chain complex $\mathcal{M}_*$ and cell complex Γ implies that, by using the *SNF* reduction algorithm, we are able to obtain the cellular homology of Γ. The homology generators of degree k (H_k) are computed through *SNF* reduction. Their geometric realization is obtained starting from the critical k-cells of V and navigating the gradient pairs. Computing the homology generators corresponds to compute, for each critical k-cell σ, the V-paths connecting σ to a critical $(k-1)$-cell.

The computation of such gradient paths starts from a critical k-cell σ. All the $(k-1)$-cells in the immediate boundary of σ are then selected and, among them, only the $(k-1)$-cells paired with a k-cell different from σ are considered. Such k-cells are inserted into a queue, and the traversal of the complex Γ continues in a breadth-first fashion until all the V-paths starting from σ have been visited. In 2D, for example, we start from a critical 2-cell (maximum) σ and, by following gradient pairs, we continue adding adjacent 2-cells until all V-paths from σ have been traversed.

The computation of the gradient paths originated from a critical cell is performed through constant time operations at each cell on the visited V-paths. In 2D, the extraction of the gradient paths starting from a critical k-cell requires time linear in the number of cells of Γ involved in a such V-path since each cell is visited at most once. As discussed in Sect. 9.5, in three dimensions and higher, visiting the gradient paths among saddles may exhibit a cubical time complexity.

9.6.2 Computing Persistent Homology

As mentioned in Sect. 9.2.4, it is possible to obtain the persistent homology of the input cell complex Γ by studying the persistent homology of a considerably smaller discrete Morse complex. Both constrained and unconstrained algorithms can be used for persistent homology computation. In the constrained approaches, the values of the function associated with the vertices of cell complex Γ naturally induce a filtration of Γ defined by the lower level sets of such function. In [52], for example, the generic element Γ_m of the filtration induced by the input function on Γ is the cell complex containing all the cells of Γ that have no vertex with a function value greater than m.

For unconstrained approaches, the point of view is different. The first difference is that, in this context, there is no a scalar function assigned to the vertices of the cell complex, and thus, there is no naturally induced filtration. Once a filtration is selected, the construction of the Forman gradient is limited by suitable constrains to obtain a filtered gradient vector field, so as to preserve persistent homology information [22, 45]. One can obtain a Forman gradient by sequences of homology-preserving operators such as removals of reduction or coreduction pairs. When using approaches based on homology-preserving operators, each removal of a pair needs to be compatible with the filtration. For example, the compatibility condition for reduction and coredution pairs requires that both the cells in a pair belong to the same subcomplex of the filtration.

A new approach to efficiently compute persistent homology is based on the notion of annotation of a simplicial complex [6, 19]. An interesting avenue of research would be to generalize the definition and the computation of the annotations to the context of chain complexes.

9.6.3 Applications

Homology and persistent homology computation have been applied in many different fields with a growing attention in the analysis of data in high dimensions where pure geometric tools are usually not sufficient. In addition, discrete Morse theory turned out to be a fundamental tool for computing boundary matrices, which are at the basis of any application involving homology and persistent homology computation.

In multivariate data analysis, persistent homology has been used to extract significant structures in arbitrary high-dimensional data sets, such as high-dimensional real-world data sets arising from research in cultural heritage [51], and multivariate point clouds from particle physics, political science and meteorology [50]. In [56], homological tools provide a criterion for certifying a coverage in sensor network analysis. In chemistry and biology, methods based on topology are used for understanding energy landscapes [43]. In astrophysics, homological information allows to study the topology of the Megaparsec Cosmic Web [60].

In [52] and [31], persistence homology is applied to the study of 3D images. As described in Sect. 9.4.1, the critical cells obtained with the algorithm defined in [52] (in the 3D case) are in one-to-one correspondence with the topological changes in the sub-level complexes and, thus, the persistent homology of the input complex corresponds to the persistent homology of the much smaller discrete Morse complex computed.

9.7 Concluding Remarks

We have reviewed algorithms for computing a Forman gradient on a cell complex and we have described how to retrieve information from it in order to compute homology, homology generators and persistent homology of the corresponding discrete Morse complex.

The Forman gradient, encoded on a regular grid or "on"? i.e., "or on a simplicial complex"? a simplicial complex, offers an effective framework for retrieving all information required, such as the Morse chain complex, the boundary maps of the Morse cells and a filtration. However, optimizing the storage requirements of such information and their efficient computation need to be investigated.

Datasets are characterized by a constantly growing number of sample points and consequently by a huge number of simplices, in particular when working in medium dimensions. In this area, it is crucial to be able to work effectively and efficiently with simplicial complexes or grids of high dimensions and large size. When dealing with such kind of data, the ratio between the number of simplices and sample points of the datasets increases exponentially.

The development of parallel approaches seems to be the most promising research trend. Taking advantage of GPU and multicore architectures, for improving the computation time without reducing memory, would be fundamental for developing interactive tools based on homology computation. In this direction, enhancing topological data structures with spatial indexes [58] could be an excellent way for handling huge datasets. Such data structures offer compact representations for simplicial complexes and infer a natural subdivision on them defining independent decompositions to be used in parallel computation. The preliminary results obtained in high dimensions using data stuctures based on the encoding of only the top simplices [29] further encourage the use of spatio-topological data structures, as mentioned above.

Acknowledgements This work has been partially supported by the US National Science Foundation under grant number IIS-1116747 and by the University of Genova through PRA 2013. The authors wish to thank Davide Bolognini, Emanuela De Negri and Maria Evelina Rossi for their helpful comments and suggestions.

References

1. Agoston, M.K.: Computer Graphics and Geometric Modeling: Mathematics. Springer, London (2005)
2. Alexandroff, P., Hopf, H.: Topologic i, vol. 1035. Springer, Berlin (1935)
3. Artin, M.: Algebra. Prentice Hall, Englewood Cliffs (1991)
4. Bendich, P., Edelsbrunner, H., Kerber, M.: Computing robustness and persistence for images. IEEE Trans. Vis. Comput. Graph. **16**(6), 1251–1260 (2010)
5. Benedetti, B., Lutz, F.H.: Random discrete Morse theory and a new library of triangulations. Exp. Math. **23**(1), 66–94 (2014)

6. Boissonnat, J.D., Dey, T.K., Maria, C.: The compressed annotation matrix: an efficient data structure for computing persistent cohomology. In: Algorithms–ESA 2013, Sophia Antipolis, pp. 695–706. Springer (2013)
7. Boltcheva, D., Canino, D., Merino Aceituno, S., Léon, J.C., De Floriani, L., Hétroy, F.: An iterative algorithm for homology computation on simplicial shapes. Comput. Aided Des. **43**(11), 1457–1467 (2011)
8. Bremer, P.T., Hamann, B., Edelsbrunner, H., Pascucci, V.: A topological hierarchy for functions on triangulated surfaces. IEEE Trans. Vis. Comput. Graph. **10**(4), 385–396 (2004)
9. Canino, D., De Floriani, L., Weiss, K.: IA*: an adjacency-based representation for non-manifold simplicial shapes in arbitrary dimensions. Comput. Graph. **35**(3), 747–753 (2011)
10. Carlsson, G., Ishkhanov, T., De Silva, V., Zomorodian, A.J.: On the local behavior of spaces of natural images. Int. J. Comput. Vis. **76**(1), 1–12 (2008)
11. Cazals, F., Chazal, F., Lewiner, T.: Molecular shape analysis based upon the Morse-Smale complex and the Connolly function. In: Proceedings of 9th Annual Symposium on Computational Geometry, pp. 351–360. ACM Press, New York (2003)
12. Cerri, A., Ferri, M., Giorgi, D.: Retrieval of trademark images by means of size functions. Graph. Models **68**(5), 451–471 (2006)
13. Chung, M.K., Bubenik, P., Kim, P.T.: Persistence diagrams of cortical surface data. In: Information Processing in Medical Imaging, pp. 386–397. Springer, Berlin/New York (2009)
14. Čomić, L., De Floriani, L., Iuricich, F.: Simplification operators on a dimension-independent graph-based representation of Morse complexes. In: Hendriks, C.L.L., Borgefors, G., Strand R. (eds.) ISMM. Lecture Notes in Computer Science, vol. 7883, pp. 13–24. Springer, Berlin/New York (2013)
15. Čomić, L., De Floriani, L., Iuricich, F., Fugacci, U.: Topological modifications and hierarchical representation of cell complexes in arbitrary dimensions. Comput. Vis. Image Underst. **121**, 2–12 (2014)
16. Connolly, M.L.: Measurement of protein surface shape by solid angles. J. Mol. Graph. **4**(1), 3–6 (1986)
17. De Floriani, L., Hui, A.: Data structures for simplicial complexes: an analysis and a comparison. In: Desbrun, M., Pottmann, H. (eds.) Proceedings of 3rd Eurographics Symposium on Geometry Processing. ACM International Conference on Proceeding Series, vol. 255, pp. 119–128. Eurographics Association, Aire-la-Ville (2005)
18. Dequeant, M.L., Ahnert, S., Edelsbrunner, H., Fink, T.M., Glynn, E.F., Hattem, G., Kudlicki, A., Mileyko, Y., Morton, J., Mushegian, A.R., et al.: Comparison of pattern detection methods in microarray time series of the segmentation clock. PLoS One **3**(8), e2856 (2008)
19. Dey, T.K., Fan, F., Wang, Y.: Computing topological persistence for simplicial maps. arXiv preprint arXiv:1208.5018 (2012)
20. Dey, T.K., Hirani, A.N., Krishnamoorthy, B., Smith, G.: Edge contractions and simplicial homology. arXiv preprint arXiv:1304.0664 (2013)
21. Dłotko, P., Kaczynski, T., Mrozek, M., Wanner, T.: Coreduction homology algorithm for regular cw-complexes. Discret. Comput. Geom. **46**(2), 361–388 (2011)
22. Dłotko, P., Wagner, H.: Simplification of complexes of persistent homology computations. Homol. Homotopy Appl. **16**(1), 49–63 (2014)
23. Edelsbrunner, H.: Algorithms in Combinatorial Geometry. Springer, Berlin (1987)
24. Edelsbrunner, H., Harer, J.: Persistent homology-a survey. Contemp. Math. **453**, 257–282 (2008)
25. Edelsbrunner, H., Letscher, D., Zomorodian, A.J.: Topological persistence and simplification. Discret. Comput. Geom. **28**(4), 511–533 (2002)
26. Fellegara, R., Iuricich, F., De Floriani, L., Weiss, K.: Efficient computation and simplification of discrete Morse decompositions on triangulated terrains. In: 22th ACM SIGSPATIAL International Symposium on Advances in Geographic Information Systems, ACM-GIS 2014, Dallas, 4–7 Nov 2014 (2014)
27. Forman, R.: Combinatorial vector fields and dynamical systems. Mathematische Zeitschrift **228**, 629–681 (1998)

28. Forman, R.: Morse theory for cell complexes. Adv. Math. **134**, 90–145 (1998)
29. Fugacci, U., Iuricich, F., De Floriani, L.: Efficient computation of simplicial homology through acyclic matching. In: Proceedings of 5th International Workshop on Computational Topology in Image Context (CTIC 2014), Timisoara (2014)
30. Ghrist, R.: Barcodes: the persistent topology of data. Bull. Am. Math. Soc. **45**(1), 61–75 (2008)
31. Günther, D., Reininghaus, J., Wagner, H., Hotz, I.: Efficient computation of 3D Morse-Smale complexes and persistent homology using discrete Morse theory. Vis. Comput. **28**(10), 959–969 (2012)
32. Gyulassy, A., Bremer, P.T., Pascucci, V.: Computing Morse-Smale complexes with accurate geometry. IEEE Trans. Vis. Comput. Graph. **18**(12), 2014–2022 (2012). doi:10.1109/TVCG.2012.209
33. Gyulassy, A., Bremer, P.T., Hamann, B., Pascucci, V.: A practical approach to Morse-Smale complex computation: scalability and generality. IEEE Trans. Vis. Comput. Graph. **14**(6), 1619–1626 (2008)
34. Gyulassy, A., Bremer, P.T., Hamann, B., Pascucci, V.: Practical considerations in Morse-Smale complex computation. In: Pascucci, V., Tricoche, X., Hagen, H., Tierny, J. (eds.) Topological Methods in Data Analysis and Visualization: Theory, Algorithms, and Applications, Mathematics and Visualization, pp. 67–78. Springer, Heidelberg (2011)
35. Gyulassy, A., Kotava, N., Kim, M., Hansen, C., Hagen, H., Pascucci, V.: Direct feature visualization using Morse-Smale complexes. IEEE Trans. Vis. Comput. Graph. **18**(9), 1549–1562 (2012)
36. Harker, S., Mischaikow, K., Mrozek, M., Nanda, V.: Discrete Morse theoretic algorithms for computing homology of complexes and maps. Found. Comput. Math. **14**(1), 151–184 (2014)
37. Harker, S., Mischaikow, K., Mrozek, M., Nanda, V., Wagner, H., Juda, M., Dłotko, P.: The efficiency of a homology algorithm based on discrete Morse theory and coreductions. In: Proceedings of 3rd International Workshop on Computational Topology in Image Context (CTIC 2010), Cádiz. Image A, vol. 1, pp. 41–47 (2010)
38. Hatcher, A.: Algebraic Topology. Cambridge University Press, Cambridge/New York (2002)
39. King, H., Knudson, K., Mramor, N.: Generating discrete Morse functions from point data. Exp. Math. **14**(4), 435–444 (2005)
40. Lewiner, T., Lopes, H., Tavares, G.: Optimal discrete Morse functions for 2-manifolds. Comput. Geom. **26**(3), 221–233 (2003)
41. Lewis, R.H., Zomorodian, A.J.: Multicore homology via Mayer Vietoris. arXiv preprint arXiv:1407.2275 (2014)
42. Lundell, A.T., Weingram, S.: The topology of CW complexes. Van Nostrand Reinhold Company, New York (1969)
43. Martin, S., Thompson, A., Coutsias, E.A., Watson, J.P.: Topology of cyclo-octane energy landscape. J. Chem. Phys. **132**(23), 234115 (2010). doi:10.1063/1.3445267
44. Milnor, J.: Morse Theory. Princeton University Press, Princeton (1963)
45. Mischaikow, K., Nanda, V.: Morse theory for filtrations and efficient computation of persistent homology. Discret. Comput. Geom. **50**(2), 330–353 (2013)
46. Mrozek, M., Batko, B.: Coreduction homology algorithm. Discret. Comput. Geom. **41**(1), 96–118 (2009)
47. Mrozek, M., Wanner, T.: Coreduction homology algorithm for inclusions and persistent homology. Comput. Math. Appl. **60**(10), 2812–2833 (2010)
48. Munkres, J.: Elements of Algebraic Topology. Advanced Book Classics. Perseus Books, New York (1984)
49. Nanda, V.: The Perseus software project for rapid computation of persistent homology. http://www.math.rutgers.edu/~vidit/perseus/index.html
50. Rieck, B., Leitte, H.: Structural analysis of multivariate point clouds using simplicial chains. Comput. Graph. Forum **33**(8), 28–37 (2014). doi:10.1111/cgf.12398
51. Rieck, B., Mara, H., Leitte, H.: Multivariate data analysis using persistence-based filtering and topological signatures. IEEE Trans. Vis. Comput. Graph. **18**(12), 2382–2391 (2012). doi:10.1109/TVCG.2012.248

52. Robins, V., Wood, P.J., Sheppard, A.P.: Theory and algorithms for constructing discrete Morse complexes from grayscale digital images. IEEE Trans. Pattern Anal. Mach. Intell. **33**(8), 1646–1658 (2011)
53. Rosenfeld, A., Kak, A.C.: Digital Picture Processing. Academic Press, London (1982)
54. Shivashankar, N., Maadasamy, S., Natarajan, V.: Parallel computation of 2D Morse-Smale complexes. IEEE Trans. Vis. Comput. Graph. **18**(10), 1757–1770 (2012)
55. Shivashankar, N., Natarajan, V.: Parallel computation of 3D Morse-Smale complexes. Comput. Graph. Forum **31**(3), 965–974 (2012)
56. de Silva, V., Ghrist, R.: Coverage in sensor networks via persistent homology. Algebr. Geom. Topol. **7**, 339–358 (2007). doi:10.2140/agt.2007.7.339
57. Wang, Y., Agarwal, P.K., Brown, P.H.E., Rudolph, J.: Coarse and reliable geometric alignment for protein docking. In: Proceedings of Pacific Symposium on Biocomputing, Hawaii, vol. 10, pp. 65–75 (2005)
58. Weiss, K., De Floriani, L., Fellegara, R., Velloso, M.: The PR-star octree: a spatio-topological data structure for tetrahedral meshes. In: GIS, Chicago, pp. 92–101 (2011)
59. Weiss, K., Iuricich, F., Fellegara, R., De Floriani, L.: A primal/dual representation for discrete Morse complexes on tetrahedral meshes. Comput. Graph. Forum **32**(3), 361–370 (2013)
60. Van de Weygaert, R., Vegter, G., Edelsbrunner, H., Jones, B.J., Pranav, P., Park, C., Hellwing, W.A., Eldering, B., Kruithof, N., Bos, E., et al.: Alpha, Betti and the megaparsec universe: on the topology of the cosmic web. In: Transactions on Computational Science XIV, pp. 60–101. Springer, Berlin/New York (2011). http://arxiv.org/abs/1306.3640
61. Zomorodian, A.J.: Topology for Computing, vol. 16. Cambridge University Press, Cambridge/New York (2005)

Chapter 10
Sparse Models for Intrinsic Shape Correspondence

Jonathan Pokrass, Alexander M. Bronstein, Michael M. Bronstein, Pablo Sprechmann, and Guillermo Sapiro

Abstract We present a novel sparse modeling approach to non-rigid shape matching using only the ability to detect repeatable regions. As the input to our algorithm, we are given only two sets of regions in two shapes; no descriptors are provided so the correspondence between the regions is not know, nor do we know how many regions correspond in the two shapes. We show that even with such scarce information, it is possible to establish very accurate correspondence between the shapes by using methods from the field of sparse modeling, being this, the first non-trivial use of sparse models in shape correspondence. We formulate the problem of *permuted sparse coding*, in which we solve simultaneously for an unknown permutation ordering the regions on two shapes and for an unknown correspondence in functional representation. We also propose a robust variant capable of handling incomplete matches. Numerically, the problem is solved efficiently by alternating the solution of a linear assignment and a sparse coding problem. The proposed methods are evaluated qualitatively and quantitatively on standard benchmarks containing both synthetic and scanned objects.

10.1 Introduction

Matching of deformable shapes is a notoriously difficult problem playing an important role in many applications [17]. Unlike rigid matching where the correspondence can be parametrized by a small number of parameters (rotation and translation of one shape w.r.t. the other [5, 10]), non-rigid matching typically uses point-wise

---

J. Pokrass • A.M. Bronstein (✉)
School of Electrical Engineering, Tel Aviv University, Tel Aviv, Israel
e-mail: bron@cs.technion.ac.il

M.M. Bronstein
Faculty of Informatics, Institute of Computational Science, University of Lugano, Lugano, Switzerland

P. Sprechmann • G. Sapiro
School of Electrical and Computer Engineering, Duke University, Durham, NC, USA

M. Breuß et al. (eds.), *Perspectives in Shape Analysis*, Mathematics and Visualization, DOI 10.1007/978-3-319-24726-7_10

representation of correspondence, which results in the number of degrees of freedom growing exponentially with the number of matched points.

Non-rigid correspondence methods try to find correspondence by minimizing some structure distortion. The structures can be point-wise (local descriptors [3, 14, 33, 37]), pair-wise (distances [6, 8, 13, 23]), or higher order [38].

In order to make the matching problem computationally feasible, it is crucial to reduce the size of the search space [34]. Most methods use a combination of point- and pair-wise structure matching in order to achieve this, and typically consist of three main components: feature detection, feature description, and regularization. Given two shapes, a *feature detector* allows to find a set of landmarks (points or regions) that are repeatable, i.e., appear (possibly with some inaccuracy) on both shapes. A *feature descriptor* then assigns to each feature a vector capturing some local geometric properties of the shape; very often, the two processes are combined into a single one. Using the descriptors, landmarks on two shapes can be matched (it has been shown [27] that under some conditions, correct landmark matching fully determines the intrinsic correspondence between the shapes). Such a matching reduces the search space size to points with similar descriptors. However, since the matching uses only local information, such correspondence can be noisy, and some kind of *regularization* based on higher-order information is needed to rule out bad or inconsistent correspondences. This information is also used to establish the correspondence between the rest of the points on the shapes. Often, the process is applied hierarchically, restricting the candidate matches to points in the proximity of the landmarks [31].

Computer graphics and geometry processing literature contains a plethora of approaches for each of the aforementioned components. Feature detection methods try to locate stable points or regions [11, 21] that are invariant under isometric deformations and robust to noise. Popular feature descriptors include the heat kernel signature (HKS) [14, 33], wave kernel signature (WKS) [3], global point signature (GPS) [30] or methods adopted from the domain of image analysis [37]. As regularization, pairwise structures such as geodesic [6, 23] or diffusion distances [8] and higher-order structures [38] have been used.

Alternatively, there have been several attempts to represent correspondences with a small set of parameters. Elad and Kimmel [13] used multidimensional scaling (MDS)-type methods to embed the intrinsic structure of the shapes into a low-dimensional Euclidean space, posing the problem of non-rigid matching as a problem of rigid matching of the corresponding embeddings ("canonical forms"). Mateus et al. [22] used spectral embeddings instead of MDS. Lipman and Funkhouser [20] embedded the shapes into a disk by means of conformal maps and represented the correspondence as a Möbius transformation.

More recently, Ovsjanikov et al. [26] introduced the functional representation of correspondences, allowing to perform a "calculus" of correspondences. In this approach, correspondence is modeled as a correspondence between functions on two shapes rather than points, and can be compactly represented in the Laplace-Beltrami eigenbasis as a matrix of coefficients of decomposition of the basis functions of the first shape in the basis of the second one. In this paper, we will be relying upon this latter representation.

10.1.1 Main Contribution

The main practical contribution of this paper is an approach for finding dense intrinsic correspondence between near-isometric shapes with very little known information: we only assume to be able to detect regions in two shapes in a repeatable enough way (i.e., that at least some regions in one shape correspond accurately enough to some other regions in another shape). No region descriptors are given, so the correspondence of the regions is unknown. The assumption of near-isometry assures that in the functional representation of [26], the unknown correspondence can be represented as a sparse matrix. The assumption of repeatable regions implies that there exists some unknown permutation that orders the regions according to their correspondence.

We formulate the problem of *permuted sparse coding*, in which we simultaneously look for the permutation and the correspondence, thereby introducing the very successful area of sparse modeling into efficient and state-of-the-art shape correspondence. We note that with the permutation fixed, our problem becomes the standard sparse coding problem; having the correspondence fixed, the problem becomes a linear assignment. This allows efficient numerical solution by alternating the two aforementioned problems and employing efficient solvers that exist for both.

Our method relies on a pretty common assumption that the shapes are nearly-isometric (though our experimental results show our approach still works even when departing from this assumption), and out of all methods we are aware of, it uses perhaps the scarcest amount of data to establish dense correspondence between the shapes. For example, sandard region detectors with high repeatability such as [21] are sufficient.

Compared to recent techniques for region-wise shape matching (see, e.g., [15, 16, 28, 36]), our approach has several important practical advantages: First, we do not use any feature descriptor. Second, most region-wise correspondence approaches require an additional step of extending the correspondence between matched regions to the rest of the points.

The rest of the paper is organized as follows. In Sect. 10.2, we overview the functional representation of correspondences, allowing to work with correspondences as algebraic structures, and state the main notions in sparse modeling. In Sect. 10.3, we formulate our problem of permuted sparse coding for establishing correspondence from a set of repeatable regions given in unknown order. We then extend the problem to the general setting where the region detection process is not perfectly repeatable. In Sect. 10.4, we describe the numerical optimization used to solve our permuted sparse coding problem. Experimental results are shown in Sect. 10.5. Finally, Sect. 10.6 discusses the limitations and possible extensions of the proposed framework and concludes the paper.

10.2 Background

10.2.1 Functional Representation of Correspondences

The direct representation of correspondences as maps between two non-Euclidean spaces limits the range of tools that can be employed for correspondence computation due to the lack of an algebraic structure. In this paper, we rely on the functional representation of correspondences introduced in [26], which overcomes this limitation. In what follows, we briefly review the main idea of such functional representations.

Let X and Y be two shapes, modeled as compact smooth Riemannian manifolds, related by a bijective correspondence $t : X \to Y$. Then, for any real function $f : X \to \mathbb{R}$, we can construct a corresponding function $g : Y \to \mathbb{R}$ as $g = f \circ t^{-1}$. The correspondence t uniquely defines a mapping between two function spaces $T : \mathscr{F}(X, \mathbb{R}) \to \mathscr{F}(Y, \mathbb{R})$, where $\mathscr{F}(X, \mathbb{R})$ denotes the space of real functions on X. Such a representation is linear, since for every pair of functions f_1, f_2 and scalars α_1, α_2,

$$\begin{aligned} T(\alpha_1 f_1 + \alpha_2 f_2) &= (\alpha_1 f_1 + \alpha_2 f_2) \circ t^{-1} \\ &= \alpha_1 f_1 \circ t^{-1} + \alpha_2 f_2 \circ t^{-1} = \alpha_1 T(f_1) + \alpha_2 T(f_2). \end{aligned} \tag{10.1}$$

Assuming that X is equipped with a basis $\{\phi_i\}_{i \geq 1}$, any $f : X \to \mathbb{R}$ can be represented as

$$f = \sum_{i \geq 1} a_i \phi_i \tag{10.2}$$

with the a_i being some representation coefficients (in case of an orthonormal basis, $a_i = \langle f, \phi_i \rangle$; in the general case, the coefficients are found by projecting the function f on the bi-orthonormal basis). Due to the linearity of T,

$$T(f) = T\left(\sum_{i \geq 1} a_i \phi_i \right) = \sum_{i \geq 1} a_i T(\phi_i) \tag{10.3}$$

If the shape Y is further equipped with a basis $\{\psi_j\}_{j \geq 1}$, then for every i there exists coefficients c_{ij} such that

$$T(\phi_i) = \sum_{j \geq 1} c_{ij} \psi_j, \tag{10.4}$$

and we can write

$$T(f) = \sum_{i,j \geq 1} a_i c_{ij} \psi_j. \tag{10.5}$$

Let us now assume finite sampling of X and Y, with m samples (for simplicity, we assume that the shapes are sampled at the same number of samples m. The extension to the case with a different number of samples is straightforward). The bases are represented as the $m \times n$ matrices $\mathbf{\Phi}$ and $\mathbf{\Psi}$ containing, respectively, n discretized functions ϕ_i and ψ_j as the columns. The functions f and $g = T(f)$ can now be represented as n-dimensional vectors $\mathbf{f} = \mathbf{\Phi a}$ and $\mathbf{g} = \mathbf{\Psi b}$ with the coefficients $\mathbf{a}$ and $\mathbf{b}$. Using this notation, Equation (10.5) can be rewritten as $\mathbf{\Psi b} = T(\mathbf{\Phi a}) = \mathbf{\Psi C}^{\mathrm{T}}\mathbf{a}$; since $\mathbf{\Psi}$ is invertible, this simply means that

$$\mathbf{b}^{\mathrm{T}} = \mathbf{a}^{\mathrm{T}}\mathbf{C}. \tag{10.6}$$

Thus, the $n \times n$ matrix $\mathbf{C}$ fully encodes the linear map T between the functional spaces, and contains the coordinates in the basis $\mathbf{\Psi}$ of the mapped elements of the basis $\mathbf{\Phi}$.

10.2.2 Point-to-Point Correspondence

Point-to-point correspondences assume that each point i on X corresponds to some point j on Y. In functional representation, this is equivalent to having $\mathbf{C}$ that makes each row of $\mathbf{\Psi C}^{\mathrm{T}}$ coincide with some row of $\mathbf{\Phi}$ [26]. In applications requiring point-to-point correspondence, given some $\mathbf{C}$, it can be converted into a point-to-point correspondence by thinking of $\mathbf{\Phi}$ and $\mathbf{\Psi}$ as n-dimensional points clouds, and orthogonal matrix $\mathbf{C}$ as a rigid alignment transformation between them. This procedure is equivalent to iterative closest point (ICP) in n dimensions [26], initialized with the given $\mathbf{C}_0$: first, for each row i of $\mathbf{\Psi C_0}^{\mathrm{T}}$, find the closest row j_i^* in $\mathbf{\Phi}$ (this operation can be performed efficiently using approximate nearest neighbor algorithms). Then, find orthonormal $\mathbf{C}$ minimizing $\sum_i \|\mathbf{\Phi}_{j_i^*} - \mathbf{\Psi C}^{\mathrm{T}}\|_2$ and set $\mathbf{C}_0 = \mathbf{C}$. This operation is repeated until convergence and can be performed efficiently over all the vertexes of X and Y using approximate nearest neighbor algorithms.

A more naive approach not imposing orthonormality of $\mathbf{C}$ is simply to map every standard Euclidean basis vector $\mathbf{e}_i$ in $\mathbb{R}^m$ representing a delta function centered at point i on X to the band-limited approximation, $\mathbf{\Psi C \Phi}^{\mathrm{T}}\mathbf{e}_i$, of the corresponding indicator function on Y. If the maximum value of the latter vector is attained at point j on Y, the correspondence between point i on X and point j on Y is established.

10.2.3 Sparse Modeling

One of the main tools that will be used in this paper are *sparse models*. In what follows, we give a very brief overview of this vast field, and refer the reader to [12] for a comprehensive treatise. The central assertion of sparse modeling is that many families of signals (and later operations as here introduced) can be represented as a sparse linear combination in an appropriate domain, usually referred to as the *dictionary*. This can be written as $\mathbf{x} \approx \mathbf{D}\mathbf{z}$, where $\mathbf{x}$ denotes the signal, $\mathbf{D}$ the dictionary, and $\mathbf{z}$ the sparse vector of representation coefficients. The dictionary is often selected to be *overcomplete*, i.e., with more columns than rows.

Finding the representation of a signal $\mathbf{x}$ in a given dictionary $\mathbf{D}$ is usually referred to as sparse representation *pursuit* or *sparse coding*. Among the variety of pursuit methods, we will focus on the so-called Lasso formulation [35] that finds $\mathbf{z}$ as the solution to the unconstrained convex program

$$\min_{\mathbf{z}} \|\mathbf{x} - \mathbf{D}\mathbf{z}\|_2^2 + \lambda \|\mathbf{z}\|_1. \tag{10.7}$$

The first term is the data fitting term, while the second term involving the ℓ_1 norm, $\|\mathbf{z}\|_1 = |z_1| + \ldots + |z_n|$, promotes a sparse solution; the parameter λ controls the relative importance of the latter. Proximal splitting methods [24] are among the most efficient and most frequently used numerical tools to solve problem (10.7); in Sect. 10.4, we present a variant of the proximal splitting algorithms for the solution of the pursuit problem arising in shape correspondence as detailed in the sequel.

In some cases, signals not admitting the simplistic model of element-wise sparsity still manifest more intricate types of *structured* sparsity. In structured sparse models, the non-zero elements of $\mathbf{z}$ come in groups or, more generally, in hierarchies of groups. A common class of structured pursuit problems can be formulated as convex programs of the form

$$\min_{\mathbf{z}} \|\mathbf{x} - \mathbf{D}\mathbf{z}\|_2^2 + \lambda \|\mathbf{z}\|_{1,2}, \tag{10.8}$$

where the $\ell_{1,2}$ norm, $\|\mathbf{z}\|_{1,2} = \|\mathbf{z}_1\|_2 + \cdot + \|\mathbf{z}_k\|_2$, assumes that the vector $\mathbf{z}$ is decomposed into k non-overlapping sub-vectors $\mathbf{z}_i$, and promotes group-wise sparse solutions (i.e., the solution will have a small number of groups with non-zero coefficients, but the sub-vectors representing each such non-zero group will be dense).

While structured sparse models enforce group structure of each representation vector independently, it is often useful to consider the structure shared by multiple vectors. *Collaborative* sparse models operate on data matrices $\mathbf{X}$, in which each column corresponds to a data vector, and assert that the patterns of non-zero coefficients are shared across the corresponding representation vectors, $\mathbf{Z}$. This is achieved by solving a pursuit problem of the form

$$\min_{\mathbf{Z}} \|\mathbf{X} - \mathbf{D}\mathbf{Z}\|_{\mathrm{F}}^2 + \lambda \|\mathbf{Z}\|_{2,1}, \tag{10.9}$$

where the first term involving the Frobenius norm serves as the data fitting term, and the second term with the $\ell_{2,1}$ norm promotes row-wise sparsity of the solution. The $\ell_{2,1}$ norm is defined as $\|\mathbf{Z}\|_{2,1} = \|\mathbf{z}_1^{\mathrm{T}}\|_2 + \cdots + \|\mathbf{z}_m^{\mathrm{T}}\|_2$, where $\mathbf{z}_i^{\mathrm{T}}$ denotes the i-th row of $\mathbf{Z}$ (note the difference from the $\ell_{1,2}$ column-wise counterpart!).

In this paper, we use formulate the shape correspondence problem using a sparse model, and use sparse modeling tools to efficiently solve it.

10.3 Sparse Modeling of Correspondences

In case the shapes X and Y are isometric and the corresponding Laplace-Beltrami operators have simple spectra (no eigenvalues with multiplicity greater than one), the harmonic bases (Laplacian eigenfunctions) have a compatible behavior, $\psi_i = T(\phi_i)$ such that $c_{ij} = \pm\delta_{ij}$. Choosing the discretized eigenfunctions of the Laplace-Beltrami operator as $\mathbf{\Phi}$ and $\mathbf{\Psi}$ causes every low-distortion correspondence being represented by a nearly diagonal, and therefore very sparse, matrix $\mathbf{C}$.

In practice, due to lack of perfect isometry and numerical inaccuracies, the diagonal structure of $\mathbf{C}$ is manifested for the first eigenfunctions corresponding to the small eigenvalues (low frequencies), and is gradually lost with the increase of the frequency (see, e.g., Fig. 10.1). However, a correspondence with low metric distortion will usually be represented by a sparse $\mathbf{C}$. We use this property to

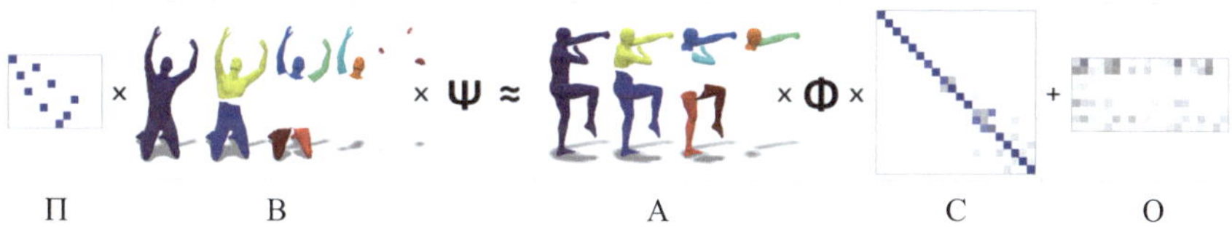

Fig. 10.1 Near isometric shape correspondence as a sparse modeling problem (see details in text): Indicator functions of repeatable regions on two shapes are detected and represented as matrices of coefficients $\mathbf{A}$ and $\mathbf{B}$ in the corresponding orthonormal harmonic bases $\mathbf{\Phi}$ and $\mathbf{\Psi}$. When the regions are brought into correspondence, the point-to-point correspondence between the shapes can be encoded by an approximately diagonal matrix $\mathbf{C}$. In the proposed procedure termed as *permuted sparse coding*, we solve $\mathbf{\Pi B} = \mathbf{AC} + \mathbf{O}$ simultaneously for an approximately diagonal $\mathbf{C}$ and the permutation $\mathbf{\Pi}$ bringing the indicator functions into correspondence. To cope with imperfectly matching regions, we relax the surjectivity of the permutation and absorb the mismatches into a row-wise sparse outlier matrix $\mathbf{O}$. For visualization purposes, the coloring of the regions is consistent as after the application of the permutation. Correspondence is shown between a synthetic TOSCA and scanned SCAPE shape

formulate the computation of correspondences in terms of a sparse representation pursuit problem.

Let us assume to have some *region* (or *feature*) *detection* process that given a shape X produces a collection of functions $\{f_i : X \to \mathbb{R}\}$ based on the intrinsic properties of the shape only. For example, the f_i's can be indicator functions of the maximally stable components (regions) of the shape [21]. Since the process is intrinsic, given a nearly isometric deformation Y or X, it should produce a collection of similar functions $\{g_j : Y \to \mathbb{R}\}$.

To simplify the presentation, let us assume that the process is perfectly *repeatable* in the sense that it finds q functions on X and Y, such that for every f_i there exists a $g_j = f_i \circ t$ related by the unknown correspondence t. We stress that the ordering of the f_i's and g_j's is *unknown*, i.e., we do not know to which g_j in Y a f_i in X corresponds. This ordering can be expressed by an unknown $q \times q$ permutation matrix $\mathbf{\Pi}$ (in Sect. 10.3.2, we consider the more general case when the number of functions detected on X and Y can be different, i.e., $\mathbf{\Pi}$ is non-square).

Representing the functions in the bases on each shape, we have $\mathbf{f}_i = \mathbf{\Phi}\mathbf{a}_i$ and $\mathbf{g}_j = \mathbf{\Psi}\mathbf{b}_j$. Since each pair of corresponding $\mathbf{f}_i$ and $\mathbf{g}_j$ shall satisfy (10.6), we can write

$$\mathbf{\Pi}\mathbf{B} = \mathbf{A}\mathbf{C}, \tag{10.10}$$

where $\mathbf{A}$ and $\mathbf{B}$ are the $q \times n$ matrices containing, respectively, $\mathbf{a}_i^{\mathrm{T}}$ and $\mathbf{b}_j^{\mathrm{T}}$ as the rows, and $\pi_{ij} = 1$ if $\mathbf{a}_i$ corresponds to $\mathbf{b}_j$ and zero otherwise.

10.3.1 Permuted Sparse Coding

Note that in relation (10.10), both $\mathbf{\Pi}$ and $\mathbf{C}$ are unknown, and solving for them is a highly ill-posed problem. However, by recalling that the correspondence we are looking for should be represented by a nearly-diagonal $\mathbf{C}$, we formulate the following problem

$$\min_{\mathbf{C},\mathbf{\Pi}} \frac{1}{2}\|\mathbf{\Pi}\mathbf{B} - \mathbf{A}\mathbf{C}\|_{\mathrm{F}}^2 + \lambda\|\mathbf{W} \odot \mathbf{C}\|_1, \tag{10.11}$$

where the minimum is sought over $n \times n$ matrices $\mathbf{C}$ (capturing the correspondence t between the shapes in the functional representation) and $q \times q$ permutations $\mathbf{\Pi}$ (capturing the correspondence between the detected regions on the shapes). The first term containing the Frobenius norm can be interpreted as the data term, while the second term, containing the weighted ℓ_1 norm promotes a sparse $\mathbf{C}$; $\odot$ denotes element-wise multiplication, and the non-negative parameter λ determines the relative importance of the penalty. Small weights w_{ij} in $\mathbf{W}$ are assigned close to the diagonal, while larger weights are selected for the off-diagonal elements. This promotes diagonal solutions.

The solution of (10.11) can be obtained using alternating minimization iterating over $\mathbf{C}$ with fixed $\mathbf{\Pi}$, and $\mathbf{\Pi}$ with fixed $\mathbf{C}$. Note that with fixed $\mathbf{\Pi}$, we can denote $\mathbf{B}' = \mathbf{\Pi}\mathbf{B}$ and reduce problem (10.11) to

$$\min_{\mathbf{C}} \frac{1}{2}\|\mathbf{B}' - \mathbf{A}\mathbf{C}\|_{\mathrm{F}}^2 + \lambda\|\mathbf{W} \odot \mathbf{C}\|_1, \tag{10.12}$$

which resembles the Lasso problem frequently employed for the pursuit of sparse representations. On the other hand, when $\mathbf{C}$ is fixed, we set $\mathbf{A}' = \mathbf{A}\mathbf{C}$, reducing the optimization objective to

$$\begin{aligned} \|\mathbf{\Pi}\mathbf{B} - \mathbf{A}'\|_{\mathrm{F}}^2 &= \\ \mathrm{tr}\left(\mathbf{B}^{\mathrm{T}}\mathbf{\Pi}^{\mathrm{T}}\mathbf{\Pi}\mathbf{B}\right) - 2\mathrm{tr}\left(\mathbf{B}^{\mathrm{T}}\mathbf{\Pi}^{\mathrm{T}}\mathbf{A}'\right) &+ \mathrm{tr}\left(\mathbf{A}'^{\mathrm{T}}\mathbf{A}'\right). \end{aligned} \tag{10.13}$$

Since $\mathbf{\Pi}$ is a permutation matrix, $\mathbf{\Pi}^{\mathrm{T}}\mathbf{\Pi} = \mathbf{I}$, and the only non-constant term remaining in the objective is the second linear term. Problem (10.11) thus becomes

$$\max_{\mathbf{\Pi}} \mathrm{tr}\left(\mathbf{\Pi}^{\mathrm{T}}\mathbf{E}\right), \tag{10.14}$$

where $\mathbf{E} = \mathbf{A}'\mathbf{B}^{\mathrm{T}} = \mathbf{A}\mathbf{C}\mathbf{B}^{\mathrm{T}}$ and the maximization is performed over permutation matrices. To make it practically solvable, we allow $\mathbf{\Pi}$ to be a double-stochastic matrix, which yields the following linear assignment problem:

$$\max_{\mathbf{\Pi}\geq\mathbf{0}} \ \mathrm{vec}(\mathbf{E})^{\mathrm{T}}\mathrm{vec}(\mathbf{\Pi}) \ \text{s.t.} \begin{cases} \mathbf{\Pi}\mathbf{1} = \mathbf{1} \\ \mathbf{\Pi}^{\mathrm{T}}\mathbf{1} = \mathbf{1}. \end{cases} \tag{10.15}$$

We refer to problem (10.11) as *permuted sparse coding*, and propose to solve it by alternating the solution of the standard sparse coding problem (10.12) and the solution of the linear assignment problem (10.15). The sparsity constraint has a regularization effect on this, otherwise extremely ill-posed, problem, and the following strong property holds:

Proposition 10.1 *The process alternating subproblems* (10.12) *and* (10.15) *converges to a local minimizer of the permuted sparse coding problem* (10.11).

Due to lack of space, the proof will be provided in an extended version of this contribution. This result means, among other, that despite the relaxation of the permutation matrix to a double-stochastic matrix in the assignment subproblem (10.15), we are actually recovering a true permutation matrix. This follows from the total unimodularity of the constraints in (10.15).

We further conjecture that when the solution of (10.12) attains a sufficiently small value of the data fitting term (the ℓ_2 term), global convergence to a unique minimizer can be guaranteed under non-restrictive technical assumptions. While we do not yet have a formal proof for this empirically observed behavior, we believe that techniques similar to [1] can be used to prove this conjecture.

10.3.2 Robust Permuted Sparse Coding

So far, we have assumed the existence of a bijective, albeit unknown, correspondence between the f_i's and the g_j's. In practice, the process detecting these functions (e.g., stable regions) is often not perfectly repeatable. In what follows, we will make a more realistic assumption that q functions f_i are detected on X, and r functions g_j detected on Y (without loss of generality, $q \leq r$), such that some f_i's have no counterpart g_j, and vice versa. This partial correspondence can be described by a $q \times r$ partial permutation matrix $\mathbf{\Pi}$ in which now some columns and rows may vanish.

Let us assume that $s \leq q$ f_i's have corresponding g_j's. This means that there is no correspondence between $r-s$ rows of $\mathbf{B}$ and $q-s$ rows of $\mathbf{A}$, and the relation $\mathbf{\Pi B} \approx \mathbf{AC}$ holds only for an unknown subset of its rows. The mismatched rows of $\mathbf{B}$ can be ignored by letting some columns of $\mathbf{\Pi}$ vanish, meaning that the correspondence is no more surjective. This can be achieved by relaxing the equality constraint $\mathbf{\Pi}^{\mathrm{T}}\mathbf{1} = \mathbf{1}$ in (10.15) replacing it with $\mathbf{\Pi}^{\mathrm{T}}\mathbf{1} \leq \mathbf{1}$. However, dropping injectivity as well and relaxing $\mathbf{\Pi 1} = \mathbf{1}$ to $\mathbf{\Pi 1} \leq \mathbf{1}$ would result in the trivial solution $\mathbf{\Pi} = \mathbf{0}$. To overcome this difficulty, we demand every row of $\mathbf{A}$ to have a matching row in $\mathbf{B}$, and absorb the $r - s$ mismatches in a row-sparse $q \times n$ outlier matrix $\mathbf{O}$ that we add to the data term of (10.11). This results in the following problem

$$\min_{\mathbf{C},\mathbf{O},\mathbf{\Pi}} \frac{1}{2}\|\mathbf{\Pi B} - \mathbf{AC} - \mathbf{O}\|_{\mathrm{F}}^2 + \lambda\|\mathbf{W} \odot \mathbf{C}\|_1 + \mu\|\mathbf{O}\|_{2,1}, \tag{10.16}$$

which we refer to as *robust permuted sparse coding*. The last term involves the $\ell_{2,1}$ norm

$$\|\mathbf{O}\|_{2,1} = \sum_{i=1}^{r} \|\mathbf{o}_i^{\mathrm{T}}\|_2, \tag{10.17}$$

which can be thought of as the ℓ_1 norm of the vector of the ℓ_2 norms of the rows $\mathbf{o}_i^{\mathrm{T}}$ of $\mathbf{O}$. The $\ell_{2,1}$ norm promotes row-wise sparsity, allowing to absorb the errors in the data term corresponding to the rows of $\mathbf{A}$ having no corresponding rows in $\mathbf{B}$; the parameter $\mu \geq 0$ controls the amount of regularization. The $q \times r$ matrix $\mathbf{\Pi}$ is searched over all injective correspondences.

As before, problem (10.16) is split into two sub-problems, one with the fixed permutation $\mathbf{\Pi}$,

$$\min_{\mathbf{C},\mathbf{O}} \frac{1}{2}\|\mathbf{B}' - \mathbf{AC} - \mathbf{O}\|_{\mathrm{F}}^2 + \lambda\|\mathbf{W} \odot \mathbf{C}\|_1 + \mu\|\mathbf{O}\|_{2,1}, \tag{10.18}$$

with $\mathbf{B}' = \mathbf{\Pi B}$, and the other one with the fixed $\mathbf{C}$,

$$\max_{\mathbf{\Pi} \geq \mathbf{0}} \mathrm{vec}(\mathbf{E})^{\mathrm{T}}\mathrm{vec}(\mathbf{\Pi}) \text{ s.t. } \begin{cases} \mathbf{\Pi 1} = \mathbf{1} \\ \mathbf{\Pi}^{\mathrm{T}}\mathbf{1} \leq \mathbf{1}, \end{cases} \tag{10.19}$$

with $\mathbf{E} = (\mathbf{AC})\,\mathbf{B}^{\mathrm{T}}$. Note that an injective correspondence is relaxed into a row-wise stochastic and column-wise sub-stochastic matrix $\mathbf{\Pi}$. Proposition 10.1 simply extends to the robust formulation as well.

10.4 Numerical Solution

The solution of the robust permuted sparse coding problem (10.16) is reduced to alternating two relatively standard optimization problems, and there exist many readily available efficient numerical tools to solve them. For the sake of completeness, we provide a concise description of the involved numerics.

Problem (10.19), being a simple linear assignment problem, is solved using the Hungarian algorithm. As an alternative, linear programming can be employed. To reduce the search space size, we disallow certain impossible permutations such as those relating regions with very distinct sizes.

In order to solve (10.18), we use the family of forward-backward splitting algorithms [24] designed for solving unconstrained optimization problems in which the cost function can be split into the sum of two terms,

$$\min_{\mathbf{x}} h_1(\mathbf{x}) + h_2(\mathbf{x}), \tag{10.20}$$

one, h_1, convex and differentiable with an α-Lipschitz continuous gradient and another, h_2, convex extended real valued and possibly non-smooth. Clearly, problem (10.18) falls in this category.

The forward-backward splitting method with fixed constant step defines a series of iterates, $\{\mathbf{x}^k\}_k$,

$$\mathbf{x}^{k+1} = \mathbf{P}_{\alpha h_2}\left(\mathbf{x}^k - \frac{1}{\alpha}\nabla h_1(\mathbf{x}^k)\right), \tag{10.21}$$

where

$$\mathbf{P}_{\alpha h_2}(\mathbf{x}) = \arg\min_{\mathbf{u}} \|\mathbf{u} - \mathbf{x}\|_2^2 + \alpha h_2(\mathbf{u}) \tag{10.22}$$

denotes the proximal operator of h_2. Many alternatives have been studied in the literature to improve the convergence rate of forward-backward splitting algorithms [4, 24]. Accelerated versions reach quadratic convergence rates (the best possible for the class of first order methods). The discussion of theses methods is beyond of the scope of this paper.

In our case, the objective comprises a quadratic function $h_1 = \|\mathbf{B}' - \mathbf{AC} - \mathbf{O}\|_{\mathrm{F}}^2$ and the non-smooth function $h_2 = \lambda\|\mathbf{W} \odot \mathbf{C}\|_1 + \mu\|\mathbf{O}\|_{2,1}$. The proximal operator splits into two operators, one in $\mathbf{C}$ and another one in $\mathbf{O}$, both having closed forms. The proximal operator corresponding to the ℓ_1 norm term is given by the weighted

input : Data $\mathbf{B}', \mathbf{A}$; parameters λ, μ; step size α.
output: Sparse matrix $\mathbf{O}$ and row-wise sparse outlier matrix $\mathbf{O}$
Initialize $\mathbf{O}^0 = \mathbf{B}'$ and $\mathbf{C}^0 = \mathbf{0}$.
for *k=1,2,...,until convergence* **do**

$$\mathbf{C}^{k+1} = \mathbf{P}_1\left((\mathbf{I} - \frac{1}{\alpha}\mathbf{A}^{\mathrm{T}}\mathbf{A})\mathbf{C}^k - \frac{1}{\alpha}\mathbf{A}^{\mathrm{T}}(\mathbf{O}^k - \mathbf{B}')\right)$$

$$\mathbf{O}^{k+1} = \mathbf{P}_2\left((1 - \frac{1}{\alpha})\mathbf{O}^k - \frac{1}{\alpha}(\mathbf{A}\mathbf{C}^k - \mathbf{B}')\right)$$

end

Algorithm 1: Forward-backward splitting method for the solution of (10.18)

soft threshold function

$$\mathbf{P}_1(\mathbf{C}) = \max\left\{|\mathbf{C}| - \frac{\lambda}{\alpha}\mathbf{W}\right\} \odot \operatorname{sign}(\mathbf{C}), \tag{10.23}$$

where the absolute value and the sign functions are applied element-wise. The i-th row of the proximal operator corresponding to the $\ell_{2,1}$ norm term is given by

$$(\mathbf{P}_2(\mathbf{O}))_i = \max\left\{\|\mathbf{o}_i^{\mathrm{T}}\|_2 - \frac{\mu}{\alpha}\right\} \frac{\mathbf{o}_i^{\mathrm{T}}}{\|\mathbf{o}_i^{\mathrm{T}}\|_2}. \tag{10.24}$$

The gradient of the quadratic data term with respect to $\mathbf{C}$ and $\mathbf{O}$ is given straightforwardly by

$$\begin{aligned} \nabla_{\mathbf{C}} h_1 &= \mathbf{A}^{\mathrm{T}}\mathbf{A}\mathbf{C} + \mathbf{A}^{\mathrm{T}}\mathbf{O} - \mathbf{A}^{\mathrm{T}}\mathbf{B}' \\ \nabla_{\mathbf{O}} h_1 &= \mathbf{O} + \mathbf{A}\mathbf{C} - \mathbf{B}'. \end{aligned} \tag{10.25}$$

The Lipschitz constant of the gradient determining the step size is bounded by the maximum eigenvalue

$$\alpha \leq \lambda_{\max}\begin{pmatrix} \mathbf{A}^{\mathrm{T}}\mathbf{A} & \mathbf{A}^{\mathrm{T}} \\ \mathbf{I} & \mathbf{A} \end{pmatrix}. \tag{10.26}$$

Plugging the above expressions together into (10.21) yields the forward-backward splitting optimization summarized in Algorithm 1.

10.5 Experimental Results

In order to evaluate our approach, we performed several experiments on data from the TOSCA [7], SHREC'11 [9] and SCAPE [2] datasets. The TOSCA set contains high-quality (10–50 K vertices) synthetic triangular meshes of humans and animals in different poses with known ground truth correspondences between them. SHREC'11 contains meshes from the TOSCA set undergoing simulated transformations. The SCAPE set contains high-resolution (12 K vertices) scans of a real human in different poses.

For each pair of shapes we calculated the MSERs using 6–10 eigenfunctions and selected regions with areas of at least 5–10 % of the total shape area, resulting in about 5–15 detected regions (see Fig. 10.1). These parameters were selected empirically for our data sets.

The segments of each shape were projected onto 20 eigenfunctions and the corresponding $\mathbf{C}$ matrix was calculated by solving the sparse coding subproblem (10.18) using an accelerated variant of the method described in Sect. 10.4. The linear assignment subproblem (10.15) was solved using the Hungarian method [19]. We initialized the permutation matrix with $\mathbf{\Pi} = \frac{1}{q}\mathbf{1}\mathbf{1}^{\mathrm{T}}$, and the correspondence matrix with $\mathbf{C} = \mathbf{0}$. We observed a rapid convergence of the alternating minimization procedure in one or two iterations (see Fig. 10.2 where for visualization purposes, $\mathbf{\Pi}$ was initialized to identity). We found that the method consistently converged to the same solution regardless of the initialization. Finally, after convergence of the alternating minimization, the resulting $\mathbf{C}$ was refined using the method described in Sect. 10.2.2.

The robustness of the method is demonstrated in Figs. 10.3, 10.4, and 10.5; correct correspondences are computed even when the shapes undergo non-isometric deformations and are contaminated by geometric or topological noise. In Fig. 10.6, we used around 45 WKS features detected on two SCAPE shapes, to demonstrate that our method works equally well with point features. Observe how robust permuted sparse coding detects and ignores features without matches, and note the

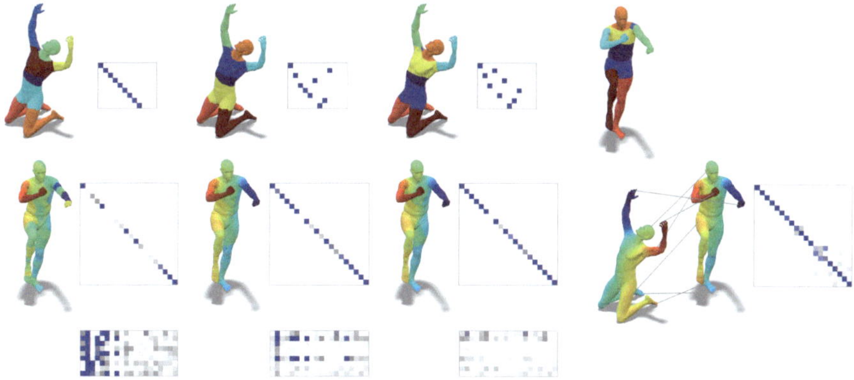

Fig. 10.2 Outer iterations of robust permuted sparse coding alternating the solution of the sparse representation purusit problem (10.18) with the linear assignment problem (10.19). Three iterations, shown left-to-right, are required to achieve convergence. Depicted are the permutation matrix $\mathbf{\Pi}$ (*first row*), the correspondence matrix $\mathbf{C}$ (*second row*), and the outlier matrix $\mathbf{O}$ (*last row*). The resulting point-to-point correspondence and the correspondence matrix $\mathbf{C}$ refined using the ICP as described in Sect. 10.2.2 are shown in the *rightmost column*

Fig. 10.3 Dense point-to-point correspondences obtained between the left TOSCA human shape and its approximate isometries. Corresponding points are marked with consistent colors. The average correspondence distortion is depicted in units of the reference shape diameter. The highest distortions are obtained on the non-isometric joints, but do not exceed 6 % of the diameter

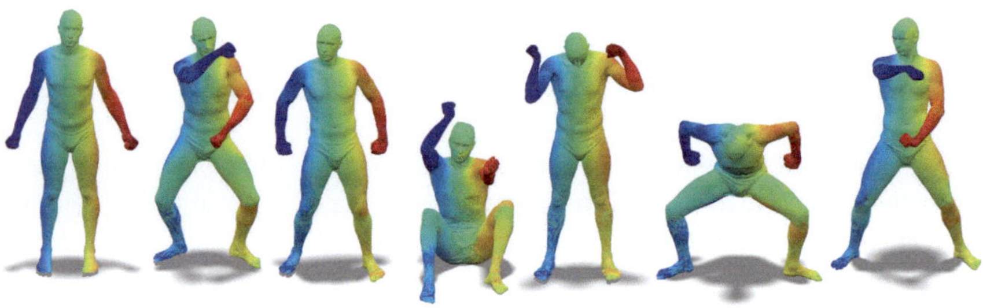

Fig. 10.4 Dense point-to-point correspondences obtained between the left SCAPE human shape and various other poses. Corresponding points are marked with consistent colors

Fig. 10.5 *First row*: point-to-point correspondences obtained between different non-isometric shapes: male and female (*left*); two strongly non-isometric deformations of the dog shape from the TOSCA set (*middle*); TOSCA and SCAPE human shapes (*right*). *Second row*: Point-to-point correspondences obtained between SHREC shapes undergoing nearly isometric deformations and (from *left* to *right*) spike noise, Gaussian noise, and topological noise in the form of large and small holes

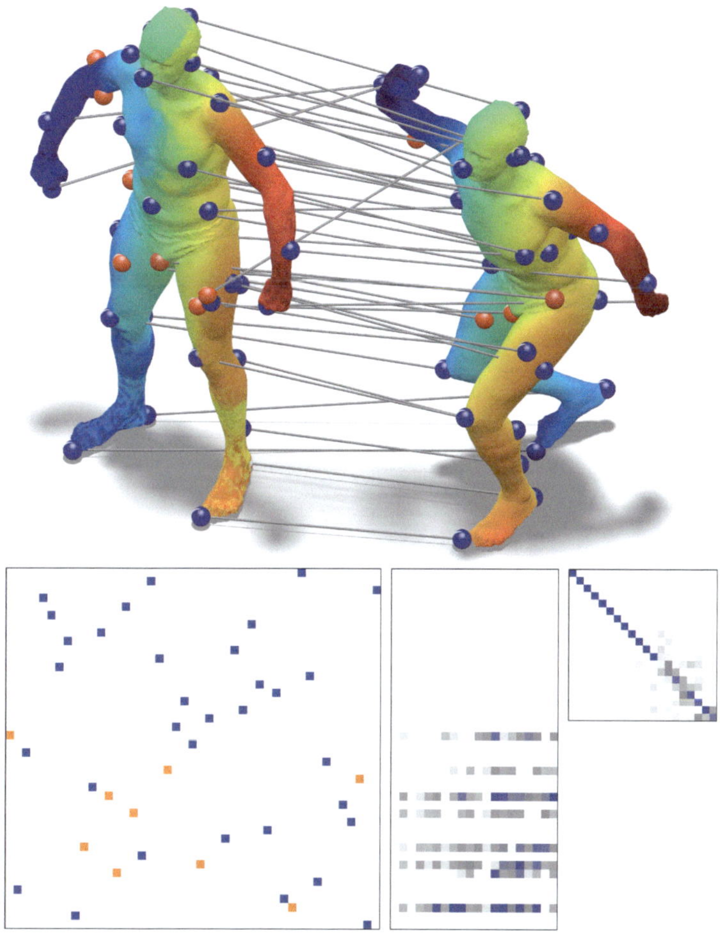

Fig. 10.6 *Top*: dense point-to-point correspondences obtained between two SCAPE human shapes using 45 and 43 WKS point features (rejected features are marked in *red*). Corresponding points are marked with consistent colors. *Bottom*, *left*-to-*right*: recovered permutation matrix $\mathbf{\Pi}$ (rejected matches are marked in *red*); outlier matrix $\mathbf{O}$; and correspondence matrix $\mathbf{C}$

effect of such outliers on the matrices $\mathbf{\Pi}$ and $\mathbf{O}$. Figure 10.7 shows a quantitative evaluation and comparison of our algorithm to other correspondence algorithms on the SCAPE data set. The evaluation was performed using the code and data from [18]. Comparison to [26] was performed in two settings: In the first setting, $k = 20$ basis functions were used with indicator functions of the detected stable regions (about ten regions per shape). In the second setting, $k = 100$ harmonics were used, and 200 wave kernel maps were automatically generated for each region, following verbatim [26]. Our method outperforms existing methods while using less information. Finally, Fig. 10.8 shows the failure of our approach for very non-isometric shapes.

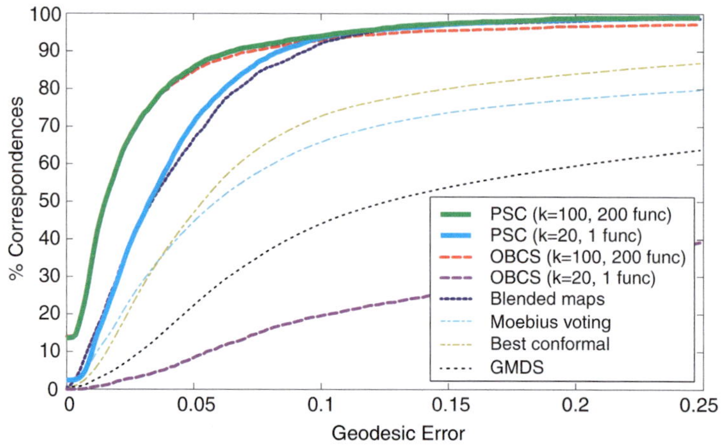

Fig. 10.7 Quantitative evaluation of the proposed permuted sparse coding (PSC) shape correspondence algorithm and its comparison to other correspondence algorithms on the SCAPE shapes using the evaluation protocol from [18]. Compared are Ovsjanikov et el. original method [26] (OBSC), and blended maps [18]

Fig. 10.8 Dense point-to-point correspondences obtained between the left TOSCA human shape and various other non-isometric shapes. The approach fails for significantly non-isometric shapes due to deviation from the diagonal form of **C**

The code used in the experiments was implemented in Matalb with parts written in C. The approximate nearest neighbor search in the ICP refinement step was accelerated using the FLANN library. The experiments were run on a 2.4 GHz Intel Xeon CPU. End-to-end execution time varied from 10 to 50 s, with the detailed breakdown summarized in Table 10.1.

Table 10.1 Average runtime (in seconds) as a function of the shape size for different stages in the proposed method: Basis – harmonic basis computation; MSER – region detection; Opt. – alternating minimization procedure; Ref. – ICP-based refinement and point-to-point correspondence computation; Tot. – total runtime

Vertices	Basis	MSER	Opt.	Ref.	Tot.
5 K	0.53	0.61	7.80	1.41	**10.35**
10 K	0.99	1.32	7.91	2.70	**12.92**
20 K	2.03	3.58	7.91	5.52	**19.04**
50 K	5.57	14.23	7.85	13.99	**41.64**

10.6 Discussion and Conclusion

In this paper, we posed the problem of finding intrinsic correspondence between near-isometric deformable shapes as a problem of sparse modeling. Given only two sets of regions in the two shapes with unknown correspondence, we are able to infer a dense correspondence between the shapes from two assumptions: that at least some of the regions in the two sets are corresponding; and that the shapes are nearly-isometric. The latter assumption implies that in functional representation in harmonic bases the unknown correspondence between the shapes is modeled as a sparse nearly-diagonal matrix; the former assumption implies that there exists an unknown permutation that reorders the regions in corresponding order. To find both the permutation and the correspondence, we formulate the novel permuted sparse coding problem and propose its efficient solution. An additional sparse coding term addressing outliers is added to the model for handling partial matching and formulated as the robust permuted sparse coding.

To the best of our knowledge, among other dense correspondence techniques, our method relies on the smallest amount of information (the ability to find some repeatable regions) and quite generic assumption (near-isometric shapes). In particular, it allows us to use only a region detector without a feature descriptor to find a high-quality correspondence between two shapes.

We note that, as in [26], we explicitly assume that the shapes are nearly isometric, and that their Laplacians have simple spectrum. This assumption assures that the Laplacian eigenbases $\mathbf{\Phi}$ and $\mathbf{\Psi}$ have a compatible behavior, and as a result $\mathbf{C}$ has a nearly-diagonal structure. If we try to relax the restriction on multiplicity, $\mathbf{C}$ will still be sparse, but with unknown sparse structure. We can still use our problem in this setting, imposing a different sparsity constraint on $\mathbf{C}$.

Relaxing the assumptions even more, we can depart from the near-isometric model, e.g. considering applications where one wishes to match shapes with roughly similar geometry but very different details (such as a horse and an elephant). In such a generic setting, the Laplacian eigenbases may differ dramatically, and thus $\mathbf{C}$ have a non-sparse structure. It is possible to incorporate the bases $\mathbf{\Phi}$ and $\mathbf{\Psi}$ as variables into our problem, and in addition to finding the permutation $\mathbf{\Pi}$ and correspondence $\mathbf{C}$ find also the bases in which $\mathbf{C}$ will have a diagonal structure. This problem is akin to dictionary learning used in the sparse modeling literature. In future research, we will study such a generalization of our framework in the hope to find correspondences between non-isometric shapes. Another possible generalization of our problem is for finding correspondence between collections of shapes [18, 25].

It is also worthwhile noting that the novel structured sparse modeling techniques introduced in [32] provide an alternative to the optimization-based pursuit by replacing the iterative proximal algorithm with a learned fixed-complexity feed-forward network. Approaching shape correspondence as a learning problem from this perspective seems a very attractive future research direction.

Finally, being purely intrinsic, the described correspondence computation algorithms are agnostic to intrinsic symmetries [29], i.e., automorphisms that do not affect the manifold metric. Incorporating extrinsic information such as the direction of the normal to the surface, or adding knowingly corresponding *seed* points [1] can resolve these ambiguities. We leave these issues for future research.

Acknowledgements Work partially supported by GIF, ISF, and BSF. M.B. is supported by the ERC Starting grant No. 307047. A.B. is supported by the ERC Starting Grant No. 335491.

References

1. Aflalo, Y., Bronstein, A., Kimmel, R.: On convex relaxation of graph isomorphism. Proc. Nat. Acad. Sci. **112**(10), 2942–2947 (2015)
2. Anguelov, D., Srinivasan, P., Koller, D., Thrun, S., Rodgers, J., Davis, J.: Scape: shape completion and animation of people. In: Proceedings of the SIGGRAPH Conference, Los Angeles (2005)
3. Aubry, M., Schlickewei, U., Cremers, D.: The wave kernel signature: a quantum mechanical approach to shape analysis. In: Proceeding of Workshop on Dynamic Shape Capture and Analysis, Barcelona (2011)
4. Beck, A., Teboulle, M.: A fast iterative shrinkage-thresholding algorithm for linear inverse problems. SIAM J. Img. Sci. **2**, 183–202 (2009)
5. Besl, P.J., McKay, N.D.: A method for registration of 3D shapes. Trans. PAMI **14**, 239–256 (1992)
6. Bronstein, A.M., Bronstein, M.M., Kimmel, R.: Generalized multidimensional scaling: a framework for isometry-invariant partial surface matching. PNAS **103**(5), 1168–1172 (2006)
7. Bronstein, A.M., Bronstein, M.M., Kimmel, R.: Numerical Geometry of Non-rigid Shapes. Springer, New York (2008)
8. Bronstein, A.M., Bronstein, M.M., Kimmel, R., Mahmoudi, M., Sapiro, G.: A Gromov-Hausdorff framework with diffusion geometry for topologically-robust non-rigid shape matching. IJCV **89**(2–3), 266–286 (2010)
9. Bronstein, M.M., Bustos, B., Darom, T., Horaud, R., Hotz, I., Keller, Y., Keustermans, J., Kovnatsky, A., Litman, R., Reininghaus, J., Sipiran, I., Smeets, D., Suetens, P., Vandermeulen, D., Zaharescu, A., Zobel, V., Boyer, E., Bronstein, A.M.: Shrec 2011: robust feature detection and description benchmark. In: EUROGRAPHICS Workshop on 3D Object Retrieval (3DOR), Llandudno (2011)
10. Chen, Y., Medioni, G.: Object modeling by registration of multiple range images. In: Proceeding of Conference on Robotics and Automation, Sacramento (1991)

11. Digne, J., Morel, J.M., Audfray, N., Mehdi-Souzani, C.: The level set tree on meshes. In: Proceeding 3DPVT, Paris (2010)
12. Elad, M.: Sparse and redundant representations: from theory to applications in signal and image processing. Springer, New York (2010)
13. Elad, A., Kimmel, R.: Bending invariant representations for surfaces. In: Proceedings of CVPR, Colorado, pp. 168–174 (2001)
14. Gebal, K., Bærentzen, J.A., Aanæs, H., Larsen, R.: Shape analysis using the auto diffusion function. Comput. Graph. Forum **28**(5), 1405–1413 (2009)
15. Golovinskiy, A., Funkhouser, T.: Consistent segmentation of 3d models. Comput. Graph. **33**(3), 262–269 (2009)
16. Huang, Q., Koltun, V., Guibas, L.: Joint shape segmentation with linear programming. TOG **30**, 125 (2011)
17. Kaick, O.V., Zhang, H., Hamarneh, G., Cohen-Or, D.: A survey on shape correspondence. Comput. Graph. Forum **20**, 1–23 (2010)
18. Kim, V.G., Lipman, Y., Funkhouser, T.: Blended intrinsic maps. TOG **30**(4), 79 (2011)
19. Kuhn, H.W.: The Hungarian method for the assignment problem. Nav. Res. Logist. Quart. **2**, 83–97 (1955)
20. Lipman, Y., Funkhouser, T.: Mobius voting for surface correspondence. ACM Trans. Graph. (Proc. SIGGRAPH) **28**(3), 72 (2009)
21. Litman, R., Bronstein, A.M., Bronstein, M.M.: Diffusion-geometric maximally stable component detection in deformable shapes. Comput. Graph. **35**(3), 549–560 (2011)
22. Mateus, D., Horaud, R., Knossow, D., Cuzzolin, F., Boyer, E.: Articulated shape matching using Laplacian eigenfunctions and unsupervised point registration. In: Proceeding CVPR, Anchorage (2008)
23. Memoli, F., Sapiro, G.: A theoretical and computational framework for isometry invariant recognition of point cloud data. Found. Comput. Math. **5**(3), 313–347 (2005)
24. Nesterov, Y.: Gradient methods for minimizing composite objective function. In: CORE Discussion Paper 2007/76, Center for Operations Research and Econometrics (CORE). Catholic University of Louvain, Louvain-la-Neuve (2007)
25. Nguyen, A., Ben-Chen, M., Welnicka, K., Ye, Y., Guibas, L.: An optimization approach to improving collections of shape maps. Comput. Graph. Forum **30**, 1481–1491 (2011)
26. Ovsjanikov, M., Ben-Chen, M., Solomon, J., Butscher, A., Guibas, L.: Functional maps: a flexible representation of maps between shapes. TOG **31**(4), 129–139 (2012)
27. Ovsjanikov, M., Mérigot, Q., Mémoli, F., Guibas, L.: One point isometric matching with the heat kernel. Comput. Graph. Forum **29**, 1555–1564 (2010)
28. Pokrass, J., Bronstein, A.M., Bronstein, M.M.: A correspondence-less approach to matching of deformable shapes. In: Proceeding SSVM, Ein-Gedi (2011)
29. Raviv, D., Bronstein, A.M., Bronstein, M.M., Kimmel, R.: Symmetries of non-rigid shapes. In: Proceeding of Workshop on Non-rigid Registration and Tracking Through Learning (NRTL), Stony Brook (2005)
30. Rustamov, R.M.: Laplace-Beltrami eigenfunctions for deformation invariant shape representation. In: Proceeding of SGP, Barcelona, pp. 225–233 (2007)
31. Sahillioglu, Y., Yemez, Y.: Coarse-to-fine combinatorial matching for dense isometric shape correspondence. Comput. Graph. Forum **32**, 177–189 (2012)
32. Sprechmann, P., Bronstein, A.M., Sapiro, G.: Learning efficient structured sparse models. In: Proceedings of ICML, Edinburgh (2012)
33. Sun, J., Ovsjanikov, M., Guibas, L.J.: A concise and provably informative multi-scale signature based on heat diffusion. In: Proceedings of SGP, Berlin (2009)
34. Tevs, A., Berner, A., Wand, M., Ihrke, I., Seidel, H.P.: Intrinsic shape matching by planned landmark sampling. Comput. Graph. Forum **30**, 543–552 (2011)
35. Tibshirani, R.: Regression shrinkage and selection via the LASSO. J. R. Stat. Soc. Ser. B **58**(1), 267–288 (1996)

36. Van Kaick, O., Tagliasacchi, A., Sidi, O., Zhang, H., Cohen, D.-Or, Wolf, L., Hamarneh, G.: Prior knowledge for part correspondence. Comput. Graph. Forum **30**, 553–562 (2011)
37. Zaharescu, A., Boyer, E., Varanasi, K., Horaud, R.: Surface feature detection and description with applications to mesh matching. In: Proceedings of CVPR, Miami (2009)
38. Zeng, Y., Wang, C., Wang, Y., Gu, X., Samaras, D., Paragios, N.: Dense non-rigid surface registration using high-order graph matching. In: Proceedings of CVPR, San Francisco (2010)

Chapter 11
Applying Random Forests to the Problem of Dense Non-rigid Shape Correspondence

Matthias Vestner, Emanuele Rodolà, Thomas Windheuser, Samuel Rota Bulò, and Daniel Cremers

Abstract We introduce a novel dense shape matching method for deformable, three-dimensional shapes. Differently from most existing techniques, our approach is general in that it allows the shapes to undergo deformations that are far from being isometric. We do this in a supervised learning framework which makes use of training data as represented by a small set of example shapes. From this set, we learn an implicit representation of a shape descriptor capturing the variability of the deformations in the given class. The learning paradigm we choose for this task is a random forest classifier. With the additional help of a spatial regularizer, the proposed method achieves significant improvements over the baseline approach and obtains state-of-the-art results while keeping a low computational cost.

11.1 Introduction

Matching three-dimensional shapes is a pervasive problem in computer vision, computer graphics and several other fields. Nevertheless, while the advances made by works such as [2, 4, 10, 14, 23, 29] have been dramatic, the problem is far from being solved.

Many methods in shape matching use a notion of similarity that is defined on a very general set of possible shapes. Due to the highly ill-posed nature of the shape matching problem, it is very unlikely that a general method will reliably find good matchings between arbitrary shapes. In fact, while many matching methods (such as methods based on metric distortion [4, 20, 22] and eigen-decomposition of the Laplacian [2, 23, 29]) mostly capture near-isometric deformations, others might consider too general deformations which are not consistent with the human

---

M. Vestner (✉) • E. Rodolà • T. Windheuser • D. Cremers
Technische Universität München, Munich, Germany
e-mail: matthias.vestner@in.tum.de

S.R. Bulò
Fondazione Bruno Kessler, Trento, Italy

M. Breuß et al. (eds.), *Perspectives in Shape Analysis*, Mathematics and Visualization, DOI 10.1007/978-3-319-24726-7_11

intuition of correspondence. In applications where the class of encountered shapes is in-between, adapting the matching methods at hand is often very tedious.

In this paper we try to bridge the gap between general shape matching methods and application-specific algorithms by taking a learning-by-examples approach.

In our scenario, we assume to have a set of training shapes which are equivalent up to some class of non-isometric deformations. Our goal is to learn from these examples how to match two shapes falling in the equivalence class represented by the training set. To this end, we treat the shape matching problem as a classification problem, where input samples are points on the shape manifold and the output class is an element of a canonical label set, which might e.g. coincide with the manifold of one of the shapes in the training set. A first contribution of this paper consists in a new random forest classifier, which can tackle this unconventional classification problem in an efficient and effective way, starting from a general parametrizable shape descriptor. Our classifier is designed in a way to randomly explore the descriptor's parametrization space and find the most discriminative features that properly recover the transformation map characterizing the shape category at hand. In this work, we consider the *wave kernel signature* (WKS) [2] as the shape descriptor. This descriptor is known to be invariant to isometric transformations, but the forest can effectively exploit it to match shapes that undergo non-rigid and non-isometric deformations.

In some sense, the output of the random forest can be seen as a new descriptor by itself that is tuned to the shapes and deformations appearing in the training set. In this respect, the proposed method is complementary to existing shape descriptors as it can improve the performance of a given descriptor [11, 12, 32]. Early attempts to apply machine learning techniques to the problem of non-rigid correspondence [25, 28] consider shapes represented by signed distance functions. We follow the intrinsic view point, considering shapes given by their boundary surface, seen as a Riemannian manifold.

One of the main benefits of our approach is the fact that the random forest classifier gives for each point on the shape an ordered set of matching candidates, hence delivering a dense point-to-point matching. Since such a descriptor does not include any spatial regularity, we propose to use a regularization technique along the lines of the *functional maps framework* [16]. We experimentally validate that the proposed learning approach improves the underlying general descriptor significantly, and it performs better than other state-of-the-art matching algorithms on equivalent benchmarks.

An earlier version of this work was published in [21].

11.1.1 Intrinsic Point Descriptors

We consider 3D shapes that are represented by their boundary surface, a two-dimensional Riemannian manifold (M, g) without boundary. A point descriptor is a function ϕ that assigns to each point on the surface an element of a metric space

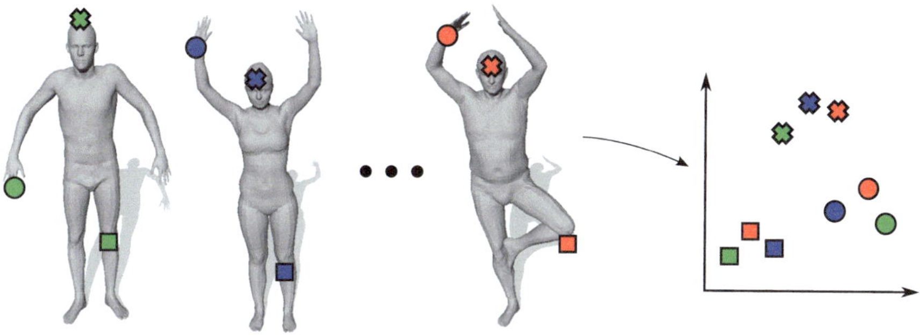

Fig. 11.1 A good point descriptor should at the same time assign similar values to corresponding points on deformed shapes and dissimilar values to non-corresponding points

D, the descriptor space. A good point descriptor should satisfy two competing properties (Fig. 11.1):

- **deformation-invariance:** it should assign similar values to corresponding points on deformed shapes
- **discriminativity:** it should well distinguish non-corresponding points

While it is in principle possible to construct a descriptor that is invariant under an arbitrary large class of deformations (e.g. the constant function), it is evident that there will always be a tradeoff between deformation-invariance and discriminativity.

The descriptors we consider are based on the spectrum of the *Laplace-Beltrami operator* $\Delta_M = -\mathrm{div}_M(\nabla_M)$. Being a symmetric operator the spectrum of Δ_M consists of real eigenvalues $\lambda_1, \lambda_2, \ldots$ and the corresponding eigenfunctions $\gamma_1, \gamma_2, \ldots$ can be chosen to be real valued and orthonormal. Moreover, Δ_M is a non-negative operator with a one-dimensional kernel and a compact pseudo-inverse, so we can order the eigenvalues $0 = \lambda_1 < \lambda_2 \leq \ldots$ and assign to each point $x \in M$ a vector $p \in \mathbb{R}^{2K}$, $p = (\lambda_1, \ldots, \lambda_K, \gamma_1(x), \ldots, \gamma_K(x))$. The Laplace Beltrami Operator is purely intrinsic as it is uniquely determined by the metric tensor $g = (g_{ij})_{i,j=1}^2$ (respectively its inverse $(g^{ij})_{i,j=1}^2$):

$$\Delta_M = \frac{1}{\sqrt{\det g}} \sum_{i,j=1}^{2} \frac{\partial}{\partial x_i} \left(g^{ij} \sqrt{\det g} \frac{\partial}{\partial x_j} \right). \tag{11.1}$$

As a consequence the eigenvalues λ_k as well as the corresponding eigenspaces do not change whenever a shape undergoes an isometric deformation. The eigenbases however are not uniquely determined, even in the case of one dimensional eigenspaces the normalized eigenvectors are only unique up to sign. Nevertheless from the representation p it is possible to construct descriptors that are invariant under isometric deformations. Given a collection $(t_i)_{i=1}^n$ of positive numbers, the

Heat Kernel Signature (HKS)

$$HKS(p) = \left(\sum_k \exp(-\lambda_k t_i) \gamma_k(x)^2 \right)_{i=1}^n \in \mathbb{R}^n \tag{11.2}$$

is a n-dimensional intrinsic point-descriptor [29]. From a physical point of view each component tells us how much heat $u(x, t_i)$ remains at point x after time t_i when the initial distribution of heat is a unit heat source at the very same point:

$$\Delta u = u_t \tag{11.3}$$

$$u(0, \cdot) = \delta_x \tag{11.4}$$

Since the class of isometric deformations includes reflections, any intrinsic descriptor will assign identical values to a point and its symmetric counterpart, whenever shapes exhibit bilateral intrinsic symmetries. Using information about the symmetry [18] or making use of extrinsic information as in [27] would overcome this problem.

From a signal processing viewpoint HKS can be seen as a collection of low-pass filters and thus it is not appropriate to localize features, see Fig. 11.2. Motivated by this observation Aubry et al. [2] introduced the Wave Kernel Signature (WKS), a descriptor where the low-pass filters are replaced by band pass filters:

$$WKS(p) = \left(\sum_k f_{(e_i, \sigma_i^2)}(\lambda_k)^2 \gamma_k(x)^2 \right)_{i=1}^n \in \mathbb{R}^n \tag{11.5}$$

Here the parameters (e_i, σ_i^2) correspond to mean and variance of the log-normal energy distributions

$$f_{(e,\sigma^2)}(\lambda) \propto \exp(-\frac{(\log e - \log \lambda)^2}{2\sigma^2}) \tag{11.6}$$

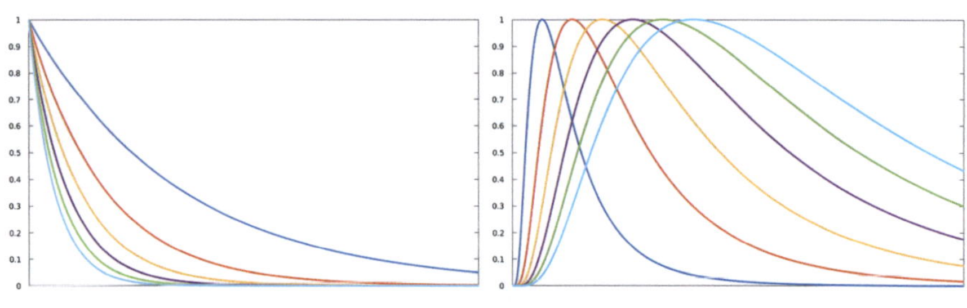

Fig. 11.2 The weighting functions of the heat kernel signature (*left*) can be seen as low-pass filters, the ones of the wave kernel signature (*right*) in contrary behave like band-pass filters

Fig. 11.3 Finding a correspondence between shapes should be feasible even if they are far from being isometric

The authors propose fixed values for the parameters (e_i, σ_i) depending on the truncated spectrum of the Laplace-Beltrami-operator. Moreover they equip the descripor with a metric related to the L^1-distance.

In this work the parameters will be learned from training data, a distance function between vector valued descriptors is unneeded since descriptors are compared component wise in a hierarchical manner (Sects. 11.2.1.1 and 11.2.1.3).

Both, HKS and WKS, are invariant under isometric deformations. However the human notion of similarity by far exceeds the class of isometries. Asking for a correspondence between an adult and a child or even an animal like a gorilla is a feasible task for us. Figure 11.3 shows examples of shapes taken from different datasets [1, 5, 19, 21] that could in principle be put into correspondence. By choosing application dependent parameters one can achieve descriptors that are less sensitive to the type of deformation one is interested in. In this work we implicitly determine optimal parameters when the deformation class is represented by a set of training shapes with known ground truth correspondence.

11.1.2 Discretized Surfaces and Operators

In practice the shapes are given as triangular meshes $M = (V_M, F_M)$. We will henceforth identify a shape M by the set of it vertices V_M. A one-to-one correspondence between two shapes can then be represented by a permutation matrix, a fuzzy correspondence, i.e. a function that assigns to each point a probability distribution over the other shape, respectively as a left-stochastic matrix. Functions defined on a shape become vectors and linear operators acting on them, e.g. the Laplace-Beltrami operator can be written as matrices. Inner products between functions are calculated via an area-weighted inner product between the vectors representing them. We chose the popular cotangent scheme [15] as the discretization of the Laplacian.

11.2 Dense Correspondence Using Random-Forests

In this work we treat the shape matching problem as a classification problem, where input samples are points on the shape and the output class is an element of a canonical label set, which might e.g. coincide with one of the shapes in the training set (the *reference shapes*). The classifier we choose is a Random forest, designed in a way to randomly explore the descriptor's parametrization space and find the most discriminative features that properly recover the transformation map characterizing the shape category at hand. In this work, we consider the *wave kernel signature* (WKS) as the parametrizable point descriptor (weak classifier). In general other choices of parametrizable descriptors, e.g. HKS, are possible. As mentioned in Sect. 11.1.1 any classifier based on isometry-invariant point descriptors can not distinguish a point from its symmetric counterpart. Thus the fuzzy outcome of the Random forest classifier has to be regularized in order to get a consistent correspondence.

11.2.1 Learning and Inference Using Random Forests

Random forests [3] are ensembles of decision trees that have become very popular in the computer vision community to solve both classification and regression problems with applications ranging from object detection, tracking and action recognition [9] to semantic image segmentation and categorization [26], and 3D pose estimation [30], to name just a few. The forest classifier is particularly appealing because its trees can be trained efficiently and techniques like bagging and randomized feature selection allow to limit the correlation among trees and thus ensure good generalization. We refer to [7] for a detailed review.

11.2.1.1 Inference

In the context of shape matching, a decision tree comprised by the forest routes a point m of a test shape M from the root of the tree to a leaf node, where a probability distribution defined on a discrete label set L is assigned to the point. The path from the root to a leaf node is determined by means of binary decision functions called *split functions* located at the internal nodes, which given a shape point return L or R depending on whether the point should be forwarded to the left or to the right with respect to the current node. According to this inference procedure, each tree $t \in \mathscr{F}$ of a forest $\mathscr{F}$ provides a posterior probability $\mathrm{P}(\ell|m,t)$ of label $\ell \in L$, given a point $m \in M$ in a test shape M (Fig. 11.4).

This probability measure is the one associated with the leaf of tree $t \in \mathscr{F}$ that the shape point would reach. The prediction of the whole forest $\mathscr{F}$ is finally obtained

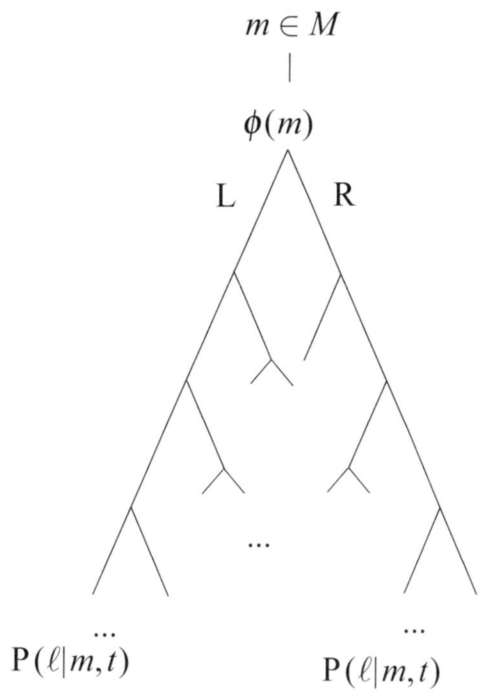

Fig. 11.4 At each inner node of a decision tree a binary split function is evaluated. Depending on the result the point m is either routed to the left or to the right. Leafs of the tree correspond to probability distributions in the label space. A random forest is a collection of mulitple decision trees

by averaging the predictions of the single trees:

$$P(\ell|m, \mathscr{F}) = \frac{1}{|\mathscr{F}|} \sum_{t \in \mathscr{F}} P(\ell|m, t) \; . \tag{11.7}$$

The outcome of the prediction over an entire shape M can be represented as a left-stochastic matrix X_M encoding the probabilistic canonical transformation, where

$$(\mathrm{X}_M)_{ij} = P\left(\ell_i|m_j, \mathscr{F}\right) \tag{11.8}$$

for each $\ell_i \in L$ and $m_j \in M$. Using Bayes' theorem we can further construct a fuzzy correspondence between two previously unseen shapes (i.e. no members of the training set).

11.2.1.2 Learning

During the learning phase, the structure of the trees, the split functions and the leaf posteriors are determined from a training set. Let $\{(R_i, T_i)\}_{i=1}^{\mathsf{m}}$ be a set of m reference shapes R_i each equipped with a canonical transformation, i.e. a bijection $T_i : R_i \to L$ between the vertex set of the reference shape and the label set L. A training set $\mathbb{T}$

for the random forest is given by the union of the graphs of the mappings T_i, i.e.

$$\mathbb{T} = \{(\boldsymbol{r}, T_i(\boldsymbol{r})) \ : \ \boldsymbol{r} \in R_i,\ 1 \leq i \leq \mathsf{m}\} \ . \tag{11.9}$$

The learning phase that creates each tree forming the forest consists in a recursive procedure that starting from the root iteratively splits the actual terminal nodes. During this process each shape point of the training set is routed through the tree in a way to partition the whole training set across the terminal nodes. The decision whether a terminal node has to be further split and how the splitting will take place is purely local as it involves exclusively the shape points that have reached that node. A terminal node typically becomes a leaf of the tree if the depth of the node exceeds a given limit, if the size of the subset of training samples reaching the node is small enough, or if the entropy of the sample's label distribution is low enough. If this is the case, then the leaf node is assigned the label distribution of subset $\mathbb{S}$ of training samples that have reached the leaf, i.e.

$$\mathrm{P}\left(\ell|\mathbb{S}\right) = \frac{|\{(\boldsymbol{r}, \ell) \in \mathbb{S}\}|}{|\mathbb{S}|} \ . \tag{11.10}$$

The probability distribution $\mathrm{P}\left(\cdot|\mathbb{S}\right)$ will become the posterior probability during inference for every shape point reaching the leaf. Consider now the case where the terminal node is split. In this case, we have to select a proper split function $\psi(r) \in \{\mathsf{L}, \mathsf{R}\}$ that will route a point r reaching the node to the left or right branch. An easy and effective strategy for guiding this selection consists in generating a finite pool Ψ of random split functions and retaining the one maximizing the information gain with respect to the label space L. The information gain $\mathrm{IG}\left(\psi\right)$ due to split function $\psi \in \Psi$ is given by the difference between the entropy of the node posterior probability defined as in (11.10) before and after having performed the split. In detail, if $\mathbb{S} \subseteq \mathbb{T}$ is the subset of the training set that has reached the node to be split and $\mathbb{S}^{\mathsf{L}}$, $\mathbb{S}^{\mathsf{R}}$ is the partition of $\mathbb{S}$ induced by the split function ψ then $\mathrm{IG}\left(\psi\right)$ is given by

$$\mathrm{IG}\left(\psi\right) = \mathrm{H}\left(\mathrm{P}\left(\cdot|\mathbb{S}\right)\right) - \mathrm{H}\left(\mathrm{P}\left(\cdot|\mathbb{S}\right)|\psi\right) \ , \tag{11.11}$$

where $\mathrm{H}\left(\cdot\right)$ denotes the entropy and

$$\mathrm{H}\left(\mathrm{P}\left(\cdot|\mathbb{S}\right)|\psi\right) = \frac{|\mathbb{S}^{\mathsf{L}}|}{|\mathbb{S}|}\mathrm{H}\left(\mathrm{P}\left(\cdot|\mathbb{S}^{\mathsf{L}}\right)\right) + \frac{|\mathbb{S}^{\mathsf{R}}|}{|\mathbb{S}|}\mathrm{H}\left(\mathrm{P}\left(\cdot|\mathbb{S}^{\mathsf{R}}\right)\right) \ . \tag{11.12}$$

Intuitively the information gain of a split function is higher, the better it seperates members belonging to different classes (see Fig. 11.5).

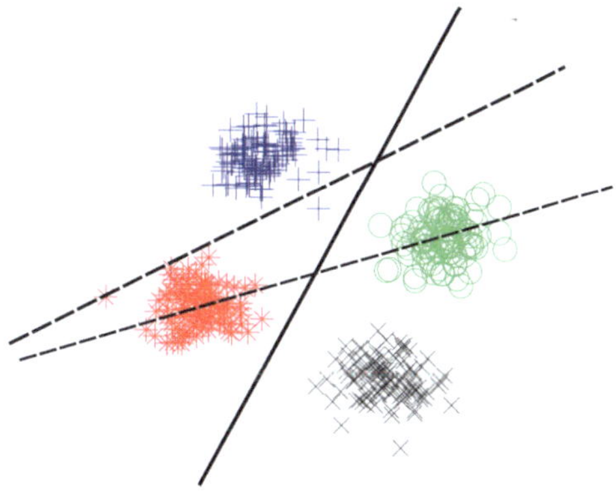

Fig. 11.5 The split function visualized as a *solid line* has the highest information gain (IG) among the three candidates

11.2.1.3 Choice of Decision Functions

During the build up of the forest the randomized training approach allows us to vary the parametrization of the shape descriptor for each point of the shape. In fact, we can in principle let the forest automatically determine the optimal discriminative features of the chosen descriptor for the matching problem at hand. In this work we have chosen the Wave Kernel Signature (WKS) but as mentioned above, in principle any parametrizable feature descriptor (e.g. HKS) can be considered. From a practical perspective, it can be shown [2] that the sum in (11.5) can be restricted to the first $\bar{k} < \infty$ components. We make explicit in (11.5) the dependency on $\bar{k}$ by writing:

$$p(m; e, \bar{k}) = \sum_{k=1}^{\bar{k}} f_e^2(\nu_k)\phi_k^2(m)\,. \tag{11.13}$$

We are now in the position of generating at each node of a tree during the training phase a pool of randomized split functions by sampling an energy level e_i, a number of eigenpairs $\bar{k}_i$ and a threshold τ_i. Accordingly, the split functions will take the form:

$$\psi_i(m) = \begin{cases} \mathsf{L} & \text{if } p(m; e_i, \bar{k}_i) > \tau_i \\ \mathsf{R} & \text{otherwise}\,. \end{cases} \tag{11.14}$$

11.2.2 *Interpretation and Regularization of the Forests Prediction*

The simplest way to infer a correspondence from a forest prediction consists in assigning each point $m \in M$ to the most likely label according to its final distribu-

Fig. 11.6 The coordinate functions from a test shape M (*standing cat*) are transferred to a reference shape R (*walking cat*) via the functional map $T_{X_{M,R}}$ induced by the forest prediction. Most of the ambiguities arise in f_x, and are due to the global intrinsic symmetry of the cat. The *first column* shows the map f_x on the test cat, while the *second* and *third columns* are obtained by mapping f_x without and with regularization respectively. The remaining four columns show the mappings of f_y and f_z *without* regularization. The symmetric ambiguities disappear as a result of the regularization process (columns (*a*)–(*c*), matches encoded by color)

tion, i.e., the label maximizing $\mathrm{P}(\ell|\boldsymbol{m}, \mathscr{F})$. If we are also given a reference shape R from the training set, the maximum a posteriori estimate of ℓ can be transformed into a point-to-point correspondence from M to R via the known bijection $T : R \to L$. Figure 11.6a, b show an example of this approach. The resulting correspondence is exact for about 50 % of the points, whereas it induces a large metric distortion on the rest of the shape. However, this is not a consequence of the particular criterion we adopted when applying the prediction. Indeed, the training process can not distinguish symmetric points and is oblivious to the underlying manifolds as it is only based on pointwise information: the correspondence estimates are taken *independently* for each point and thus the metric structure of the test shape is not taken into account during the regression. Nevertheless, as we shall see, the predicted distributions carry enough information that can be exploited to obtain a consistent matching.

11.2.2.1 Functional Maps

Multiplying X_M (as defined in (11.8)) from the left with the permutation matrix associated to the known bijection $T : L \to R$ between the label space L and a reference shape R gives raise to another left-stochastic matrix $X_{M,R}$. As pointed out in [16] this (fuzzy) correspondence $X_{M,R}$ can be interpreted as a linear map $T_{X_{M,R}} : L^2(M) \to L^2(R)$. In Fig. 11.6 (first 7 columns) we use such a construction to map the coordinate functions $f_i : M \to \mathbb{R}$ (where $i \in \{x, y, z\}$) to scalar functions on R. Specifically, we plot $\boldsymbol{f}_i$ and their reconstructions $\boldsymbol{g}_i = T_{X_{M,R}}\boldsymbol{f}_i$. Note that the reference shape is axis-aligned, so that the x coordinates of its points grow from the right side (blue) to the left side of the model (red).

As in [16] from now on we consider $T_{X_{M,R}}$ in the truncated harmonic bases on the respective shapes and by that dramatically reduce the size of the problem. Since the LB-eigenfunctions are chosen to form orthonormal bases, the norms considered in the following section are invariant under this basis-transform. For simplicity we will still denote the associated matrix by $X_{M,R}$.

11.2.2.2 Metric Distortion Using Functional Maps

The plots we show in Fig. 11.6 tell us that most of the error in the correspondence arises from the (global) intrinsic symmetries of the shape. As mentioned previously, this is to be expected since the training process does not exploit any kind of structural information about the manifolds.

This suggests the possibility to regularize the prediction by introducing metric constraints on the correspondence. Specifically, we consider an objective of the form

$$E(\mathtt{X}) = c(\mathtt{X}_{M,R}, \mathtt{X}) + \rho(\mathtt{X}) \,, \tag{11.15}$$

where $\mathtt{X}$ is a correspondence between shapes M and R. The first term (or *cost*) ensures closeness to the prediction given by the forest, while the second term is a regularizer giving preference to geometrically consistent solutions.

A functional map is assumed to be geometrically consistent if it approximately preserves distance maps. Suppose for the moment we are given a sparse collection of matches $O \subset M \times R$. Then for each $(\boldsymbol{p}, \boldsymbol{q}) \in O$ we can define the two distance maps $d_p : M \to \mathbb{R}$ and $d_q : R \to \mathbb{R}$ as

$$d_p(\boldsymbol{x}) = d_M(\boldsymbol{p}, \boldsymbol{x}) \,, \qquad d_q(\boldsymbol{y}) = d_R(\boldsymbol{q}, \boldsymbol{y}) \,. \tag{11.16}$$

With these definitions, we can express the regularity term $\rho(\mathtt{C})$

$$\rho(\mathtt{C}) = \sum_{(\boldsymbol{p},\boldsymbol{q})\in O} \omega_{pq} \| X_{M,R} \boldsymbol{d}_p - \boldsymbol{d}_q \|_2^2 \,, \tag{11.17}$$

with weights $\omega_{pq} \in [0, 1]$ (Fig. 11.7).

In order for the regularization to work as expected, the provided collection of matches should constrain well the solution, in the sense that it should help to

Fig. 11.7 In the regularization step first a coarse subsampling of the shape is constructed via Euclidean farthest point sampling (dots on the left shape). In the small set of predicted matches O (cross product of dots on the two shapes) a sparse correspondence is obtained using an l^1 constrained optimazation technique. We expect a consistent correspondence to approximately preserve the distance maps d_p

disambiguate the intrinsic symmetries of the shape. For example, matches along the tail of the cat would bring little to no information on what solution to prefer. In practice, we can seek for a few matches that cover the whole shape and be as accurate as possible. To this end, we generate evenly distributed samples $V_{\text{fps}} \subset M$ on the test shape via farthest point sampling [13] by using the extrinsic Euclidean metric. Then, we construct a matching problem restricted to the set of *predicted* matches

$$O = \{(\boldsymbol{m}, \boldsymbol{r}) \in V_{\text{fps}} \times R \mid (\mathrm{X}_{M,R})_{\boldsymbol{rm}} > 0\} . \tag{11.18}$$

In practice this set is expected to be small, since the prediction given by the forest is very sparse and we select around 50 farthest samples per test shape ($\approx$0.2 % of the total number of points on the adopted datasets). This results in a small matching problem that we solve via game-theoretic matching [20], a ℓ_1-regularized technique that allows to obtain sparse, yet very accurate solutions in an efficient manner. Once a sparse set of matches is obtained, we solve (11.15) as the weighted least-squares problem

$$\min_{\mathrm{X}} \|\mathrm{X}_{M,R} - \mathrm{X}\|_F^2 + \sum_{(\boldsymbol{p},\boldsymbol{q}) \in O} \omega_{\boldsymbol{pq}} \|\mathrm{X}\boldsymbol{d}_{\boldsymbol{p}} - \boldsymbol{d}_{\boldsymbol{q}}\|_2^2 , \tag{11.19}$$

where $\omega_{\boldsymbol{pq}} \in [0, 1]$ are weights (provided by the game-theoretic matcher) giving a measure of confidence for each match $(\boldsymbol{p}, \boldsymbol{q}) \in O$. Figure 11.6c shows the result of the regularization performed using 25 sparse matches (indicated by small spheres).

Notice that the distance between functional maps is yet not well understood. The authors of [6] suggest to replace the Frobenius norm in (11.19) with a regularized l^0 norm of the vector of singular values:

$$\|A\|_{\varepsilon} = \sum_i \frac{\sigma(A)_i^2}{\sigma(A)_i^2 + \varepsilon} \tag{11.20}$$

Assuming the shapes to be (nearly) isometric one can expect the Laplace Beltrami operators on the shapes to commute with the functional map, i.e. (in the harmonic bases):

$$X\Lambda_M = \Lambda_R X \tag{11.21}$$

where Λ_M and Λ_R are the diagonal matrices of the singular values. A measure of deviation from (11.21) can be used as an alternative regularity cost.

11.2.3 Experimental Results

In all our experiments we used the WKS as pointwise descriptor for the training process. As in [16], we limited the size of the bases on the shapes to the first 100 eigenfunctions of the Laplace-Beltrami operator, computed using the cotangent scheme [15].

11.2.3.1 Comparison with Dense Methods

In this set of experiments we compare with the state of the art techniques in (dense) non-rigid shape matching, namely the functional maps pipeline [16], blended intrinsic maps (BIM) [10], and the coarse-to-fine combinatorial approach of [24]. We perform these comparisons on the TOSCA high-resolution dataset [5]. The dataset consists of 80 shapes belonging to different classes, with resolutions ranging in 4–52K points. Shapes within the same class have the same connectivity and undergo nearly-isometric deformations. Ground-truth point mapping among shapes from the same class is available. In particular, given a predicted map $f : M \to N$ and the corresponding ground-truth $g : M \to N$, we define the *error* of f as

$$\varepsilon(f, g) = \sum_{m \in M} d_N(f(\boldsymbol{m}), g(\boldsymbol{m})), \tag{11.22}$$

where d_N is the geodesic metric on N, normalized by $\sqrt{Area(N)}$ to allow inter-class comparisons. Similarly, we define the average (pointwise) geodesic error as $\frac{\varepsilon(f,g)}{|M|}$.

Although the methods considered in these experiments do not rely on any prior learning, the comparison is still meaningful as it gives an indication of the level of accuracy that our approach can attain in this class of problems. The experiments were designed on the same benchmark and following a procedure similar to the one reported in [10, 16]. Specifically, for each model M of a class (e.g., the class of dogs), we randomly picked other 6 models from the same class (not including M), and trained a random forest with them (thus, we only considered classes with at least 6 shapes). Then we predicted a dense correspondence for M according to the technique described in Sect. 11.2.2.

We show the results of this experiment in Fig. 11.8 (right). Each curve depicts the percentage of matches that attain an error below the threshold given on the x-axis. Our method (red line) detects 90 % correct correspondences within a geodesic error of 0.05. Almost all correct matches are detected within an error of 0.1. This is compatible with and even improves the results given by the other methods on the same data. Note that our training process only makes use of pointwise information (namely, the WKS); in contrast, the functional maps pipeline (blue line) adopts several heuristics (WKS preservation constraints in addition to orthogonality of $\mathtt{C}$, region-wise features, etc.) in order to constrain the solution optimally [16]. Upon

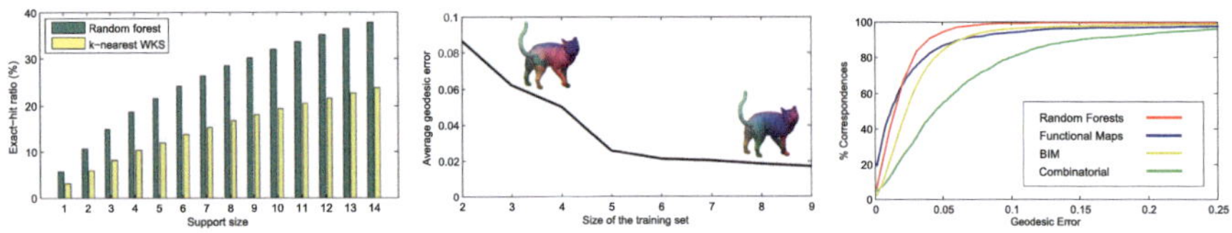

Fig. 11.8 *Left*: Fraction of exact matches predicted by a random forest vs. maximum support size of the probability distributions on a test shape. The forest was trained with 9 shapes. *Middle*: Sensitivity to number of shapes used in the training set. Note how the correspondence predicted using little training data (*top-left* model) is only partially regularized. *Right*: Comparison with the state-of-the-art methods on nearly-isometric shapes (TOSCA). Symmetric correspondences are considered correct solutions for all methods

visual inspection, we observed that most of the errors in our method were due to the poor choice of points made in the regularization step. This is analogous to what is reported for the BIM method [10]. Typically, we observed that around 20 well-distributed points are sufficient to obtain accurate results.

11.2.4 Sensitivity to Training Parameters

We performed a sensitivity analysis of our method with respect to the parameters used in the training process, namely the size of the training set and the number of trees in the forest. In these experiments we employed the cat models from the TOSCA dataset (28K vertices) with the corresponding ground-truth.

In Fig. 11.8 (middle) we plot the average geodesic error obtained by a test shape (depicted along the curve) as we varied the number of shapes in the training set. The geodesic error of the correspondence stabilizes when at least 6 shapes are used for training. This means that only 6 samples per label are sufficient in order to determine an optimal parametrization of the nearly-isometric deformations occurring on the shape. This result contrasts the common setting in which random forests are trained with copious amounts of data [8, 30], making the approach rather practical when only limited training data is available.

Figure 11.8 (left) shows the change in accuracy as we increase the number of trees in the forest. Note that increasing the number of trees directly induces a larger support of the probability distributions over L. In other words, each point of the test shape receives more candidate matches if the forest is trained with more trees (see Eq. (11.7)). The hit ratio in the bar plot is defined as the fraction of *exact* predictions given by the forest over the entire test shape. We compare the results with the hit ratio obtained by looking for k-nearest neighbors in WKS descriptor space, with k equal to the maximum support size employed by the forest at each level. From this plot we see that the forest predictions are twice as accurate as WKS predictions for equal support sizes. In particular, random forest predicts the *exact* match for almost half (around 14K points) of the shape when trained with 15 trees.

Fig. 11.9 Comparison between our method and an approach based on WKS affinity using shapes from the dataset of Vlasic et al. Columns one to four show the predicted and regularized solutions for both approaches. The last three columns show how the indicator function at one point gets functionally mapped to a second shape, by using the (non-regularized) X obtained from the forest, and by X_{WKS}

Finally, in Fig. 11.9 we show a qualitative comparison between our method and an approach based on WKS. The rationale of this experiment is to show that the prediction given by the forest gives better results than what can be obtained without prior learning within the same pipeline (i.e., prediction followed by regularization). Specifically, for each point in one shape we construct a probability distribution on the other shape based on a measure of descriptor affinity in WKS space. We then estimated a functional map $\mathrm{C}_{\mathrm{WKS}}$ from the resulting set of constraints, and plotted a final correspondence before and after regularization.

11.2.5 Learning Non-isometric Deformations

In this section we consider a scenario in which the shapes to be matched may undergo more general (i.e., far from isometric) deformations. Examples of such deformations include local and global changes in scale, topological changes, resampling, partiality, and so forth. Until now, few methods have attempted to tackle this class of problems. Most dense approaches [10, 16, 17, 24] are well-defined in the quasi-isometric and conformal cases only; instances of inter-class matching were considered in [10], but the success of the method depends on the specific choice of (usually hand-picked) feature points used in the subsequent optimization. Sparse methods considering the general setting from a metric perspective [4, 20, 22] attempt to formalize the problem by using the language of quadratic optimization, leading to difficult and highly non-convex formulations. An exception to the general trend was given in [31], where the matching is formulated as a linear program in the product space of manifolds. The method allows to obtain dense correspondences for more general deformations, but it assumes consistent topologies and is computationally expensive ($\sim$2 h to match around 10K vertices). Another recent approach [11] attempts to model deviation from isometry in the framework of functional maps, by seeking compatible harmonic bases among two shapes. However, it relies on a (sparse) set of matches being given as input and it shares with [31] the high computational cost.

Fig. 11.10 Example of dense shape matching using random forests under non-isometric deformations. Shapes in the shaded area are a subset of the training set. The forest is trained with wave kernel descriptors and consists of 80K training classes with 19 samples per class. Matches are encoded by color

As described in Sect. 11.2, the forest does not contain any explicit knowledge of the type of deformations it is asked to parametrize. This means that, in principle, one could feed the learning process with training data coming from any collection of shapes, with virtually no restrictions on the transformations that the shapes are allowed to undergo. Clearly, an appropriate choice of the pointwise descriptor should be made in order for the forest to provide a concise and discriminative model. To test this scenario, we constructed a synthetic dataset consisting of 8 high-resolution (80K vertices) models of a kid under different poses (quasi-isometries), and 11 additional models of increasingly corpulent variants of the same kid (local scale deformations) with a fixed pose (see Fig. 11.10). The shapes have equal number of points and point-to-point ground-truth is available. We test the trained random forest with a plump kid having a previously unseen pose.

Note that the result is reasonably accurate if we keep in mind the noisy setting: the forest was trained with WKS descriptors, which are originally designed for quasi-isometric deformations, and thus not expected to work well in the more general setting [12]. Despite being just a qualitative evaluation, this experiment demonstrates the generality of our approach. The matching process we described can still be employed in general non-rigid scenarios if provided with limited, yet sufficiently discriminative training data.

11.3 Conclusions

In this article, we showed how the random forest learning paradigm can be employed for problems of dense correspondence among deformable 3D shapes. To our knowledge, this is among the first attempts at introducing a statistical learning view on this family of problems. The effectiveness of our approach is demonstrated on a standard benchmark, where we obtain comparable results with respect to the state of the art, and very low prediction times for shapes with tens of thousands of vertices. The approach is flexible in that it provides a means to model deformations which are far from isometric, and it consistently obtains high predictive performance on all tested scenarios.

References

1. Anguelov, D., Srinivasan, P., Koller, D., Thrun, S., Rodgers, J., Davis, J.: Scape: shape completion and animation of people. In: ACM Transactions on Graphics (TOG), vol. 24, pp. 408–416. ACM, New York (2005)
2. Aubry, M., Schlickewei, U., Cremers, D.: The wave kernel signature: a quantum mechanical approach to shape analysis. In: ICCV Workshops, Barcelona (2011)
3. Breiman, L.: Random forests. In: Machine Learning, vol. 45. Springer, Berlin/Heidelberg (2001)
4. Bronstein, A.M., Bronstein, M.M., Kimmel, R.: Generalized multidimensional scaling: a framework for isometry-invariant partial surface matching. PNAS **103**(5), 1168–1172 (2006)
5. Bronstein, A.M., Bronstein, M.M., Kimmel, R.: Numerical Geometry of Non-Rigid Shapes. Springer, New York (2008). Incorporated, 1 edition
6. Corman, É., Ovsjanikov, M., Chambolle, A.: Supervised descriptor learning for non-rigid shape matching. In: European Conference on Computer Vision, pp 283–298 (2014)
7. Criminisi, A., Shotton, J., Konukoglu, E.: Decision Forests: A Unified Framework for Classification, Regression, Density Estimation, Manifold Learning and Semi-supervised Learning. Foundations and Trends in Computer Graphics and Vision. Now, Boston (2012)
8. Fanelli, G., Gall, J., Van Gool, L.: Real time head pose estimation with random regression forests. In: CVPR, Colorado Springs (2011)
9. Gall, J., Yao, A., Razavi, N., Van Gool, L., Lempitsky, V.: Hough forests for object detection, tracking, and action recognition. PAMI **33**(11), 2188–2202 (2011)
10. Kim, V.G., Lipman, Y., Funkhouser, T.: Blended intrinsic maps. In: SIGGRAPH 2011, Vancouver (2011)
11. Kovnatsky, A., Bronstein, M.M., Bronstein, A.M., Glashoff, K., Kimmel, R.: Coupled quasi-harmonic bases. Comput. Graph. Forum **32**(2pt4), 439–448 (2013)
12. Litman, R., Bronstein, A.M.: Learning spectral descriptors for deformable shape correspondence. TPAMI **36**(1), 170–180 (2013)
13. Mémoli, F.: Gromov-Wasserstein distances and the metric approach to object matching. Found. Comput. Math. **11**, 417–487 (2011)
14. Mémoli, F., Sapiro, G.: A theoretical and computational framework for isometry invariant recognition of point cloud data. Found. Comput. Math. **5**(3), 313–347 (2005)
15. Meyer, M., Desbrun, M., Schröder, P., Barr, A.H.: Discrete differential-geometry operators for triangulated 2-manifolds. In: Proceedings of VisMath, Berlin (2002)
16. Ovsjanikov, M., Ben-Chen, M., Solomon, J., Butscher, A., Guibas, L.: Functional maps: a flexible representation of maps between shapes. ACM Trans. Graph. **31**(4), 30:1–30:11 (2012)
17. Ovsjanikov, M., Mérigot, Q., Mémoli, F., Guibas, L.: One point isometric matching with the heat kernel. Comput. Graph. Forum **29**(5), 1555–1564 (2010)
18. Ovsjanikov, M., Sun, J., Guibas, L.: Global intrinsic symmetries of shapes. In: Computer Graphics Forum, vol. 27, pp. 1341–1348. Wiley Online Library (2008)
19. Pickup, D., Sun, X., Rosin, P., Martin, R., Cheng, Z., Lian, Z., Aono, M., Ben Hamza, A., Bronstein, A., Bronstein, M., et al.: Shrec14 track: shape retrieval of non-rigid 3d human models. 3DOR **4**(7), 8 (2014)

20. Rodolà, E., Bronstein, A.M., Albarelli, A., Bergamasco, F., Torsello, A.: A game-theoretic approach to deformable shape matching. In: CVPR, Providence (2012)
21. Rodolá, E., Rota Bulo, S., Windheuser, T., Vestner, M., Cremers, D.: Dense non-rigid shape correspondence using random forests. In: CVPR, Columbus, pp. 4177–4184 (2014)
22. Rodolà, E., Torsello, A., Harada, T., Kuniyoshi, T., Cremers, D.: Elastic net constraints for shape matching. In: ICCV, Sydney (2013)
23. Rustamov, R.M.: Laplace-beltrami eigenfunctions for deformation invariant shape representation. In: SGP, Barcelona. Eurographics Association (2007)
24. Sahillioğlu, Y., Yemez, Y.: Coarse-to-fine combinatorial matching for dense isometric shape correspondence. Comput. Graph. Forum **30**(5), 1461–1470 (2011)
25. Schölkopf, B., Steinke, F., Blanz, V.: Object correspondence as a machine learning problem. In: ICML, Bonn, pp. 776–783. ACM (2005)
26. Shotton, J., Johnson, M., Cipolla, R.: Semantic texton forests for image categorization and segmentation. In: CVPR, Anchorage (2008)
27. Shtern, A., Kimmel, R.: Matching lbo eigenspace of non-rigid shapes via high order statistics. arXiv preprint arXiv:1310.4459 (2013)
28. Steinke, F., Blanz, V., Schölkopf, B.: Learning dense 3d correspondence. In: Advances in Neural Information Processing Systems, pp. 1313–1320. MIT Press, Cambridge/London (2006)
29. Sun, J., Ovsjanikov, M., Guibas, L.: A concise and provably informative multi-scale signature based on heat diffusion. In: SGP, Berlin. Eurographics Association (2009)
30. Taylor, J., Shotton, J., Sharp, T., Fitzgibbon, A.: The vitruvian manifold: inferring dense correspondences for one-shot human pose estimation. In: CVPR, Pittsburgh (2012)
31. Windheuser, T., Schlickewei, U., Schmidt, F.R., Cremers, D.: Geometrically consistent elastic matching of 3d shapes: a linear programming solution. In: ICCV, Barcelona (2011)
32. Windheuser, T., Vestner, M., Rodola, E., Triebel, R., Cremers, D.: Optimal intrinsic descriptors for non-rigid shape analysis. In: BMVC, Nottingham. BMVA Press (2014)

Chapter 12
Accelerating Deformable Part Models with Branch-and-Bound

Iasonas Kokkinos

Abstract Deformable Part Models (DPMs) play a prominent role in current object recognition research, as they rigorously model the shape variability of an object category by breaking an object into parts and modelling the relative locations of the parts. Still, inference with such models requires solving a combinatorial optimization task. In this chapter, we will see how Branch-and-Bound can be used to efficiently perform inference with such models. Instead of evaluating the classifier score exhaustively for all part locations and scales, such techniques allow us to quickly focus on promising image locations. The core problem that we will address is how to compute bounds that accommodate part deformations; this allows us to apply Branch-and-Bound to our problem. When comparing to a baseline DPM implementation, we obtain exactly the same results but can perform the part combination substantially faster, yielding up to tenfold speedups for single object detection, or even higher speedups for multiple objects.

12.1 Introduction

In computer vision the term 'shape' is used in a strict sense to refer to explicit geometric information, such as contours that correspond to surface boundaries, and in a broader sense to describe whatever is unaffected by appearance changes. The treatment of shape in terms of contours was the main theme of geometric 3D recognition [19, 33, 36, 53] before the advent of statistical techniques at the beginning of the previous decade. Shape has hence been used in high-level vision in its second sense, through features such as Shape Context [1], Scale-Invariant Feature Transforms (SIFT) [34] or Histograms-of-Gradients (HOG) [5], which describe shape in terms of distributions on invariant features, such as gradient histograms, or Convolutional Neural Networks [15, 16, 28, 37], which learn transformation-robust object representations. While such features provide a robust description of shape to

I. Kokkinos (✉)
Center for Visual Computing, Centrale-Supélec and INRIA-Saclay, Grande Voie des Vignes, 92295 Chatenay-Malabry, France
e-mail: iasonas.kokkinos@ecp.fr

M. Breuß et al. (eds.), *Perspectives in Shape Analysis*, Mathematics and Visualization, DOI 10.1007/978-3-319-24726-7_12

object detection tasks, a more explicit represention of shape is desireable in tasks which require a more detailed object description, such as pose estimation.

Such a representation is currently most successfully provided by deformable part models (DPM), defined in terms of a set of parts that deform with respect to each other. Such models have been shown to largely outperform rigid detectors on challenging benchmarks when trained discriminatively [10], and have become a standard in object detection and pose estimation research [50, 52]. At the heart of these models lies the optimization of a merit function with respect to the part displacements. In this work we take the merit function for granted, using the discriminatively trained models of [10], and focus on the computational efficiency of the optimization problem.

The most common detection algorithm used in conjunction with DPMs relies on the Generalized Distance Transform (GDT) algorithm [11], whose complexity is linear in the image size. Despite the algorithm's striking efficiency this approach still needs to thoroughly evaluate the object score everywhere in the image, which can become time demanding. In this work we introduce bounding-based techniques, which extend to part-based models the Branch-and-Bound (BB) and Cascaded Detection (CD) techniques used for Bag-of-Word classifiers in [29, 30] respectively. For this we exploit and adapt the Dual Tree (DT) data structure of [18] to provide the bounds required by BB/CD; we originally presented this technique in [24], but the current chapter provides a more thorough presentation and evaluation.

Our method is fairly generic; it applies to any star-shaped graphical model involving continuous variables, and pairwise potentials expressed as separable, decreasing binary potential kernels. We evaluate our technique using the mixture-of-deformable part models of [10]. Our algorithm delivers *exactly the same* results, but is substantially faster. We also develop a multiple-object detection variation of the system, where all object hypotheses are inserted in the same priority queue. If our task is to find the best (or k-best) object hypotheses in an image this can result in more than a 100-fold speedup. These speedups refer to the part combination process, after the unary part scores have been computed.

This chapter is structured as follows: after briefly covering prior work in Sect. 12.2, in Sect. 12.3 we first describe the cost function used in DPMs, and then motivate the use of bounding-based techniques for efficient object detection. In Sect. 12.4 we start with a high-level description of BB and CD in a general setting, and then proceed to describe the details of their implementation for detection with DPMs: in Sect. 12.4.3 we describe how we bound the DPM score and in Sect. 12.4.3.3 we describe how we keep the computation of the bound tractable. Qualitative results are provided throughout the text; we provide systematic experimental results on the Pascal VOC dataset in Sect. 12.5.

12.2 Previous Work on Efficient Detection

Cascade Detection (CD) algorithms were introduced in the beginning of the previous decade in the context of boosting [47] and coarse-to-fine detection [14] and have led to a proliferation of computer vision applications. However these works deal with 'monolithic' object models, i.e. there is no notion of deformable parts in the representation. Incorporating parts can make detection more challenging, since combinatorial optimization problems emerge.

The combinatorics of matching have been extensively studied for rigid objects [19], while [35] used A^* for detecting object instances. For categories, recent works [4, 13, 27, 31, 39, 43] have focused on reducing the high-dimensional pose search space during detection by initially simplifying the cost function being optimized, mostly using ideas similar to A^* and coarse-to-fine processing. In the recent work of [10] thresholds pre-computed on the training set are used to prune computation and result in substantial speedups compared to GDTs. However this approach requires tuning thresholds using the training set and comes only with approximate guarantees.

A line of work which brought new ideas into detection has been based on Branch-and-bound (BB). Even though BB was studied at least as early as [20], it was typically considered to be appropriate only for geometric matching/instance-based recognition. A most influential paper has been the Efficient Subwindow Search (ESS) technique of [29], where an upper bound of a bag-of-words classifier score delivers the bounds required by BB. Later [32] combined Graph-Cuts with BB for object segmentation, while in [30] a Cascaded Detection (CD) system for efficient detection was devised by introducing a minor variation of BB.

Our work is positioned with respect to these works as follows: unlike existing BB/CD works [29–32], we use the DPM cost and thereby accommodate parts in a rigorous energy minimization framework. And unlike the pruning-based works [4, 10, 13, 39], we do not make any approximations or assumptions about when it is legitimate to stop computation; our method is exact.

We obtain the bounds required by BB/CD by adapting the Dual Tree data structure of [18], originally developed in the context of nonparametric density estimation. To the best of our knowledge, Dual Trees have been minimally used in object detection; we are only aware of the work in [21] which used Dual Trees to efficiently generate particles for Nonparametric Belief Propagation. Here we show that Dual Trees can be used for part-based detection, which is related conceptually, but entirely different technically.

A considerable body of work has been developed around the efficient approximation of the part scores of DPMs [6, 7, 25, 26, 40–42, 44, 46]. These can be understood as complementary to the work presented here, in the sense that we consider that the part scores have been computed, and tackle the remaining combinatorial problem of 'assembling' the object parts. We actually deal with the approximations incurred by fast part computation in [25, 26] and show that they can be seamlessly integrated into the bounding-based framework presented here.

12.3 Object Detection with DPMs

The state $\mathbf{x}$ of a general DPM, e.g. [12, 49] encodes the object's putative configurations in terms of P position vectors, $x_p, p = 1, \ldots P$:

$$\mathbf{x} = \{x_1, \ldots, x_P\}, \quad x_p \in [1, K] \times [1, L], \tag{12.1}$$

where x_p can correspond to any of the $K \times L = N$ image pixels. For the most general graph topology N^P part combinations would need to be considered, severely raising the computational cost of DPMs. Coming up with algorithms of a smaller complexity is thus crucial for fast object detection in the presence of deformations.

Star-shaped DPMs [8–10] take a step in this direction, by constraining the model's topology so that a single part is designated as the 'root' node of a graph, and the remaining parts as the leaf nodes. All leaf nodes $p = 2, \ldots, P$ are connected exclusively with the root node $p = 1$, i.e. we have a star-shaped graphical model.

If the root node is placed at x, the score for a part p being placed at x' is given by $m_p(x', x) = U_p(x') + B_p(x', x)$, where the unary term $U_p(x')$ measures the fidelity of the image around position x' to the appearance model of the p-th part and the pairwise term $B_p, (\mathbf{x}_p, \mathbf{x}'_p)$ measures the geometric consistency of the positions of part p with respect to the root's position.

In particular, in [9] the appearance term $U_p(x') = \langle w_p, H(x') \rangle$ is formed as the inner product of a HOG feature $H(x')$ at x' with a discriminant w_p for p. This captures the local fidelity of the image to the appearance model of part p. The pairwise terms constrain the relative location x' of each part p w.r.t. the location x of the root in terms of a quadratic function of the form:

$$B_p(x', x) = -\left(x' - x - \mu_p\right)^T I_p \left(x' - x - \mu_p\right), \tag{12.2}$$

where $I_p = \mathrm{diag}(H_p, V_p)$ is a diagonal 'precision' matrix, μ_p is the nominal relative location vector, and for the root node, $p = 1$, we consider:

$$B_1(x', x) = \begin{cases} -\infty, & x' \neq x \\ 0, & x' = x \end{cases} \tag{12.3}$$

for convenience, practically ensuring that the 'root' part is pinned at position x. We can view the expression in Eq. 12.2 as related to the log-likelihood of the relative locations under a diagonal-covariance Gaussian model.

A star-shaped DPM scores a configuration $\mathbf{x} = (x_1, \ldots, x_P)$ by summing the merit of its parts:

$$M(\mathbf{x}) = \sum_{p=1}^{P} m_p(x_p, x_1). \tag{12.4}$$

To decide if a location x can serve as the root of an object, we maximize over all configurations that place the root at x:

$$S(x) \doteq \max_{\mathbf{x}:x_1=x} M(\mathbf{x}) \overset{Eq.\,12.4}{=} \max_{\mathbf{x}} \sum_{p=1}^{P} m_p(x_p, x) \tag{12.5}$$

$$= \sum_{p=1}^{P} m_p(x), \quad \text{where} \quad m_p(x) \doteq \max_{x'} U_p(x') + B_p(x', x). \tag{12.6}$$

To go from Eqs. 12.5 to 12.6 we use the fact that $M(\mathbf{x})$ factorizes over x_p; $m_p(x)$ serves as notation for the 'messages' being sent from the part nodes to the root node, and is obtained by eliminating x' from $m_p(x_p, x)$. The part-to-root message passing described by Eq. 12.6 is identical to the leaf-to-parent message passing equations of the Max-Product algorithm [23] if we use the logarithm of the probabilities.

The flow of computation of this algorithm is illustrated in Fig. 12.1. As one can see, the part scores have sharply peaked responses, but tend to provide many false positives, while the result of message passing (left-to-right transition) and

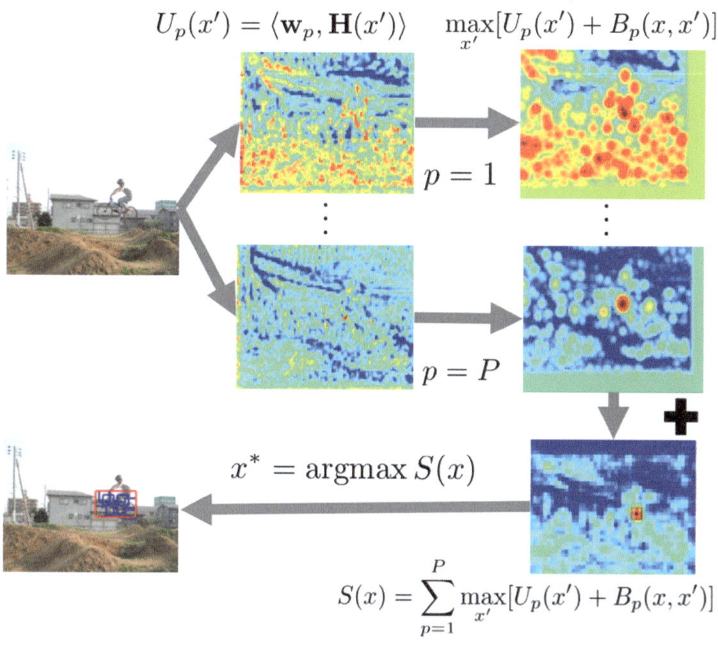

Fig. 12.1 Pipeline of object detection with star-shaped Deformable Part Models: the image features are filtered with a set of templates, providing part-specific unary terms. These are used to pass messages regarding the object's position to the root, where messages are summed to compute the overall score. From the maximum of this score we can obtain the best-scoring object hypothesis, as well as the position of the parts that support it

summation (top-to-bottom transition), performed at the root node provides a well-localized estimate of the object's position.

12.3.1 Complexity of Object Detection with Star-Shaped DPMs

During detection our goal is to identify either (a) $M^* = \{\arg\max_x S(x)\}$, or (b) $M^\theta = \{x : S(x) \geq \theta\}$. We will refer to case (a) as **first-best detection** and (b) as **threshold-based detection**. Case (a) is encountered commonly in pose estimation, or during latent SVM training, when maximizing over latent variables. Case (b) corresponds to the common setup for detection, where all image positions scoring above a threshold are used as object hypotheses.

A naive approach to solve both of those cases is to consider all possible values of x, evaluate $S(x)$ on them and then recover the solutions. The complexity of this would be $O(PN^2)$, where $N = |\{x\}|$ is the cardinality of the set of possible locations considered (Eq. 12.6 suggests doing N maximizations per point, and we have N points and P parts).

But due to the particular form of the pairwise term in Eq. 12.2, the maximization within each summand $m_p(x)$ in Eq. 12.6 lends itself to efficient computation in batch mode for all values of x using a Generalized Distance Transform (GDT) [11], in time $O(N)$. So the standard approach taken so far is to maximize each summand separately with GDTs and then add up the scores at all image locations to obtain the overall object score; this yields an overall complexity of $O(PN)$. Even though the $O(PN)$ complexity achieved with GDTs is remarkably fast (requiring 1–2 s for multi-scale processing of VGA-sized images), the N factor can still slow things down for large images. This motivates an approach to detection that can potentially operate with a complexity that is sublinear in the number of pixels – which can only be accomplished if we can somehow 'skip' unpromising pixel positions. This is implemented in a rigorous, fail-proof manner with bounding-based techniques, as described below.

12.4 Bounding-Based Detection with DPMs

Our approach to accelerating detection starts from the observation that if we use a fixed threshold for detection, e.g. -1 for an SVM classifier, then the GDT-based approach outlined above can be wasteful. In particular it treats equally all image locations, even when we can quickly realize that some of them score far below the threshold. This is illustrated in Fig. 12.2: in (a) we show the part-root configuration that gives the maximum score, and in (b) the score of a bicycle model from [10] over the whole image domain. The tiny part of the image scoring above a conservative threshold of -1 is encircled by a black contour in (b).

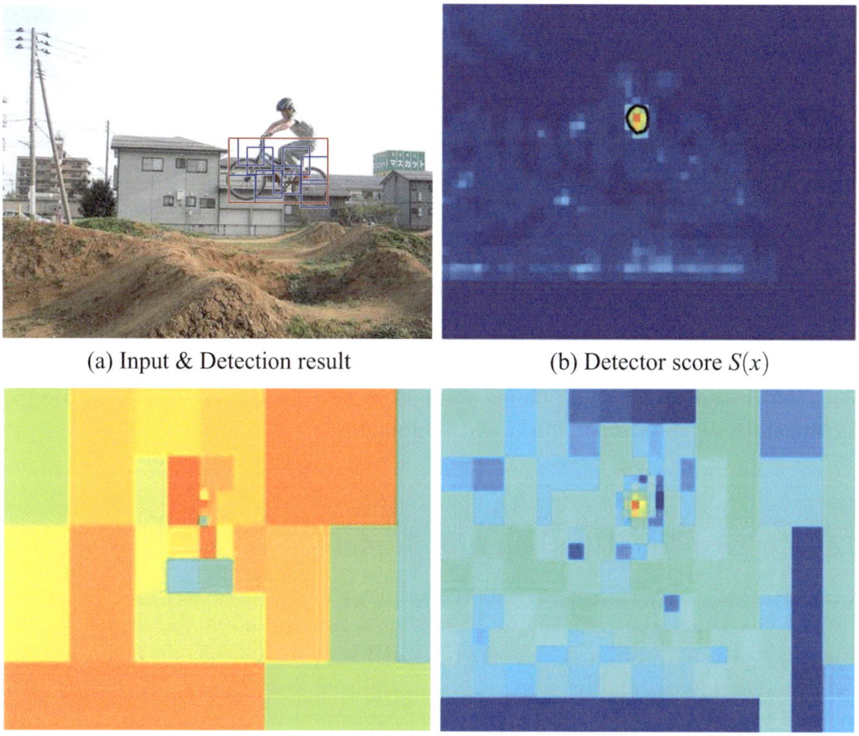

(a) Input & Detection result

(b) Detector score $S(x)$

(c) Branch and Bound for $\{\arg\max_x S(x)\}$

(d) Cascaded Detection for $\{x : S(x) \geq -1\}$.

Fig. 12.2 Motivation for a bounding-based approach (note that the classifier is designed to 'fire' on the *top-left* corner of the object's bounding box): standard part-based models evaluate a classifier's score $S(x)$ over the whole image domain. Typically only a tiny portion of the image domain should have large scores – in (**b**) we draw a *black* contour around $\{x : S(x) > -1\}$ for an SVM-based classifier. Our algorithm ignores large intervals with low $S(x)$ by upper bounding their values, and postponing their exploration in favor of more promising ones. In (**c**) we show as heat maps the upper bounds of the intervals visited by our algorithm until the strongest location was explored, and in (**d**) of the intervals visited until all locations x with $S(x) > -1$ were explored

Our approach instead speeds up detection by upper bounding the score of the detector within *intervals* of x. These bounds can be rapidly obtained using low-cost operations, as will be detailed in the following. Having a bound allows us to use a coarse-to-fine strategy that starts from an interval containing all possible object locations and then gradually subdivides it to refine the bounds on promising sub-intervals, while avoiding the exploration of less promising ones.

This is demonstrated in Fig. 12.2c, d where we show as heat maps the upper bounds of the intervals visited by our approach for first-best and threshold-based detection respectively. The parts of the image where the heat maps are more fine-grained correspond to image locations that seemed promising and were explored at a finer level. Coarse-grained parts correspond to intervals whose upper bound was low, and the refinement of the bound was therefore avoided.

Even though the number of operations performed by our bounding-based approach is image-dependent, we can say that it is roughly *logarithmic in the image size*, since our approach recursively subdivides the explored intervals (the best-case complexity of our algorithm is $O(|M|P \log N)$). So rescaling an image by a factor of 2 will require roughly two more iterations for our algorithm, while for the GDT-based computation it will require four times the original number of operations (since we now have four times as many pixels).

We now make these high-level ideas more concrete by first describing Branch-and-Bound and Cascaded Detection, which respectively address the first-based and threshold-based detection problems outlined in Sect. 12.3.1, and then get into the technical details involved in the bound computation.

12.4.1 First-Best Detection with Branch and Bound

Branch and Bound (BB) can be used a generic maximization algorithm for non-convex or even non-differentiable functions. BB searches for the interval containing the function's maximum by using a prioritized search strategy; the priority of an interval is determined by the function's upper bound within it. The operation of BB for the maximization of a function over a domain X_0 is illustrated in Fig. 12.3: BB finds the maximum of a function by using a prioritized search strategy over intervals; at each step branching first takes place, where an interval – X_0, here – is split into two subintervals, X_1, X_2. Then bounding takes place, where the value of the function is upper bounded within each of the new intervals. This upper bound serves as a priority and dictates which interval is explored next.

The main hurdle in devising a BB algorithm is coming up with a bound that is relatively tight and also easy to compute. In Fig. 12.3 a parabola is used to upper bound a complex, non-concave function; the interval's priority can then be rapidly estimated by constructing an analytical upper bound on the parabola's value.

More concretely, if the function we want to maximize is $S(x)$, BB requires that we are able to construct an upper bound of this function's value within an interval.

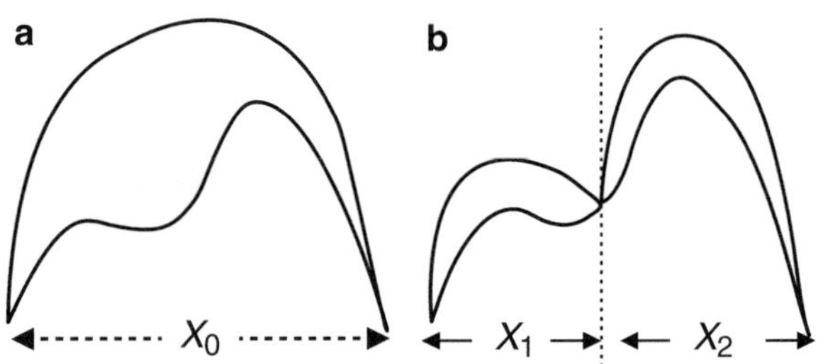

Fig. 12.3 Illustration of how Branch-and-Bound proceeds to maximize a complex, non-concave function within an interval by branching and bounding the function within intervals. Please see text for details

With a slight abuse of notation we introduce:

$$S(X) \doteq \max_{x \in X} S(x), \tag{12.7}$$

i.e. we 'overload' function symbols to take intervals as arguments. Denoting the upper bound to function S as $\overline{S}$ the requirement is that:

$$\overline{S}(X) \geq S(X) = \max_{x \in X} S(x) \quad \forall X, \quad \overline{S}(\{x\}) = S(x), \tag{12.8}$$

i.e. on a singleton our bound should be tight.

With such a bounding function at our disposal, BB searches for the maximum of a function using prioritized search over intervals, as illustrated by the pseudocode in Table 12.1. Starting from an interval corresponding to all possible object locations (X_0) the algorithm splits it into subintervals and uses the upper bounds of the latter as priorities in search. At each step the algorithm visits the most promising subinterval and the algorithm terminates when the first singleton interval, say x, is popped. This is guaranteed to be a global maximum: since the bound is tight for singletons, we know that the solutions contained in the remaining intervals of the priority queue will score below or equal to x, since the upper bound of their scores is at most $\overline{S}(\{x\}) = S(x)$.

Table 12.1 Pseudocode for Brand-and-Bound (BB) and Cascaded Detection (CD). Both algorithms use a KD-tree for the image domain, where the root node, X_0, corresponding to an interval for the whole image domain and the leaves to singletons (pixels). BB starts from the root interval and performs prioritized search to find the interval containing the best configuration. CD starts from the root node and performs a Center-Left-Right traversal of the tree to return all singletons scoring above a fixed threshold

Branch-and-Bound	Cascaded Detection
$M^* = BB(X_0, \overline{S})$	$M^\theta = CD(X, \overline{S}, \theta)$
INITIALIZE: $\mathcal{Q} = \{(X_0, \overline{S}(X_0))\}$	**if** $\overline{S}(X) < \theta$ **then**
while 1 **do**	RETURN {}
$X = \text{Pop}[\mathcal{Q}]$	**end if**
if Singleton[X] **then**	**if** Singleton[X] **then**
RETURN X { // First singleton: best X}	RETURN X { //Singleton with score $\geq \theta$}
end if	**end if**
$[X_1, X_2] = \text{Branch}[X]$	$[X_1, X_2] = \text{Branch}[X]$
Push[$\mathcal{Q}$, $(X_1, \overline{S}(X_1))$], Push[$\mathcal{Q}$, $(X_2, \overline{S}(X_2))$]	$M^\theta = CD(X_1, \overline{S}, \theta) \cup CD(X_2, \overline{S}, \theta)$
end while	RETURN M^θ

12.4.2 Threshold-Based Detection: Cascaded Detection

The BB algorithm described above is appropriate if we search for the first-best (or k-best) scoring configuration(s). This is typically the case for tasks such as training or pose estimation. But for detection we typically want to find all object locations that score above a threshold θ. To accommodate this in [24] we proposed to use prioritized search, but stop when the popped interval scores below θ. This will return all singletons scoring above θ indeed, but it is more efficient to use a cascaded detection algorithm similar to [30], which avoids the overhead of inserting/removing elements from a priority queue and is also easy to parallelize.

In particular, our adaptation of the algorithm in [30] uses a tree of intervals, with the root corresponding to the whole domain and the leaves to singletons (single pixels). The algorithm, described in pseudocode in Table 12.1, starts from the root and recursively traverses the tree in a center-left-right manner. At the center we check if the upper bound of the current node is above threshold. If it is not, we return an empty set, meaning that none of the node's children can contain an object above threshold. Otherwise, if the node is singleton, we return the actual location. Finally if the node is non-singleton we recurse to its left and right children (subintervals), and return the union of their outputs.

12.4.3 Bounding the DPM Score

Having given a high-level description of BB/CD we describe in this subsection how we compute the bounds and in the following one how we organize the computation.

The main operation required by both algorithms is to compute 'cheap' upper bounds of the DPM score function $S(x)$ within an interval X. From Eq. 12.6 we have that $S(x) = \sum_p m_p(x)$ and we are now concerned with forming an upper bound for the quantity $S(X) = \max_{x \in X} \sum_p m_p(x)$. We can upper bound $S(X)$ as follows:

$$\overline{S}(X) \doteq \sum_p \overline{m}_p(X) \geq \sum_p m_p(X) = \sum_p \max_{x \in X} m_p(x) \geq \max_{x \in X} \sum_p m_p(x) = S(X), \tag{12.9}$$

where $\overline{m}_p(X)$ are upper bounds on the value of $m_p(x)$ within X – we describe these below. On the left we have the construction of our upper bound and on the right the quantity we wanted to bound in the first place. The first inequality stems from the fact that $\overline{m}_p(X)$ is an upper bound for $m_p(X)$, the next equality from the definition of the 'overloaded' notation for $m(X)$. The second inequality stems from the fact that $\max_{x \in X} f(x) + \max_{x \in X} g(x) \geq \max_{x \in X} f(x) + g(x)$ for any two functions f, g, and any interval X.

We clarify that the maximization showing up here is over the interval X for which the upper bound is computed; it is not the maximization implicit in the definition of the messages in Eq. 12.6.

As we will focus on the individual summands $\overline{m}_p(X)$, we omit the p subscript. Based on Eq. 12.6, $\overline{m}(X)$ should satisfy:

$$\overline{m}(X) \geq m(X) \overset{Eq.\,12.7}{=} \max_{x \in X} m(x) \overset{Eq.\,12.6}{=} \max_{x \in X} \left[\max_{x' \in X'} m(x', x) \right], \tag{12.10}$$

where X and X' do not need to be identical (by the definition of Eq. 12.6 X' is the whole image domain). We now proceed to describe how we compute the relevant bounds efficiently.

12.4.3.1 Dual Trees and Domain Paritioning

We decompose the computation of the upper bound in Eq. 12.10 into smaller parts by using the partitions $X = \cup_{d \in D} X_d$, $X' = \cup_{s \in S} X_s$ as illustrated in Fig. 12.4. We call points contained in X' the source locations and points in X the domain locations, with the intuition that the points in X' contribute to a score in X. Making reference to Fig. 12.4, the 'domain' intervals-d could be the letters and the 'source' intervals could be the numbers.

For a given partition of X, X' we can rewrite $m(X)$ in Eq. 12.10 as:

$$m(X) = \max_d \max_{x \in X_d} \max_s \max_{x' \in X_s} m(x', x) = \max_d \max_s \mu_d^s, \quad \text{where} \tag{12.11}$$

$$\mu_d^s \doteq \max_{x \in X_d} \max_{x' \in X_s} m(x', x). \tag{12.12}$$

The quantity μ_d^s quantifies the maximal contribution of any source-interval point X_s to any domain-interval point X_d; and $m(X)$ expresses the maximal contribution

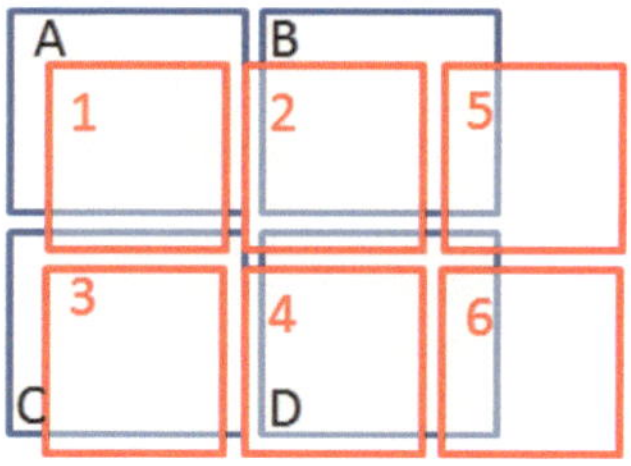

Fig. 12.4 We rely on a partition of the 'source' (*red*) and 'domain' (*blue*) points to derive rapidly computable bounds of their 'interactions'. This could indicate for example that points lying in square 6 cannot have a large effect on points in square A, and therefore we do not need to go to a finer level of resolution to exactly estimate their interactions

that any point within any source-interval can have to any point within any domain-interval.

In order to compute $m(X)$ we have at our disposal a range of partitions for the domain and source points to choose from, represented using separate KD-trees (hence the 'Dual Tree' term). As we illustrate in Fig. 12.6 and further detail in Sect. 12.4.3.3, we start from coarse partitions of X, X' and iteratively refine and prune both. To describe how exactly this takes place we first provide bounds for the associated terms.

12.4.3.2 Bounding the Appearance and Geometric Terms

Based on Eq. 12.12 and the definition of $m(x', x)$ we can upper bound μ_d^s as follows:

$$\mu_d^s = \max_{x \in X_d} \max_{x' \in X_s'} \left(U(x') + B(x', x) \right) \leq \max_{x' \in X_s'} U(x') + \max_{x \in X_d} \max_{x' \in X_d} B(x', x) \doteq \overline{\mu}_d^s, \tag{12.13}$$

where again we use the fact that $\max_{x \in X} f(x) + \max_{x \in X} g(x) \geq \max_{x \in X} (f(x) + g(x))$.

For reasons that will become clear in Sect. 12.4.3.3, we also need to lower bound the quantity

$$\lambda_d^s = \min_{x \in X_d} \max_{x' \in X_s} \left(U(x') + B(x', x).\right) \tag{12.14}$$

This provides the weakest contribution to a domain point in X_d by any source point in X_s. To bound λ_d^s we have two options:

$$\underline{\lambda}_{d,1}^s = \max_{x' \in X_s} U(x') + \min_{x \in X_d} \min_{x' \in X_s} B(x', x) \leq \lambda_d^s, \tag{12.15}$$

$$\underline{\lambda}_{d,2}^s = \min_{x' \in X_s'} U(x') + \min_{x \in X_d} \max_{x' \in X_s} B(x', x) \leq \lambda_d^s. \tag{12.16}$$

The first bound corresponds intuitively to placing the point of X_s with the best unary score, say x_b to the worst location within X_s and then evaluating the support that it lends to the 'hardest' point of X_s. This is a lower bound since x_b will actually be in at least as good a position with respect to the hardest point. The second bound corresponds to taking the point of X_s with the worst unary score, say x_w and placing it at the location in X_s that supports the hardest point of X_d. This again is a lower bound since in practice the point of X_s supporting the hardest point in X_d will have at least as good a unary score as x_w does.

We combine these two bounds into a single and tighter lower bound as:

$$\underline{\lambda}_d^s = \max\left(\underline{\lambda}_{d,1}^s, \underline{\lambda}_{d,2}^s\right). \tag{12.17}$$

In [24] we had used only the first bound. Computing Eq. 12.17 requires some additional operations, but the bound is tighter and accelerates detection by a factor of 10–20 %.

We can rapidly compute the terms involved in the bounds of Eqs. 12.13, 12.14, 12.15, and 12.16. First, the appearance-based terms, $\max_{x \in X_s} U(x)$ and $\min_{x \in X_s} U(x)$, can be computed with fine-to-coarse max-/min-imization through the KD-tree data structures. The overall complexity of computing all of the relevant terms turns out to be linear in the image size but with a particularly low constant, equal to the cost of the max/min operation.

Second, the geometric terms $\min_{x \in X_d} \max_{x' \in X_s} B(x', x)$, $\max_{x \in X_d} \max_{x' \in X_d} B(x', x)$ can be rapidly computed by exploiting the fact that X_d and X_s are rectangular. For clarity's sake we now abandon the x notation for coordinates and switch to horizonal and vertical coordinates, (h, v). Making reference to Fig. 12.5, we consider two 2D intervals, one for the domain-node X_d and one for the domain-node X_s; X_d is centered at (h_d, v_d), and has an horizontal/vertical half-range of η_d/ν_d, while for X_s the respective quantities are $(h_s, v_s), \eta_s, \nu_s$. Using the (h, v) notation, we can write the pairwise term between two points, say $x \in X_d, x' \in X_s$ as:

$$\mathscr{G}_{x,x'} = -H(h - h')^2 - V(v - v')^2 \tag{12.18}$$

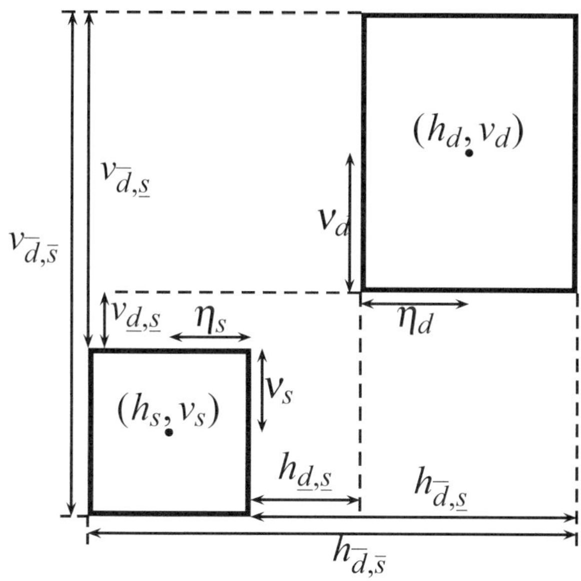

Fig. 12.5 Illustration of the terms involved in the geometric bound computations of Eqs. 12.24, 12.25, and 12.26. The d/s subscript indicates quantities relevant to the domain/source intervals respectively (we want to bound the score within the domain interval, using contributions from the source interval)

where H, V are the diagonal elements of the precision matrix showing up in Eq. 12.2; we omit the effect of the means μ in Eq. 12.2 for simplicity, but they can be trivially incorporated in what follows.

Since the pairwise cost is separable in the horizontal and vertical dimensions, we can use distributivity to break the max-/min-imization operations along separate axes. In particular, we have to compute:

$$\mathscr{G}_{\bar{d},\bar{s}} \doteq \max_{x \in X_d} \max_{x' \in X_s} \mathscr{G}_{x,x'} = \max_{h \in X_d^h} \max_{h' \in X_s^h} -H(h-h')^2 + \max_{v \in X_d^v} \max_{v' \in X_s^v} -V(v-v')^2 \tag{12.19}$$

$$= -Hh^2_{\underline{d},\underline{s}} - Vv^2_{\underline{d},\underline{s}}, \tag{12.20}$$

$$\text{where} \quad h_{\underline{d},\underline{s}} \doteq \min_{h \in X_d^h} \min_{h' \in X_s^h} |h-h'|, \quad v_{\underline{d},\underline{s}} \doteq \min_{v \in X_d^v} \min_{v' \in X_s^c} |v-v'| \tag{12.21}$$

where we use $\bar{i}, \underline{i}$ to indicate respectively that we are max-/min-imizing with respect to the points belonging to domain i, and denote by X^v, X^h the projections of a 2D interval X on the horizontal and vertical axes respectively. Similarly we get:

$$\mathscr{G}_{\underline{d},\bar{s}} \doteq \min_{x \in X_d} \max_{x' \in X_s} \mathscr{G}_{x,x'} = -Hh^2_{\bar{d},\underline{s}} - Vv^2_{\bar{d},\underline{s}}, \quad \mathscr{G}_{\underline{d},\underline{s}} \doteq \min_{x \in X_d} \min_{x' \in X_s} \mathscr{G}_{x,x'} = -Hh^2_{\bar{d},\bar{s}} - Vv^2_{\bar{d},\bar{s}},$$

$$\textit{where} \quad h_{\bar{d},\underline{s}} \doteq \max_{h \in X_d^h} \min_{h' \in X_s^h} |h-h'|, \quad v_{\bar{d},\underline{s}} \doteq \max_{v \in X_d^v} \min_{v' \in X_s^v} |v-v'| \tag{12.22}$$

$$h_{\bar{d},\bar{s}} \doteq \max_{h \in X_d^h} \max_{h' \in X_s^h} |h-h'|, \quad v_{\bar{d},\bar{s}} \doteq \max_{v \in X_d^v} \max_{v' \in X_s^v} |v-v'| \tag{12.23}$$

For the particular configuration shown in Fig. 12.5 we have:

$$h_{\bar{d},\bar{s}} = (h_d + \eta_d) - (h_s + \eta_s), v_{\bar{d},\bar{s}} = (v_d + \nu_d) - (v_s + \nu_s), \tag{12.24}$$

$$h_{\bar{d},\underline{s}} = (h_d - \eta_d) - (h_s + \eta_s), v_{\bar{d},\underline{s}} = (v_d - \nu_d) - (v_s + \nu_s), \tag{12.25}$$

$$h_{\underline{d},\underline{s}} = (h_d + \eta_d) - (h_s - \eta_s),$$
$$v_{\underline{d},\underline{s}} = (v_d + \nu_d) - (v_s - \nu_s) \tag{12.26}$$

If we consider all possible relative placements of the two rectangles we obtain the following forms for the horizontal coordinate:

$$h_{\bar{d},\bar{s}} = \lceil |h_d - h_s| + (\eta_d - \eta_s) \rceil \tag{12.27}$$

$$h_{\bar{d},\underline{s}} = \lceil |h_d - h_s| - (\eta_d + \eta_s) \rceil \tag{12.28}$$

$$h_{\underline{d},\underline{s}} = |h_d - h_s| + (\eta_d + \eta_s) \tag{12.29}$$

where $\lceil \cdot \rceil \doteq \max(\cdot, 0)$; similar expressions are used for the vertical coordinate after substituting v, ν for h, η respectively.

12.4.3.3 Dual Recursion and Supporter Pruning

We now describe how to control the complexity of maximizing over d and s in Eq. 12.11. The range of d and s will scale inversely with the cardinality of the intervals X_s, X_d, meaning that as the bounds get finer a larger number of terms will be involved; in the limit of singletons X_s, X_d we have a quadratic complexity in the number of pixels. We now describe how we use a coarse-to-fine algorithm to quickly prune the range of s involved for every d *without sacrificing accuracy*.

For this we use a Dual Recursion algorithm akin to the one originally introduced for Dual Trees by [18]. An illustration of how the algorithm works is provided in Fig. 12.6: starting from the root and going to the leaves, we recursively prune the range of source (s) intervals that should be used to bound the value at any domain (d) interval. In particular we 'descend' simultaneously on the source and domain trees; at the beginning (top) the root node of the source tree is used to bound the score of the root node of the domain tree and at the end the leaves of the source tree are used to compute the exact score of the leaves of the domain tree.

We use a recursive algorithm to limit the number of operations involved until getting to the leaves. Consider that in Eq. 12.11 we know that only a set of 'supporter' intervals $\mathscr{S}_d = \{s_i\}$ should be used in the bound computation relevant to a domain node-interval d. This means that all other source intervals cannot contribute something to any of the points contained in d. To reduce the number of operations when refining these domain and source intervals there are two observations that allow us to speed things up.

First, the children (sub-intervals) of d need to use only the children (sub-intervals) of $\mathscr{S}$, i.e. $\mathscr{S}_d \subset \cup_{\mathscr{S}_{\mathrm{pa}(d)}}\{\mathrm{ch}(s_i)\}$, where pa, ch denote the parent and child operators. If any other points were necessary, these should have been included in the domain $\mathscr{S}_d$, by the definition of the 'supporter' intervals. Second, we can remove some elements of $\cup_{\mathscr{S}_{\mathrm{pa}(d)}}\{\mathrm{ch}(s_i)\}$ when forming $\mathscr{S}_d$, if we know that these cannot contribute to the optimal score at a domain node. This requires combining $\overline{\mu}_d^s$ and the lower bounds of $\underline{\lambda}_d^s$, and relies on the following rationale, illustrated in Fig. 12.7: consider that a node d has supporters m, n, o. If two nodes n and m support a node d and their bounds are related by $\overline{\mu}_d^n < \underline{\lambda}_d^m$, the descendants of interval n can be ignored from the following maximization. This is intuitively so because the bounds become tighter as the intervals become smaller, namely lower bounds increase and upper bounds decrease.

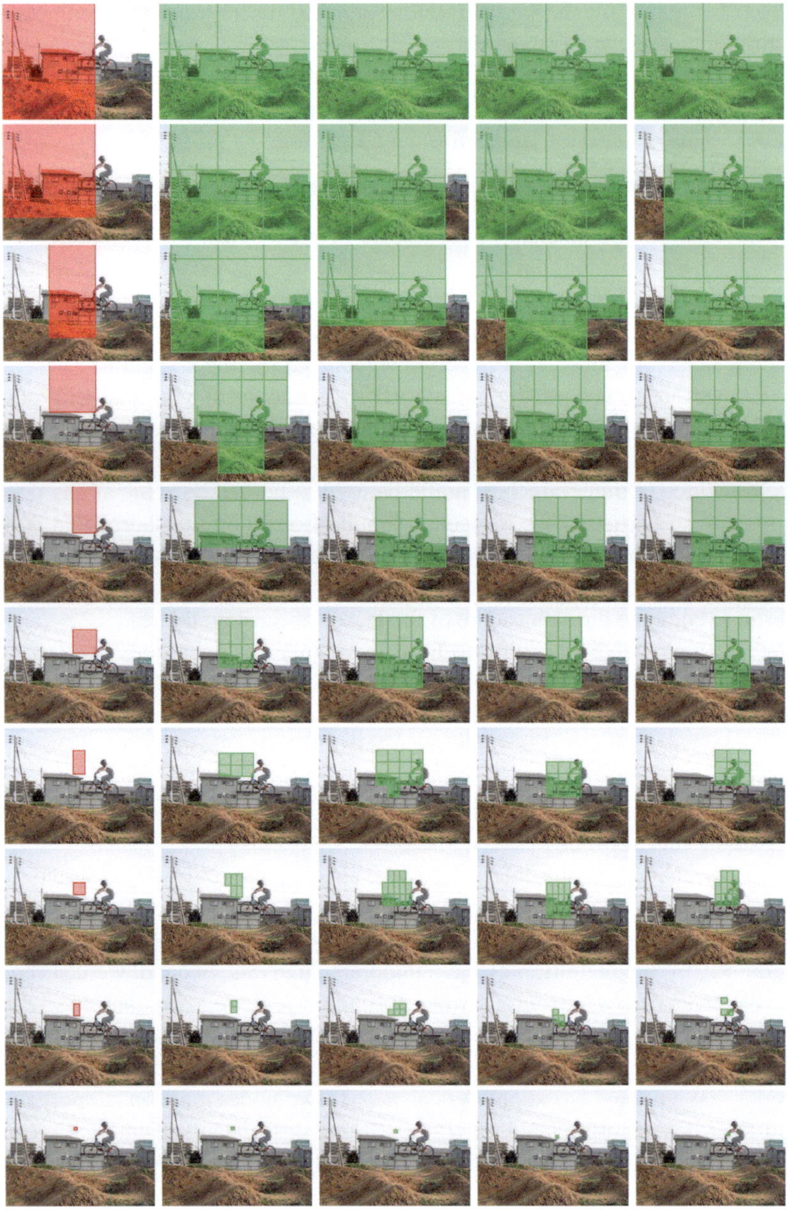

Fig. 12.6 Illustration of supporter pruning. The *left column* illustrates the succession of domain intervals that leads to the optimal object configuration. The *next four columns* illustrate the associated 'supporters' of that interval for four distinct object parts. Our algorithm starts at the top with a large interval that is supported by equally large intervals. On the way the domain and supporter intervals get refined. For each part the supporter intervals are also pruned in every step, making the overall optimization tractable. At the bottom row the domain interval is a singleton, and is supported by a single, and singleton, supporter interval. This indicates the optimal part placement for the given domain interval

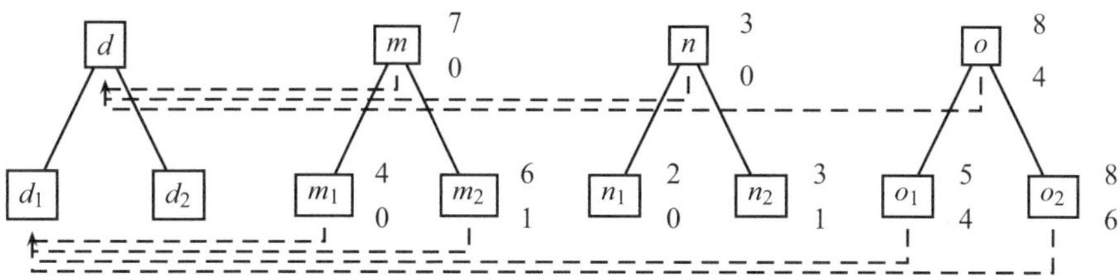

Fig. 12.7 Supporter pruning: source nodes $\{m, n, o\}$ are among the possible supporters of domain-node l. Their upper and lower bounds (shown as numbers to the right of each node) are used to prune them. Here, the upper bound for n (3) is smaller than the maximal lower bound among supporters (4, from o): this implies the upper bound of n's children contributions to l's children (shown here for l_1) will not surpass the lower bound of o's children. We can thus safely remove n from the supporters. Please see text for details

Below we provide a more concrete stament of this result, while making reference to Fig. 12.7: denote by s_1, s_2 the two children of a supporter node s, with s being one of the three right-most nodes and by d_1, d_2 the two children of node d, on the left. We have that

$$\overline{\mu}_d^s \geq \mu_d^s \geq \mu_{d_j}^{s_i} \quad \forall i \in \{1,2\}, \forall j \in \{1,2\} \tag{12.30}$$

The first inequality holds from the fact that $\overline{\mu}$ is an upper bound to μ. The second inequality holds because according to Eq. 12.12, $\mu_d^s \doteq \max_{x \in X_s} \max_{x' \in X_d} m(x, x')$ while $X_{s'} \subset X_s, X_{d'} \subset X_d$; so maximizing a function over a smaller domain will lead to a smaller quantity. In words, Eq. 12.30 tells us that the contribution $\mu_{d_j}^{s_i}$ of any child of s to any child of d cannot be larger than the upper bound $\overline{\mu}_d^s$ to the contribution of s to d.

We also have that:

$$\underline{\lambda}_d^s \leq \lambda_d^s \leq \lambda_{d_i}^s = \max(\lambda_{d_i}^{s_1}, \lambda_{d_i}^{s_2}), \quad i \in \{1,2\}. \tag{12.31}$$

The first inequality holds from the fact that $\underline{\lambda}_d^s$ lower bounds λ_d^s. The second from the definition of $\lambda_d^s = \min_{x \in X_d} \max_{x' \in X_s} m(x, x')$ in Eq. 12.14, and the fact that $X_{d_i} \subset X_d$: since for $\lambda_{d_i}^s$ we are minimizing over a smaller set, it follows that $\lambda_{d_i}^s \geq \lambda_d^s$. Finally the last equality stems from the definition of λ_d^s and the fact that $X_s = X_{s_1} \cup X_{s_2}$. In words, Eq. 12.31 tells us that if we include both children, s_1, s_2 of s as potential supporters of a child d_i of d, the support at the worst point of d_i will be at least equal to $\underline{\lambda}_d^s$.

Having obtained the expressions in Eqs. 12.30 and 12.31 for an arbitrary source node s, we now turn to how these expressions can be used in order to prune the children of a node, say n, in the light of the support delivered by another node, say

m. The condition for doing this is that $\overline{\mu}_d^n \leq \underline{\lambda}_d^m$. If this holds, then it follows that

$$\mu_{d_j}^{n_i} \leq \max(\lambda_{d_j}^{m_1}, \lambda_{d_j}^{m_2}), \quad j \in \{1,2\}, i \in \{1,2\} \tag{12.32}$$

This tells us that within the domain interval d_j any point will be getting a support from m_1, m_2 that will be at least as good as the best support it can get from n_1 or n_2. Therefore the intervals n_1, n_2 do not need to be considered anymore, and node n can be safely pruned – this is illustrated in Fig. 12.6 by the lack of connections between n_1, n_2 and the children of d.

Concisely, we prune the children of supporter l to node d if $\overline{\mu}_d^l < \max_{j \in \mathscr{S}_d} \underline{\lambda}_d^j$. This allows us to keep the maximization over d in Eq. 12.11 manageable at any point. In practice less than 15 supporters are typically involved at any point of the computation, as also shown in Fig. 12.6.

12.5 Results: Application to Deformable Object Detection

To estimate the merit of BB we first compare with the mixtures-of-DPMs developed and distributed in [17]. We directly extend the Branch-and-Bound technique that we developed for a single DPM to deal with multiple scales and mixtures ('ORs') of DPMs [10, 51], by inserting all object hypotheses into the same queue. In the Cascaded Detection case we simply do a for-loop over scales and components.

Our technique delivers the same results as [17]: other than differences due to floating/double point arithmetic the results are identical. We therefore do not provide any detection performance curves, but only timing results.

Coming to time efficiency we compare the results of the original DPM mixture model and our implementation, using 1200 images from the Pascal dataset and the models of [17] for all 20 object categories. As a first experiment we consider the standard detection scenario where we want to detect all objects in an image that score above a certain threshold. We show in Fig. 12.8a how the threshold affects the

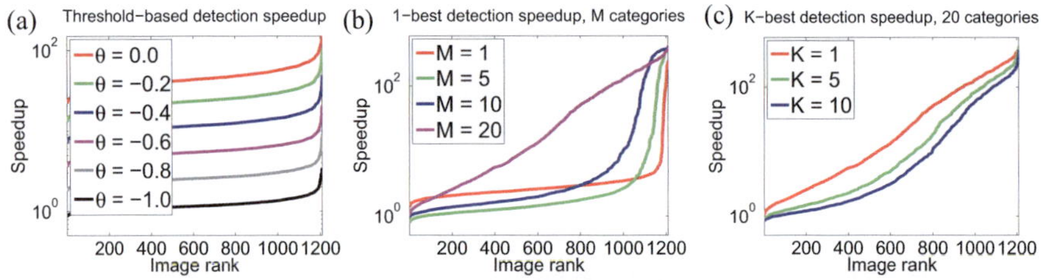

Fig. 12.8 (**a**) Single-object speedup of Cascaded Detection over GDTs on images from the Pascal dataset, (**b, c**) Multi-object speedup. Please see text for details

speedup we obtain: for a conservative threshold the speedup is typically tenfold, but as we become more aggressive it doubles.

As a second application, we consider the problem of identifying the 'dominant' object present in the image, i.e. the category that gives the largest score. Typically simpler models, like bag-of-words classifiers are applied to this problem, based on the understanding that part-based models can be time-consuming, therefore applying a large set of models to an image would be impractical.

Our claim is that Branch-and-Bound allows us to pursue a different approach, where in fact having more object categories can *increase* the speed of detection, if we leave the unary potential computation aside. Specifically, our approach can be directly extended to the multiple-object detection setting; as long as the scores computed by different object categories are commensurate, they can all be inserted in the same priority queue. In our experiments we observed that we can get a response with less computation per model by introducing more models. The reason for this is that including into our object repertoire a model giving a large score helps BB stop; otherwise BB keeps searching for another object.

The plots in Fig. 12.8b, c show systematic results for this experiment on the Pascal dataset. We compare the time that would be required by GDT to perform detection of all multiple objects considered in Pascal, to that of a model simultaneously exploring all categories. In (b) we show how finding the first-best result is accelerated as the number of objects (M) increases; while in (c) we show how increasing the 'k' in 'k-best' affects the speedup. For small values of k the gains become more pronounced. Of course if we use Cascaded Detection the speedup does not change for multiple categories when compared to plot (a), since essentially the objects do not 'interact' in any way (we do not use nonmaximum suppression). But as we turn to the best-first problem, the speedup becomes dramatic, and can often be more than 100-fold.

We note that the timings in these plots refer to the 'message passing' part implemented with GDT and not the computation of unary potentials, which is common for both models, or the KD-tree construction, which is linear in the image size.

A more thorough breakdown of all costs can be found in Table 12.2. These results are obtained by summing over all 20 categories, and averaging over 1200 images from the Pascal VOC dataset; we report mean and standard deviation.

The first two rows report the cost of computing unary terms – these costs are common to the two methods being compared. The first method uses the BLAS-accelerated implementation of inner products provided in [17], while the second method uses the FFT in conjunction with the patchwork technique of [7] in order to accelerate computations. The method of [7] has a clear advantage.

The next row reports the time required to construct the KD-trees for the part and root intervals, alongside with the associated fine-to-coarse max-/min-imization operations; these are unique to our method, and of linear complexity, but we observe that their overall cost is negligible with respect to the overall computation costs.

Table 12.2 Timings for the treatment of all 20 categories per image on a single core, in seconds. Please see text for details

	Our work	GDT-DPMs [17]
Unary terms (BLAS)	23.20 ± 1.49	23.20 ± 1.49
Unary terms (FFT) [7]	9.20 ± 1.21	9.20 ± 1.21
KD-trees	1.72 ± 0.21	0.00 ± 0.00
Detection, $\theta = 0.0$	0.25 ± 0.07	10.74 ± 1.02
Detection, $\theta = -.2$	0.47 ± 0.12	10.74 ± 1.02
Detection, $\theta = -.4$	0.93 ± 0.22	10.74 ± 1.02
Detection, $\theta = -.6$	1.95 ± 0.42	10.74 ± 1.02
Detection, $\theta = -.8$	4.17 ± 0.84	10.74 ± 1.02
Detection, $\theta = -1$	9.14 ± 1.79	10.74 ± 1.02
Detection, 1-best	0.41 ± 0.08	10.74 ± 1.02
Detection, 5-best	0.47 ± 0.09	10.74 ± 1.02
Detection, 10-best	0.48 ± 0.10	10.74 ± 1.02

The next six rows compare the cost of Cascaded Detection for a range of thresholds, with the linear-complexity GDT.[1] The last three rows compare the cost of Branch-and-Bound for K-best detection with GDT.

Apart from the clear relative improvements with respect to GDTs, we observe that by combining the FFT-based unary term computation with our Branch-and-Bound implementation of DPM inference we can detect 20 objects in potentially less than 10 s per image; these computation costs can be easily reduced further with multi-threaded computation and of course also by porting part of the computation to GPUs.

We are working on incorporating such aspects into our existing implementation, which is available from http://cvn.ecp.fr/personnel/iasonas/dpms.html

12.6 Conclusions and Discussion

In this work we have introduced Dual-Tree Branch-and-Bound for efficient part-based detection. We have used Dual Trees to compute upper bounds on the cost function of a part-based model and thereby derived Branch-and-Bound and Cascaded Detection algorithms for detection. Our algorithm is exact and makes no approximations, delivering identical results with the DPMs used in [10], but substantially smaller time. Further, we have shown that the flexibility of prioritized search allows us to consider new tasks, such as multiple-object detection, which yielded speedups by two orders of magnitude or more in certain cases.

The work presented in this chapter focuses on the combinatorial optimization problem related to the 'assembly' of a deformable object from its parts; we have

[1] In our comparisons we use the original- and faster-GDT algorithm of [11] instead of the one provided in [17].

thus only treated one of the many aspects of deformable part modelling. In parallel works we have been exploring complementary aspects, including (i) the acceleration of the part score computations [25, 26], see also [6, 7, 40–42, 44, 46] (ii) extensions of Dual-Tree Branch-and-Bound to 3D [3] (iii) the treatment of better training criteria and richer graph topologies for shape segmentation with DPMs [2] (iv) the incorporation of segmentation information in DPMs [45] and, most recently (v) the use of Deep Convolutional Neural Network (DCNN) features in DPMs [37, 38] – see also [16, 48] for parallel works. These advances hint at the breadth of problems accommodated by DPMs, and shape modelling in general. We anticipate that properly treating the combinatorial structure of the optimization problem underlying DPMs will help further fuel progress across all these problems.

Acknowledgements I thank the two anonymous reviewers and Stefan Kinauer for feedback that helped improve this manuscript. I am grateful to the authors of [7, 10, 22, 29, 30] for making their code available. This work was funded by grants ANR-10-JCJC-0205 and FP7-Reconfig.

References

1. Belongie, S., Malik, J., Puzicha, J.: Shape matching and object recognition using shape contexts. IEEE Trans. PAMI **24**(4), 509–522 (2002)
2. Boussaid, H., Kokkinos, I.: Fast and exact: ADMM-based discriminative shape segmentation with loopy part models. In: CVPR, Columbus (2014)
3. Boussaid, H., Kokkinos, I., Paragios, N.: Rapid mode estimation for 3D brain MRI tumor segmentation. In: Energy Minimization Methods in Computer Vision and Pattern Recognition, Lund (2013)
4. Chen, Y., Zhu, L., Lin, C., Yuille, A.L., Zhang, H.: Rapid inference on a novel AND/OR graph for object detection, segmentation and parsing. In: Proceedings of NIPS, Vancouver (2007)
5. Dalal, N., Triggs, B.: Histograms of oriented gradients for human detection. In: Proceedings of CVPR, San Diego (2005)
6. Dean, T., Ruzon, M., Segal, M., Shlens, J., Vijayanarasimhan, S., Yagnik, J.: Fast, accurate detection of 100,000 object classes on a single machine. In: Proceedings of CVPR, Portland (2013)
7. Dubout, C., Fleuret, F.: Exact acceleration of linear object detectors. In: ECCV (3), Florence (2012)
8. Felzenszwalb, P., Huttenlocher, D.: Pictorial structures for object recognition. Int. J. Comput. Vis. **61**(1), 55–79 (2005)
9. Felzenszwalb, P., McAllester, D., Ramanan, D.: A discriminatively trained, multiscale, deformable part model. In: Proceedings of CVPR, Anchorage (2008)
10. Felzenszwalb, P.F., Girshick, R.B., McAllester, D.A.: Cascade object detection with deformable part models. In: Proceedings of CVPR, San Francisco (2010)
11. Felzenszwalb, P.F., Huttenlocher, D.P.: Distance transforms of sampled functions. Technical report, Cornell CS (2004)
12. Fergus, R., Perona, P., Zisserman, A.: Object class recognition by unsupervised scale-invariant learning. In: Proceedings of CVPR, Madison (2003)
13. Ferrari, V., Marin-Jimenez, M.J., Zisserman, A.: Progressive search space reduction for human pose estimation. In: Proceedings of CVPR, Anchorage (2008)
14. Fleuret, F., Geman, D.: Coarse-to-fine face detection. Int. J. Comput. Vis. **41**(1/2), 85–107 (2001)

15. Girshick, R., Donahue, J., Darrell, T., Malik, J.: Rich feature hierarchies for accurate object detection and semantic segmentation. In: Proceedings of CVPR, Columbus (2014)
16. Girshick, R., Iandola, F., Darrell, T., Malik, J.: Deformable part models are convolutional neural networks. arXiv preprint arXiv:1409.5403 (2014)
17. Girshick, R.B., Felzenszwalb, P.F., McAllester, D.: Discriminatively trained deformable part models, release 5. http://people.cs.uchicago.edu/~rbg/latent-release5/
18. Gray, A.G., Moore, A.W.: Nonparametric density estimation: toward computational tractability. In: SIAM International Conference on Data Mining, San Francisco (2003)
19. Grimson, W.E.L.: Object Recognition by Computer: The Role of Geometric Constraints. MIT Press, Cambridge, MA (1990). ISBN:0-262-07130-4. http://dl.acm.org/citation.cfm?id=102900
20. Huttenlocher, D., Klanderman, G., Rucklidge, W.: Comparing images using the Hausdorff distance. IEEE Trans. PAMI **15**(9), 850–863 (1993)
21. Ihler, A., Sudderh, E., Freeman, W., Willsky, A.: Efficient multiscale sampling from products of Gaussian mixtures. In: Proceedings of NIPS, Vancouver (2003)
22. Ihler, A., Sudderth, E., Freeman, W., Willsky, A.: Efficient sampling of Gaussian distributions. In: Proceedings of NIPS, Vancouver (2004)
23. Jordan, M.: Graphical models. Stat. Sci. **19**, 140–155 (2004)
24. Kokkinos, I.: Rapid deformable object detection using dual-tree branch-and-bound. In: Proceedings of NIPS, Granada (2011)
25. Kokkinos, I.: Bounding part scores for rapid detection with deformable part models. In: 2nd Parts and Attributes Workshop, in Conjunction with ECCV 2012, Florence (2012)
26. Kokkinos, I.: Shufflets: shared mid-level parts for fast multi-category detection. In: ICCV – International Conference on Computer Vision, Sydney (2013)
27. Kokkinos, I., Yuille, A.: Inference and learning with hierarchical shape models. Int. J. Comput. Vis. **93**, 201–225 (2011)
28. Krizhevsky, A., Sutskever, I., Hinton, G.: ImageNet classification with deep convolutional neural networks. In: Proceedings of NIPS, Lake Tahoe (2012)
29. Lampert, C., Blaschko, M., Hofmann, T.: Beyond sliding windows: object localization by efficient subwindow search. In: Proceedings of CVPR, Anchorage (2008)
30. Lampert, C.H.: An efficient divide-and-conquer cascade for nonlinear object detection. In: Proceedings of CVPR, San Francisco (2010)
31. Lehmann, A., Leibe, B., Gool, L.V.: Fast PRISM: branch and bound hough transform for object class detection. Int. J. Comput. Vis. **94**(2), 175–197 (2011)
32. Lempitsky, V., Blake, A., Rother, C.: Image segmentation by branch-and-mincut. In: Proceedings of ECCV, Marseille (2008)
33. Lowe, D.: Perceptual Organization and Visual Recognition. Kluwer, Boston (1985)
34. Lowe, D.: Object recognition from local scale-invariant features. In: Proceedings of ICCV, Kerkyra (1999)
35. Moreels, P., Maire, M., Perona, P.: Recognition by probabilistic hypothesis construction. In: Proceedings of ECCV, Prague, p. 55 (2004)
36. Mundy, J.L., Zisserman, A. (eds.): Geometric invariance in computer vision. MIT Press, Cambridge (1992)
37. Savalle, P.-A., Tsogkas, S., Papandreou, G., Kokkinos, I.: Deformable part models with CNN features. In: 3rd Parts and Attributes Workshop, ECCV, Zurich (2014)
38. Papandreou, G., Kokkinos, I., Savalle, P.A.: Untangling local and global deformations in deep convolutional networks for image classification and sliding window detection. arXiv (2014)
39. Pedersoli, M., Vedaldi, A., Gonzàlez, J.: A coarse-to-fine approach for fast deformable object detection. In: Proceedings of CVPR, Colorado Springs (2011)
40. Pirsiavash, H., Ramanan, D.: Steerable part models. In: CVPR, Providence (2012)
41. Sadeghi, M.A., Forsyth, D.A.: Fast template evaluation with vector quantization. In: NIPS, Lake Tahoe (2013)
42. Sadeghi, M.A., Forsyth, D.A.: 30 hz object detection with DPM V5. In: ECCV, Zurich (2014)

43. Sapp, B., Toshev, A., Taskar, B.: Cascaded models for articulated pose estimation. In: Proceedings of ECCV, Heraklion (2010)
44. Song, H.O., Zickler, S., Althoff, T., Girshick, R.B., Fritz, M., Geyer, C., Felzenszwalb, P.F., Darrell, T.: Sparselet models for efficient multiclass object detection. In: Proceedings of ECCV, Florence (2012)
45. Trulls, E., Tsogkas, S., Kokkinos, I., Sanfeliu, A., Moreno-Noguer, F.: Segmentation-aware deformable part models. In: 2014 IEEE Conference on Computer Vision and Pattern Recognition, CVPR 2014, Columbus, 23–28 June, pp. 168–175 (2014)
46. Vedaldi, A., Zisserman, A.: Sparse kernel approximations for efficient classification and detection. In: Proceedings of CVPR, Providence (2012)
47. Viola, P., Jones, M.: Rapid object detection using a boosted cascade of simple features. In: IEEE Conference on Computer Vision and Pattern Recognition (CVPR), Kauai (2001)
48. Wan, L., Eigen, D., Fergus, R.: End-to-end integration of a convolutional network, deformable parts model and non-maximum suppression. arXiv (2014)
49. Weber, M., Welling, M., Perona, P.: Unsupervised learning of models for recognition. In: Proceedings of ECCV, Dublin (2000)
50. Yang, Y., Ramanan, D.: Articulated human detection with flexible mixtures of parts. IEEE Trans. Pattern Anal. Mach. Intell. **35**(12), 2878–2890 (2013)
51. Zhu, S.C., Mumford, D.: Quest for a stochastic grammar of images. Found. Trends Comput. Graph. Vis. **2**(4), 259–362 (2007)
52. Zhu, X., Ramanan, D.: Face detection, pose estimation, and landmark localization in the wild. In: 2012 IEEE Conference on Computer Vision and Pattern Recognition, Providence, 16–21 June, pp. 2879–2886 (2012)
53. Zisserman, A., Forsyth, D.A., Mundy, J.L., Rothwell, C.A., Liu, J., Pillow, N.: 3D object recognition using invariance. Artif. Intell. **78**, 239–288 (1995)

Part III
Deformable Shape Modeling

Chapter 13
Non-rigid Shape Correspondence Using Surface Descriptors and Metric Structures in the Spectral Domain

Anastasia Dubrovina, Yonathan Aflalo, and Ron Kimmel

Abstract Finding correspondence between non-rigid shapes is at the heart of three-dimensional shape processing. It has been extensively addressed over the last decade, but efficient and accurate correspondence detection still remains a challenging task. *Generalized Multidimensional Scaling* (GMDS) is an approach that finds correspondence by mapping one shape into another, while attempting to preserve distances between pairs of corresponding points on the two shapes. A different approach consists in detecting correspondence between shapes by matching their pointwise surface descriptors. Recently, the *Spectral GMDS* (SGMDS) approach was introduced, according to which the GMDS was re-formulated in the natural spectral domain of the shapes. Here, we propose a method that combines matching based on geodesic distances and pointwise surface descriptors. Following SGMDS, in the proposed solution the entire problem is translated into the spectral domain, resulting in efficient correspondence computation. Efficiency and accuracy of the proposed method are demonstrated by comparing it to state-of-the-art approaches, using a standard correspondence benchmark.

13.1 Introduction

Shape matching is an important component of various three-dimensional shape processing tasks. When rigid shape matching is considered, the problem is reduced to the space of rigid transformations with six degrees of freedom, for which there exist several efficient solutions [3, 10, 18, 40]. However, non-rigid shape matching is challenging to formulate as a low dimensional optimization problem. When formulated as a problem of detection of point-to-point correspondence between shapes, represented by sampled surfaces, for instance, the size of the space of its possible solutions is exponential in the number of samples. Even when it is restricted

---

A. Dubrovina (✉) • Y. Aflalo • R. Kimmel
Technion, Israel Institute of Technology, Haifa, Israel
e-mail: nastyad@cs.technion.ac.il; aflalo@cs.technion.ac.il; ron@cs.technion.ac.il

M. Breuß et al. (eds.), *Perspectives in Shape Analysis*, Mathematics and Visualization, DOI 10.1007/978-3-319-24726-7_13

to the space of isometric transformations, which we focus on in this work, the problem remains a combinatorial one, if the continuity of the matching is ignored.

A common approach to shape matching consists in minimizing a certain measure of dissimilarity between shapes, which is based on robust shape properties, remaining approximately invariant under the transformation relating the shapes. In one line of approaches, the shapes were treated as metric spaces, and the discrepancy between them was measured using the Gromov-Hausdorff distance [17, 26] and its variants. Memoli and Shapiro [41, 42] first suggested to treat sampled shapes within the Gromov-Hausdorff framework. Bronstein et al. [12–14] introduced a numerical method, the GMDS, approximating the Gromov-Hausdorff distance by embedding one shape into another. The GMDS framework can be applied with geodesic [14, 42], diffusion [9, 15, 19], or commute time distances [16].

In order to detect a meaningful initial solution for the minimization problem, feature point detectors and descriptors, such as the spin images [30], heat kernel signature (HKS) [24, 62] and heat kernel maps (HKM) [45], global point signature (GPS) [56], wave kernel signature (WKS) [7], and scale-space representation [64], were employed [5, 22, 28, 53, 66]. To avoid correspondence ambiguities, as in the case of intrinsically symmetric shapes, higher order structures were employed in [61, 65]. Still, the direct consequence of this correspondence problem formulation is the large number of dimensions involved, which makes the problem intractable for large number of potentially corresponding points, or dense point matching. Some of the previous approaches addressed this problem by adopting hierarchical solvers and iterative refinement techniques (e.g., [14, 53, 58, 60, 63]), while others reduced the problem complexity by searching for correspondence between shape segments, instead of point-to-point matching [6, 29, 49].

Mapping shapes into domains other than the original 3D Euclidean space can help resolving some of the difficulties mentioned above. A different embedding domain may reduce the matching complexity significantly, if in this new domain the transformation between shapes can be modelled with a small number of parameters. Such approaches include embedding shapes into a flat Euclidean domain, by means of multidimensional scaling (MDS) [11, 23, 67], or by exploiting the eigenspace of the Laplace-Beltrami operators (LBO) of the shapes [19, 39, 57]. In these domains, isometric transformation between the shapes becomes a rigid one, and can be detected using the aforementioned algorithms for rigid alignment [3, 10, 18, 40]. In [37, 38], the shapes were conformally embedded into disks, and the transformation between them was modelled by a six parameter Möbius transform. The results obtained by Lipman and Funkhouser [38] and Lipman and Daubechies [37] were further improved by Kim et al. [33], where a set of locally good conformal maps were tailored into a globally consistent matching. However, in these approaches, the matching complexity alleviation comes at the cost of possible embedding errors and ambiguities, such as the possibly unbounded conformal factor in Lipman and Funkhouser [38], or the sign ambiguity [39, 57], which decreases the quality of the matching.

Spectral domain has been widely adopted for shape analysis and processing, and in particular – for shape matching. The eigenspace of the Laplace-Beltrami operator, commonly used for this goal [36], was exploited for surface descriptor computation [7, 22, 24, 57, 62], shape descriptor computation [54], shape flatenning [19, 39], distances computation on surfaces [9, 16, 19], spectral mesh compression [31], etc. Lately, various attempts were made to translate the matching problem into the spectral domain. Ovsjanikov et al. [44] introduced the notion of *functional maps*, where, instead of point-wise surface correspondence, they considered correspondence between spaces of functions defined on the two shapes. They showed that any transformation relating the shapes, point-to-point correspondence in particular, could by translated into a linear relationship between their corresponding Laplace-Beltrami eigenspaces, and could be parameterized by a matrix relating these eigenspaces. For computing functional maps, a number of matching regions, or feature points, denoted in Ovsjanikov et al. [44] as functional constraints, was required. Pokrass et al. [50] used maximally stable extremal regions (MSER) [20, 49] for that goal, and suggested a method for simultaneous functional map and region correspondence estimation. Shtern and Kimmel [60] used normalized spectral kernels as functional constraints, achieving state-of-art matching results. To match non-isometric shapes, which do not have compatible Laplace-Beltrami eigenspaces, Kovnatsky et al. [35] suggested constructing a common approximated eigenbase, using joint diagonalization. Ovsjanikov et al. [46] extended the functional maps approach [44] for matching symmetric shapes. There, the matching ambiguity was solved by performing the matching in an appropriate quotient space, where the symmetry was factored out. Rodola et al. [55] combined the functional maps approach with random forests classifier. The classifier was used to produce a dense fuzzy correspondence, which was then regularized using Ovsjanikov et al. [44], applied with geodesic distance-based functional constraints.

13.1.1 Contributions

In this paper, which extends our previous work [1], we consider the L_2 version of the Gromov-Hausdorff framework, augmented with surface descriptor preservation term. We cast the corresponding optimization problem into the spectral domain, using eigenbasis of the Laplace-Beltrami operator, which was proven in to be optimally tailored for representing smooth functions on manifolds [2]. We thus obtain a problem formulation, generalizing the previous spectral matching approaches [1, 44]. The resulting optimization problem is formulated in the standard least squares form, and solved analytically. In addition, in the proposed problem formulation, the distance preservation term allows us to achieve higher matching accuracy than the previous approaches, while maintaining comparable computation cost. The proposed method was evaluated using the popular TOSCA [12] and SCAPE [4] non-rigid shape datasets, and the Princeton correspondence benchmark [32].

13.2 Correspondence Problem Formulation

Let us consider the shape correspondence problem, which consists in searching for the best point to point assignment between two given shapes, S and Q. We assume that each shape is endowed with a distance measure, $d_S : S \times S \to \mathbb{R}^+ \cup \{0\}$ and $d_Q : Q \times Q \to \mathbb{R}^+ \cup \{0\}$, and a set of pointwise k-dimensional surface descriptors $h_S : S \to \mathbb{R}^k$ and $h_Q : Q \to \mathbb{R}^k$. Both the distance measures and the surface descriptors are assumed to be approximately invariant with respect to the transformation relating the shapes S and Q. Then, given a pair of shapes, a discrete point-to-point assignment between them can be defined through an indicator function $\mathscr{C} : S \times Q \to \{0, 1\}$, such that

$$\mathscr{C}(s, q) = \begin{cases} 1, & s \in S \text{ corresponds to } q \in Q, \\ 0, & \text{otherwise.} \end{cases} \tag{13.1}$$

The correspondence problem can then be formulated as a search for an assignment $\mathscr{C}$ that introduces the smallest possible distortion into surface descriptors and metric structures of the shapes, as it was previously suggested in [21, 22, 52]. However, when applied to smooth metric spaces like shapes, modelled by two-dimensional surfaces embedded into $\mathbb{R}^3$, for instance, this is a combinatorial hard problem that ignores their continuous nature.

In this work, we consider a continuous weak form of the above. Instead of the binary assignment $\mathscr{C} : S \times Q \to \{0, 1\}$, we employ a continuous fuzzy correspondence function $p : S \times Q \to \mathbb{R}^+$. We define the space of valid correspondences as all mappings $p(s, q)$ for which the following holds

$$\int_S p(s, q) da_s = 1, \ \forall q \in Q, \qquad \int_Q p(s, q) da_q = 1, \ \forall s \in S. \tag{13.2}$$

For $p(s, q)$ describing pointwise correspondence between S and Q, and any pair of corresponding points $s_0 \in S, q_0 \in Q$,

$$p(s_0, q) = \delta(q - q_0), \ \ p(s, q_0) = \delta(s - s_0).$$

Here, the delta function $\delta(s)$ is defined in a classical sense, such that $\int_S \delta(s) da_s = 1$ and $\int_S f(s)\delta(s - s_0) ds = f(s_0)$, for any continuous function $f : S \to \mathbb{R}$. In a more general setting, for any $q_0 \in Q, s_0 \in S$, $p(s, q_0)$ could be interpreted as the probability that $s \in S$ corresponds to q_0, and $p(s_0, q)$ – as the probability that $q \in Q$ corresponds to s_0.

Given a one-dimensional descriptor $h_S : S \to \mathbb{R}$, its mapping to Q is defined as

$$\int_S h_S(s)p(s,q)da_s. \tag{13.3}$$

For a k-dimensional descriptor, the above transformation is applied separately on each of the k descriptor components. The surface description distortion, introduced by the mapping p, is defined as

$$\int_Q \left\| \int_S h_S(s)p(s,q)da_s - h_Q(q) \right\|^2 da_q. \tag{13.4}$$

Similarly, given a pair of points $s \in S$ and $q \in Q$, the distance between s and the mapping of q to S is defined as

$$\int_S d_S(s,s')p(s',q)da_{s'}, \tag{13.5}$$

and the distance between q and the mapping of s to Q is defined as

$$\int_Q d_Q(q',q)p(s,q')da_{q'}. \tag{13.6}$$

Then, the metric distortion, introduced by p, can be defined as

$$\int_{S\times Q} \left(\int_S d_S(s,s')p(s',q)da_{s'} - \int_Q d_Q(q',q)p(s,q')da_{q'} \right)^2 da_s da_q. \tag{13.7}$$

Discrete setting In practice, we detect correspondences between shapes represented by triangulated meshes. In this discrete setting, the correspondence between the shapes is given by a matrix $\mathbf{P}$, which represents a sampled function $p(s,q)$. We further denote by $\mathbf{A}_S$ and $\mathbf{A}_Q$ diagonal area element matrices, where $(\mathbf{A}_S)_{ii} \approx da_{s_i}$. Given a triangulated surface, an area element da_{s_i} about a specific vertex $s_i \in S$ is approximated by the area of the Voronoi cell about that vertex, as described in Pinkall and Polthier [48]. The symmetric inter-geodesic distance matrices are denoted by $\mathbf{D}_S$ and $\mathbf{D}_Q$, such that $(\mathbf{D}_S)_{ij} = d_S(s_i,s_j)$, that is, the geodesic distance between points $s_i \in S$ and $s_j \in S$.

The above distortion measures are rewritten in matrix notation, as follows: the metric distortion (13.7) reads

$$\|\mathbf{P}\mathbf{A}_S\mathbf{D}_S - \mathbf{D}_Q\mathbf{A}_Q\mathbf{P}\|^2_{S,Q}, \tag{13.8}$$

where $\|\mathbf{F}\|^2_{S,Q}$ is the L_2 square norm of the function $F : Q \times S \to \mathbb{R}$ discretized by the matrix $\mathbf{F}$, defined as $\|\mathbf{F}\|^2_{S,Q} = \text{trace}\left(\mathbf{F}^T\mathbf{A}_Q\mathbf{F}\mathbf{A}_S\right)$. Here, to account for different distribution of sampled points on the two shapes, an inner product of a pair of functions $\mathbf{F}, \mathbf{G}$, defined on S, is computed as $\langle \mathbf{F}, \mathbf{G} \rangle_S = \mathbf{F}^T \mathbf{A}_S \mathbf{G}$.

The descriptor distortion (13.4) reads

$$\|\mathbf{P}\mathbf{A}_S\mathbf{H}_S - \mathbf{H}_Q\|_Q^2. \tag{13.9}$$

where $\|\mathbf{F}\|_Q^2 = \text{trace}\left(\mathbf{F}^T\mathbf{A}_Q\mathbf{F}\right)$. To make the descriptor distortion symmetric, we also add to it the following symmetric term

$$\|\mathbf{H}_S - \mathbf{P}^T\mathbf{A}_Q\mathbf{H}_Q\|_S^2, \tag{13.10}$$

where $\|\mathbf{F}\|_S^2 = \text{trace}\left(\mathbf{F}^T\mathbf{A}_S\mathbf{F}\right)$. When the correspondence is *orthonormal* in the sense $\mathbf{P}^T\mathbf{A}_Q\mathbf{P}\mathbf{A}_S = \mathbf{P}\mathbf{A}_S\mathbf{P}^T\mathbf{A}_Q = \mathbf{I}$, where $\mathbf{I}$ is the identity matrix, Equations (13.9) and (13.10) are equivalent.

The constraints (13.2) in matrix notation take the form

$$\mathbf{P}\mathbf{A}_S\mathbb{1} = \mathbb{1}, \qquad \mathbf{P}^T\mathbf{A}_Q\mathbb{1} = \mathbb{1}, \tag{13.11}$$

where $\mathbb{1}$ is a vector of ones. Thus, we require that a valid correspondence $\mathbf{P}$ is a weighted doubly stochastic matrix.

Finally, the metric and the descriptor distortion measures can be combined into a single optimization problem

$$\begin{aligned} \min_{\mathbf{P}} \quad & \|\mathbf{P}\mathbf{A}_S\mathbf{D}_S - \mathbf{D}_Q\mathbf{A}_Q\mathbf{P}\|_{S,Q}^2 + \mu\left(\|\mathbf{P}\mathbf{A}_S\mathbf{H}_S - \mathbf{H}_Q\|_Q^2 + \|\mathbf{H}_S - \mathbf{P}^T\mathbf{A}_Q\mathbf{H}_Q\|_S^2\right) \\ \text{s.t.} \quad & \mathbf{P}\mathbf{A}_S\mathbb{1} = \mathbb{1}, \quad \mathbf{P}^T\mathbf{A}_Q\mathbb{1} = \mathbb{1}, \end{aligned} \tag{13.12}$$

which we solve to obtain the optimal matching. In order to guarantee convergence to significant solutions, shape descriptors could be augmented with a sparse initial point to point correspondence between shapes. The latter could be represented either as the Dirac delta functions, or Gaussians centred at the corresponding points.

13.3 Problem Formulation in the Spectral Domain

The size of the optimization problem (13.12) is determined by the number of shapes' vertices. Specifically, for shapes S and Q with $|S|$ and $|Q|$ vertices respectively, the size of the correspondence matrix $\mathbf{P}$ is $|Q| \times |S|$. Recently, several papers [1, 44, 50] showed that the computational complexity can be reduced by formulating the matching problem in the spectral domain associated with the shapes. For approximately isometric shape matching, a natural spectral domain is given by the eigendecomposition of the Laplace-Beltrami operator (LBO). There exist several discretizations of the Laplace-Beltrami operator for triangulated meshes. Here, we employ the cotangent-weight scheme suggested in [43, 48]. According to it, a triangulated surface LBO is given by $\mathbf{L} = \mathbf{A}^{-1}\mathbf{W}$, where again $(\mathbf{A})_{ii} \approx da_{s_i}$, and $\mathbf{W}$ is the sparse matrix of cotangent weights. Let us denote by $\boldsymbol{\Phi}_S$ the matrix whose columns are the eigenvectors $\{\phi_i^S\}$ of the discrete Laplace-Beltrami operator of the shape S, and by $\boldsymbol{\Lambda}_S$ – the diagonal matrix of their associated eigenvalues $\{\lambda_i^S\}$. The Laplace-Beltrami eigendecomposition is then posed as a generalized eigenvalue problem

$$\mathbf{W}_S\phi_i^S = \lambda_i^S \mathbf{A}_S \phi_i^S. \tag{13.13}$$

Both W and A are symmetric matrices, and A is positive semi-definite, thus the generalized eiegnvalues λ_i^S are real. Their corresponding eigenfunctions are orthonormal with respect to the weighted inner product $\langle \phi_i^S, \phi_j^S \rangle_S = (\phi_i^S)^T \mathbf{A}_S \phi_j^S = \delta_{ij}$, where δ_{ij} is the Kronecker delta function. It follows that

$$\boldsymbol{\Phi}_S^T \mathbf{A}_S \boldsymbol{\Phi}_S = \mathbf{I}. \tag{13.14}$$

A function $f : S \to \mathbb{R}$ is expressed in the spectral domain of S as

$$f(s) = \sum_i \langle f, \phi_i^S \rangle_S \phi_i^S(s) = \sum_i a_i \phi_i^S(s), \tag{13.15}$$

where $a_i = \langle f, \phi_i^S \rangle_S$ are the spectral decomposition coefficients.

Correspondence matrix $\boldsymbol{P}$ Given the fuzzy correspondence $p(s, q) : S \times Q \to [0, 1]$, we first express it in the spectral domain of S

$$p(s, q) = \sum_i \langle p(s, q), \phi_i^S(s) \rangle_S \phi_i^S(s) = \sum_i \alpha_i(q) \phi_i^S(s).$$

For each i, $\alpha_i(q)$ is a scalar function defined on Q, and thus it in turn can be expressed in the spectral domain of Q as

$$\alpha_i(q) = \sum_j \left\langle \alpha_i(q), \phi_j^Q(q) \right\rangle_Q \phi_j^Q(q) = \sum_j \alpha_{ij} \phi_j^Q(q).$$

We now combine the last two expressions into a single spectral representation of the correspondence $p(s, q)$ in spectral domains of both S and Q

$$p(s, q) = \sum_i \sum_j \alpha_{ij} \phi_j^Q(q) \phi_i^S(s). \tag{13.16}$$

In matrix notation, the above translates to

$$\mathbf{P} = \boldsymbol{\Phi}_Q \boldsymbol{\alpha} \boldsymbol{\Phi}_S^T, \tag{13.17}$$

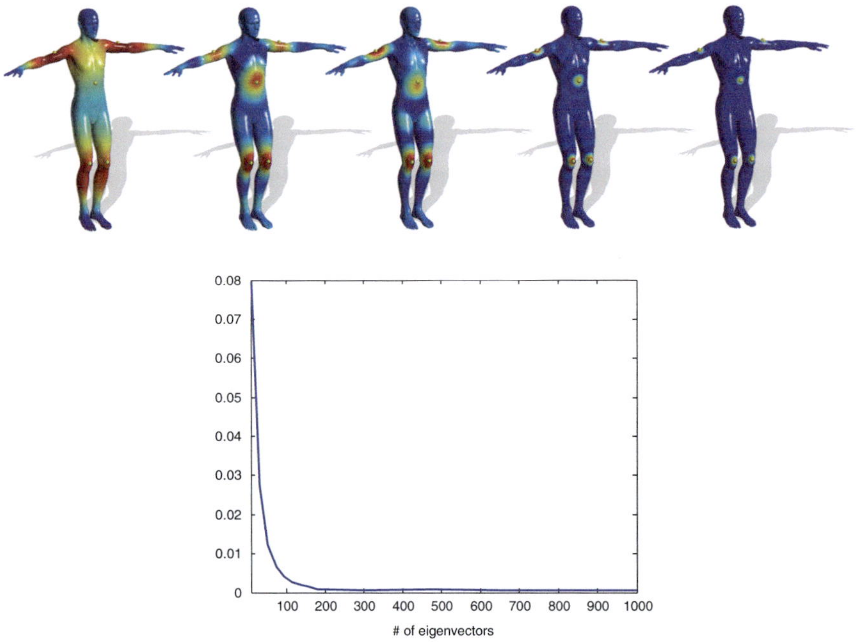

Fig. 13.1 *Top*: Mapping 5 surface-points (indicated by *yellow spheres*) to their own location, using (from left to right) 10, 50, 100, 500 and 1000 eigenvectors of the Laplace-Beltrami operator, respectively. *Bottom*: Geodesic distance error between surface points and their mapping to themselves using 10–1000 eigenfunctions, averaged over 50 points randomly sampled from S

where the entries of the matrix $\boldsymbol{\alpha}$ are $\{\boldsymbol{\alpha}\}_{ij} = \alpha_{ij}$. Thus, knowing $\boldsymbol{\alpha}$ allows us to compute $\mathbf{P}$ – a fact that we will exploit in our spectral matching formulation, which will be described in the section.

Next, we study the effect of truncating the number of the eigenvectors used in the spectral representation, to only a few leading eigenvectors. For this, consider a mapping from a shape to itself, that is $S = Q$, so that the mapping is given by $\mathbf{P} = \boldsymbol{\Phi}_S \boldsymbol{\alpha} \boldsymbol{\Phi}_S^T$, and $\boldsymbol{\alpha} = \mathbf{I}$. Figure 13.1 illustrates how truncating the eigenbasis to a varying number of eigenvectors affects the location of the surface points mapped to themselves using such mapping $\mathbf{P}$. Specifically, if the original locations of the points are given by delta functions, after the mapping we obtain filtered delta functions, as shown in Fig. 13.1 (top). The accuracy of the mapping, measured by the sum of geodesic distances between the original delta function locations and maxima of their filtered versions, and weighted by $\sqrt{\mathbf{A}_S}$, is shown in Fig. 13.1 (bottom).

From these experiments it follows that using the leading $m = 100$ eigenfunctions allows faithful representation of the correspondence. Furthermore, truncating the eigenbasis to m eigenvectors allows us to reduce the size of the matching problem: instead of searching for the correspondence matrix $\mathbf{P}$ of size $|Q| \times |S|$, we now need to compute the matrix $\boldsymbol{\alpha}$, relating the bases $\boldsymbol{\Phi}_S$ and $\boldsymbol{\Phi}_Q$, of size $m \times m$. Let us now translate the rest of the correspondence problem ingredients into the spectral domain.

The double stochastic conditions Equation (13.11) in the spectral domain take the form

$$\boldsymbol{\Phi}_Q \boldsymbol{\alpha} \boldsymbol{\Phi}_S^T \mathbf{A}_S \mathbb{1} = \mathbb{1} \quad \text{and} \quad \boldsymbol{\Phi}_S \boldsymbol{\alpha}^T \boldsymbol{\Phi}_Q^T \mathbf{A}_Q \mathbb{1} = \mathbb{1}.$$

We further denote $\boldsymbol{\eta}_S = \boldsymbol{\Phi}_S^T \mathbf{A}_S \mathbb{1}$ and $\boldsymbol{\eta}_Q = \boldsymbol{\Phi}_Q^T \mathbf{A}_Q \mathbb{1}$, and re-write the above as

$$\boldsymbol{\alpha} \boldsymbol{\eta}_S = \boldsymbol{\eta}_Q \quad \text{and} \quad \boldsymbol{\alpha}^T \boldsymbol{\eta}_Q = \boldsymbol{\eta}_S. \tag{13.18}$$

Distances and descriptors Given a d-dimensional descriptor $\mathbf{H}_S : S \to \mathbb{R}^d$, its spectral representation in the eigenspace of S is

$$\mathbf{H}_S = \boldsymbol{\Phi}_S \boldsymbol{\delta}_S, \tag{13.19}$$

where $\boldsymbol{\delta}_S = \boldsymbol{\Phi}_S^T \mathbf{A}_S \mathbf{H}_S$. In a continuous setting, this reads

$$\delta_{ij} = \int_S (H_S(s))_j \phi_i(s) da_s,$$

where $(H_S(s))_j$ is the j-th element of the descriptor H_S at $s \subset S$.

The spectral representation of the distance $d_S : S \times S \to \{\mathbb{R}^+, 0\}$ is a special case of Equation (13.16), for $Q = S$, and is given by

$$d(s, s') = \sum_i \sum_j \beta_{ij} \phi_j^S(q) \phi_i^S(s), \tag{13.20}$$

with the coefficients β_{ij} computed using

$$\beta_{ij} = \left\langle \left\langle d(s, s'), \phi_i^S(s) \right\rangle_S, \phi_j^S(s') \right\rangle_S = \int_{S \times S} d_S(s, s') \phi_i^S(s) \phi_j^S(s') da_s da_{s'}.$$

In matrix formulation, the above becomes

$$\mathbf{D}_S = \boldsymbol{\Phi}_S \boldsymbol{\beta}_S \boldsymbol{\Phi}_S^T,$$

with

$$\boldsymbol{\beta}_S = \boldsymbol{\Phi}_S^T \mathbf{A}_S \mathbf{D}_S \mathbf{A}_S \boldsymbol{\Phi}_S.$$

When, instead of n, $m \ll n$ eigenvectors of the LBO are used, we obtain an approximate distance measure, that we denote by $\tilde{\mathbf{D}}_S$.

When matching shapes with large number of vertices, the spectral representation of $\mathbf{D}_S$ can be efficiently approximated even without knowing all inter-geodesic distances. Instead, only distances computed between a small set of sampled points, usually up to 5 % of the total number of shape points, may be used to estimate the spectral distance representation $\boldsymbol{\beta}_S$. The estimation is based on the fact that distances, computed from the sampled points to the rest of the surface, capture the global structure of the shape, and the smooth local structure can be interpolated from it, using the leading eigenfunctions of the Laplace-Beltrami operator. Here, the spectral representation $\boldsymbol{\beta}$ was obtained by minimizing the bi-harmionic equation, used for distance interpolation [59].

13.4 Correspondence in Spectral Domain

We now have all the necessary ingredients to re-formulate the complete correspondence problem (13.12) in the spectral domain. Using

$$\mathbf{P}\mathbf{A}_S\mathbf{D}_S = \boldsymbol{\Phi}_Q\boldsymbol{\alpha}\boldsymbol{\Phi}_S^T\mathbf{A}_S\boldsymbol{\Phi}_S\boldsymbol{\beta}_S\boldsymbol{\Phi}_S^T = \boldsymbol{\Phi}_Q\boldsymbol{\alpha}\boldsymbol{\beta}_S\boldsymbol{\Phi}_S^T,$$

and, similarly,

$$\tilde{\mathbf{D}}_Q\mathbf{A}_Q\mathbf{P} = \boldsymbol{\Phi}_Q\boldsymbol{\beta}_Q\boldsymbol{\alpha}\boldsymbol{\Phi}_S^T,$$

the first term of the objective function in Equation (13.12) is translated into the spectral domain by

$$\|\mathbf{P}\mathbf{A}_S\mathbf{D}_S - \mathbf{D}_Q\mathbf{A}_Q\mathbf{P}\|_{S,Q} = \|\boldsymbol{\Phi}_Q\left(\boldsymbol{\alpha}\boldsymbol{\beta}_S - \boldsymbol{\beta}_Q\boldsymbol{\alpha}\right)\boldsymbol{\Phi}_S^T\|_{S,Q}.$$

In the above derivation, we used the orthonormality of the LBO eigenfunctions, namely $\boldsymbol{\Phi}_S^T\mathbf{A}_S\boldsymbol{\Phi}_S = \mathbf{I}$, $\boldsymbol{\Phi}_Q^T\mathbf{A}_Q\boldsymbol{\Phi}_Q = \mathbf{I}$. We can show that $\|\boldsymbol{\Phi}_Q\mathbf{F}\boldsymbol{\Phi}_S^T\|_{S,Q} = \|\mathbf{F}\|_F^2$, so that the distance distortion terms reads

$$\|\mathbf{P}\mathbf{A}_S\tilde{\mathbf{D}}_S - \tilde{\mathbf{D}}_Q\mathbf{A}_Q\mathbf{P}\|_{S,Q} = \|\boldsymbol{\alpha}\boldsymbol{\beta}_S - \boldsymbol{\beta}_Q\boldsymbol{\alpha}\|_F^2. \tag{13.21}$$

Similarly, for the descriptor distortion term we have

$$\mathbf{P}\mathbf{A}_S\mathbf{H}_S = \boldsymbol{\Phi}_Q\boldsymbol{\alpha}\boldsymbol{\Phi}_S^T\mathbf{A}_S\boldsymbol{\Phi}_S\boldsymbol{\delta}_S = \boldsymbol{\Phi}_Q\boldsymbol{\alpha}\boldsymbol{\delta}_S,$$
$$\mathbf{P}^T\mathbf{A}_Q\mathbf{H}_Q = \boldsymbol{\Phi}_S\boldsymbol{\alpha}^T\boldsymbol{\Phi}_Q^T\mathbf{A}_Q\boldsymbol{\Phi}_Q\boldsymbol{\delta}_Q = \boldsymbol{\Phi}_S\boldsymbol{\alpha}^T\boldsymbol{\delta}_Q.$$

We can also show that $\|\boldsymbol{\Phi}_Q\mathbf{F}\|_Q = \|\mathbf{F}\|_F^2$, so that the descriptor distortion term reads

$$\begin{aligned}\|\mathbf{PA}_S\mathbf{H}_S - \mathbf{H}_Q\|_Q^2 + \|\mathbf{H}_S - \mathbf{P}^T\mathbf{A}_Q\mathbf{H}_Q\|_S^2 =& \|\boldsymbol{\Phi}_Q(\boldsymbol{\alpha}\boldsymbol{\delta}_S - \boldsymbol{\delta}_Q)\|_Q^2 + \|\boldsymbol{\Phi}_S(\boldsymbol{\delta}_S - \boldsymbol{\alpha}^T\boldsymbol{\delta}_Q)\|_S^2 \\ =& \|\boldsymbol{\alpha}\boldsymbol{\delta}_S - \boldsymbol{\delta}_Q\|_F^2 + \|\boldsymbol{\delta}_S - \boldsymbol{\alpha}^T\boldsymbol{\delta}_Q\|_F^2.\end{aligned} \tag{13.22}$$

Finally, we re-write the optimization problem (13.12) in the spectral domain, as follows

$$\begin{aligned}&\min_{\boldsymbol{\alpha}} \ \|\boldsymbol{\alpha}\boldsymbol{\beta}_S - \boldsymbol{\beta}_Q\boldsymbol{\alpha}\|_F^2 + \mu\left(\|\boldsymbol{\alpha}\boldsymbol{\delta}_S - \boldsymbol{\delta}_Q\|_F^2 + \|\boldsymbol{\delta}_S - \boldsymbol{\alpha}^T\boldsymbol{\delta}_Q\|_F^2\right) \\ &\text{s.t.} \\ &\qquad \boldsymbol{\alpha}\boldsymbol{\eta}_S = \boldsymbol{\eta}_Q, \quad \text{and} \quad \boldsymbol{\alpha}^T\boldsymbol{\eta}_Q = \boldsymbol{\eta}_S.\end{aligned} \tag{13.23}$$

Relation to functional maps We started the analysis above by formulating the problem of pointwise shape correspondence in terms of the correspondence matrix $\mathbf{P}$. We then obtained spectral problem formulation (13.23) by translating each of the problem's components into the spectral domain, where the unknown is the matrix $\boldsymbol{\alpha}$, relating between eigenspaces of the shapes S and Q. In the functional maps approach [44], similar problem formulation was derived based on the requirement of preservation of functions mapped from one shape to another, and the spectral correspondence matrix $\boldsymbol{\alpha}$ was termed the *functional map*.

Pointwise correspondence computation For each vertex $s \in S$, we can compute its corresponding vertex on Q by mapping an indicator function $\mathbb{1}_s$, defined in S and centred at the vertex s, to Q, using the obtained spectral mapping $\boldsymbol{\alpha}$

$$\mathbf{P}\mathbb{1}_s = \boldsymbol{\Phi}_Q\boldsymbol{\alpha}\boldsymbol{\Phi}_S^T\mathbf{A}_S\mathbb{1}_s.$$

The vertex $q \in Q$ corresponding to s is given by

$$q = \operatorname*{argmax}_{\tilde{q}\in Q} \ (\mathbf{P}\mathbb{1}_s)(\tilde{q}).$$

Alternatively, the pointiwse correspondence between the shapes may be computed as suggested in Ovsjanikov et al. [44], by directly comparing the spectral representations of indicator functions defined on the two shapes.

13.4.1 Double Stochasticity Constraints

The first eigenvalue of the Laplace-Beltrami operator is $\lambda_0 = 0$, with the corresponding constant eigenvector $\phi_1 = \tau(1\,1\,1\,1\,1 \cdots 1)^T$. The eigenbasis $\boldsymbol{\Phi}$ is orthonormal with respect to the inner product defined on the shape. In particular,

$\phi_1^T \mathbf{A} \phi_1 = 1$, and therefore the constant τ is

$$\tau = \left(\sum_i \mathbf{A}_{ii} \right)^{-1/2} .$$

Let us recall that $\boldsymbol{\eta}_S$ and $\boldsymbol{\eta}_Q$ are defined as $\boldsymbol{\eta}_S = \boldsymbol{\Phi}_S^T \mathbf{A}_S \mathbb{1}$ and $\boldsymbol{\eta}_Q = \boldsymbol{\Phi}_Q^T \mathbf{A}_Q \mathbb{1}$. Again, using the orthonormality of $\boldsymbol{\Phi}$, that is $\phi_j^T \mathbf{A} \phi_1 = 0, \forall j \neq 1$, we obtain

$$\eta = \boldsymbol{\Phi}^T \mathbf{A} \mathbb{1} = \boldsymbol{\Phi}^T \mathbf{A} \, \tau^{-1} \phi_1 = \tau^{-1} (1 \; 0 \; 0 \cdots \; 0)^T .$$

For approximately isometric shapes, we can assume that $\tau_S \cong \tau_Q$, that is the total area of the shape is preserved by an isometric transformation. Therefore, the constraints $\boldsymbol{\alpha} \boldsymbol{\eta}_S = \boldsymbol{\eta}_Q, \boldsymbol{\alpha}^T \boldsymbol{\eta}_Q = \boldsymbol{\eta}_S$ become

$$\boldsymbol{\alpha} \begin{pmatrix} 1 \\ 0 \\ \vdots \\ 0 \end{pmatrix} = \begin{pmatrix} 1 \\ 0 \\ \vdots \\ 0 \end{pmatrix} \quad \text{and} \quad \boldsymbol{\alpha}^T \begin{pmatrix} 1 \\ 0 \\ \vdots \\ 0 \end{pmatrix} = \begin{pmatrix} 1 \\ 0 \\ \vdots \\ 0 \end{pmatrix} ,$$

or, equivalently,

$$\boldsymbol{\alpha} = \begin{pmatrix} 1 & 0 & \cdots & 0 \\ 0 & \alpha_{22} & \cdots & \alpha_{2m} \\ \vdots & \vdots & \vdots & \vdots \\ 0 & \alpha_{m2} & \cdots & \alpha_{mm} \end{pmatrix} .$$

Note that the above constraint preserves the constant eigenvectors of the two shapes.

13.4.2 Unconstrained Problem Formulation

Let us define a matrix $\tilde{\boldsymbol{\alpha}}$, such that

$$\tilde{\boldsymbol{\alpha}} = \begin{pmatrix} \alpha_{22} & \cdots & \alpha_{2m} \\ \vdots & \vdots & \vdots \\ \alpha_{m2} & \cdots & \alpha_{mm} \end{pmatrix} \quad \text{and} \quad \boldsymbol{\alpha} = \begin{pmatrix} 1 & 0 \\ 0 & \tilde{\boldsymbol{\alpha}} \end{pmatrix} . \tag{13.24}$$

Similarly, let us define $\beta^{11}, \boldsymbol{\beta}^{(1)}, \boldsymbol{\beta}^{(2)}, \tilde{\boldsymbol{\beta}}$, such that

$$\boldsymbol{\beta} = \begin{pmatrix} \beta^{11} & \boldsymbol{\beta}^{(1)} \\ \boldsymbol{\beta}^{(2)} & \tilde{\boldsymbol{\beta}} \end{pmatrix}, \tag{13.25}$$

where $\beta^{11} \in \mathbb{R}, \boldsymbol{\beta}^{(1)} \in \mathbb{R}^{1\times(N-1)}, \boldsymbol{\beta}^{(2)} \in \mathbb{R}^{(N-1)\times 1}, \tilde{\boldsymbol{\beta}} \in \mathbb{R}^{(N-1)\times(N-1)}$, and N is the number of points on the shape. In addition, let us define

$$\boldsymbol{\delta} = \begin{pmatrix} \boldsymbol{\delta}^{(1)} \\ \tilde{\boldsymbol{\delta}} \end{pmatrix}, \tag{13.26}$$

where $\boldsymbol{\delta}^{(1)} \in \mathbb{R}^{1\times N}$, $\tilde{\boldsymbol{\delta}} \in \mathbb{R}^{(N-1)\times N}$. Thus, matrix products $\boldsymbol{\alpha}\boldsymbol{\beta}_S$, $\boldsymbol{\beta}_Q\boldsymbol{\alpha}$ can be written as

$$\boldsymbol{\alpha}\boldsymbol{\beta}_S = \begin{pmatrix} \beta_S^{11} & \boldsymbol{\beta}_S^{(1)} \\ \tilde{\boldsymbol{\alpha}}\boldsymbol{\beta}_S^{(2)} & \tilde{\boldsymbol{\alpha}}\tilde{\boldsymbol{\beta}}_S \end{pmatrix}, \quad \boldsymbol{\beta}_Q\boldsymbol{\alpha} = \begin{pmatrix} \beta_Q^{11} & \boldsymbol{\beta}_Q^{(1)}\tilde{\boldsymbol{\alpha}} \\ \boldsymbol{\beta}_Q^{(2)} & \tilde{\boldsymbol{\beta}}_Q\tilde{\boldsymbol{\alpha}} \end{pmatrix}, \tag{13.27}$$

and $\boldsymbol{\alpha}\boldsymbol{\delta}_S$, $\boldsymbol{\alpha}^T\boldsymbol{\delta}_Q$ can be written as

$$\boldsymbol{\alpha}\boldsymbol{\delta}_S = \begin{pmatrix} \boldsymbol{\delta}_S^1 \\ \tilde{\boldsymbol{\alpha}}\tilde{\boldsymbol{\delta}}_S \end{pmatrix}, \quad \boldsymbol{\alpha}^T\boldsymbol{\delta}_Q = \begin{pmatrix} \boldsymbol{\delta}_Q^1 \\ \tilde{\boldsymbol{\alpha}}^T\tilde{\boldsymbol{\delta}}_Q \end{pmatrix}. \tag{13.28}$$

The optimization problem (13.23) then reads

$$\min_{\tilde{\boldsymbol{\alpha}}} \; \|\tilde{\boldsymbol{\alpha}}\tilde{\boldsymbol{\beta}}_S - \tilde{\boldsymbol{\beta}}_Q\tilde{\boldsymbol{\alpha}}\|_F^2 + \|\boldsymbol{\beta}_S^{(1)} - \boldsymbol{\beta}_Q^{(1)}\tilde{\boldsymbol{\alpha}}\|^2 + \|\tilde{\boldsymbol{\alpha}}\boldsymbol{\beta}_S^{(2)} - \boldsymbol{\beta}_Q^{(2)}\|^2 + \\ \mu\left(\|\tilde{\boldsymbol{\alpha}}\tilde{\boldsymbol{\delta}}_S - \tilde{\boldsymbol{\delta}}_Q\|_F^2 + \|\tilde{\boldsymbol{\delta}}_S - \tilde{\boldsymbol{\alpha}}^T\tilde{\boldsymbol{\delta}}_Q\|_F^2\right). \tag{13.29}$$

Finally, to obtain $\boldsymbol{\alpha}$, one can solve the above problem numerically, for instance, using the PBM optimization toolbox by M. Zibulevsky [8], and construct $\boldsymbol{\alpha}$ from $\tilde{\boldsymbol{\alpha}}$ using Equation (13.24). Another possibility is to formulate (13.29) as a standard least squares problem, and solve it analytically – as demonstrated in the next section. Optionally, the least squares solution may be followed by the post-processing iterative refinement [44]. While Ovsjanikov et al. [44] also suggested using least squares to compute functional maps, they employed only the descriptor preservation term, and did not enforce bi-stochastisity of the correspondence.

13.4.3 Alternative Formulation as a Standard Least Squares Problem

Denote $\bar{\boldsymbol{\alpha}} = \text{vec}\,(\tilde{\boldsymbol{\alpha}}) = \tilde{\boldsymbol{\alpha}}(:)$. Then

$$\begin{aligned}
\text{vec}\left(\tilde{\boldsymbol{\alpha}}\tilde{\boldsymbol{\beta}}_S\right) &= (\tilde{\boldsymbol{\beta}}_S^T \otimes \mathbf{I})\,\bar{\boldsymbol{\alpha}},\\
\text{vec}\left(\tilde{\boldsymbol{\beta}}_Q\tilde{\boldsymbol{\alpha}}\right) &= (\mathbf{I} \otimes \tilde{\boldsymbol{\beta}}_Q)\,\bar{\boldsymbol{\alpha}},\\
\text{vec}\left(\tilde{\boldsymbol{\alpha}}\tilde{\boldsymbol{\delta}}_S\right) &= (\tilde{\boldsymbol{\delta}}_S^T \otimes \mathbf{I})\,\bar{\boldsymbol{\alpha}},\\
\text{vec}\left(\tilde{\boldsymbol{\alpha}}^T\tilde{\boldsymbol{\delta}}_Q\right) &= \boldsymbol{\Pi}\,((\mathbf{I} \otimes \tilde{\boldsymbol{\delta}}_S^T)\,\bar{\boldsymbol{\alpha}}),
\end{aligned} \tag{13.30}$$

where $\otimes$ denotes Kronecker tensor product, and $\boldsymbol{\Pi}$ is a matrix that satisfies the relationship $\text{vec}\,(\tilde{\boldsymbol{\alpha}}) = \boldsymbol{\Pi}\ \text{vec}\left(\tilde{\boldsymbol{\alpha}}^T\right)$. By further denoting

$$\begin{aligned}
\mathbf{M}_\beta &= \tilde{\boldsymbol{\beta}}_S^T \otimes\ \mathbf{I} - \mathbf{I} \otimes \tilde{\boldsymbol{\beta}}_Q,\\
\mathbf{M}_{\delta_S} &= \tilde{\boldsymbol{\delta}}_S^T \otimes \mathbf{I},\ \ \mathbf{M}_{\delta_Q} = \boldsymbol{\Pi}\,(\mathbf{I} \otimes \tilde{\boldsymbol{\delta}}_S^T),\\
\bar{\boldsymbol{\delta}}_Q &= \text{vec}\left(\tilde{\boldsymbol{\delta}}_Q\right),\ \ \bar{\boldsymbol{\delta}}_S = \text{vec}(\tilde{\boldsymbol{\delta}}_S),\\
\mathbf{M}_{\beta_S} &= \boldsymbol{\beta}_S^{(2)} \otimes \mathbf{I},\ \ \mathbf{M}_{\beta_Q} = \mathbf{I} \otimes \boldsymbol{\beta}_Q^{(1)},\\
\bar{\boldsymbol{\beta}}_S &= \text{vec}\left(\boldsymbol{\beta}_S^{(1)}\right),\ \ \bar{\boldsymbol{\beta}}_Q = \text{vec}\left(\boldsymbol{\beta}_Q^{(2)}\right),
\end{aligned} \tag{13.31}$$

we reformulate (13.29) as

$$\begin{aligned}
\min_{\bar{\alpha}}\ & \|\mathbf{M}_\beta\bar{\boldsymbol{\alpha}}\|^2 + \|\mathbf{M}_{\beta_Q}\bar{\boldsymbol{\alpha}} - \bar{\boldsymbol{\beta}}_S\|^2 + \|\mathbf{M}_{\beta_S}\bar{\boldsymbol{\alpha}} - \bar{\boldsymbol{\beta}}_Q\|^2\\
& + \mu\left(\|\mathbf{M}_{\delta_S}\bar{\boldsymbol{\alpha}} - \bar{\boldsymbol{\delta}}_Q\|^2 + \|\mathbf{M}_{\delta_Q}\bar{\boldsymbol{\alpha}} - \bar{\boldsymbol{\delta}}_S\|^2\right).
\end{aligned} \tag{13.32}$$

This is a standard least squares problem, for which the optimal $\bar{\boldsymbol{\alpha}}$ is computed analytically.

13.5 Experimental Results

To evaluate the proposed method we used two publicly available datasets – TOSCA [12] and SCAPE [5]. The first dataset, TOSCA, contains 90 synthetic human and animal shapes, with known point-to-point correspondences between shapes in the

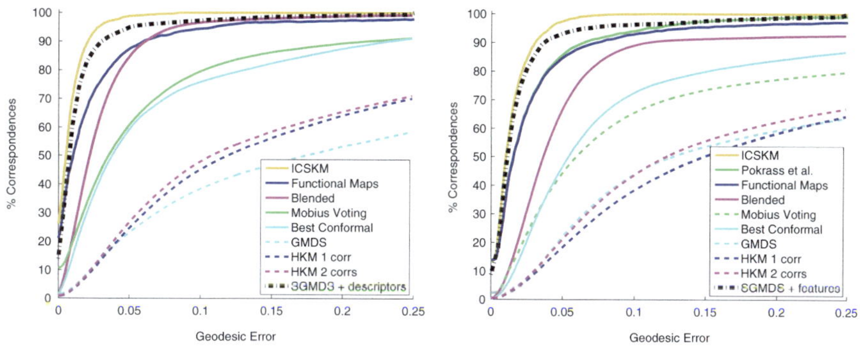

Fig. 13.2 Quantitative evaluation of the proposed approach (SGMDS + features) on all the shapes from the TOSCA (*left*) and SCAPE (*right*) datasets, using the evaluation protocol from the Princeton benchmark [33]

same class (cats, dogs, humans, etc.). The number of vertices of the shapes in this dataset varies between approximately 4000 and 50,000. The second dataset, SCAPE, contains scans of real human bodies in different poses.

In our experiments, we first sub-sampled the shapes, to obtain approximately 5 % of the shapes' vertices, using the farthest point sampling method [25, 27], and pre-computed geodesic distances between them using the fast marching method [34]. These distance were then used to compute the spectral representations $\boldsymbol{\beta}_S, \boldsymbol{\beta}_Q$. The Wave Kernel Signatures [7] were used as point-wise surface descriptors.

We implemented the proposed method in Matlab, with time consuming parts of the code implemented as Mex files in C++. All the experiments were executed on a 2.7 GHz Intel Core i7 laptop with 16 GB RAM. The solution of the least squares problem (13.32) was implemented using Matlab sparse matrix support. In our experiments we used up to 100 eigenvectors of the LBO, for which the computation time of least squares problem (13.32) was of order of 40 s.

In Fig. 13.2, we compare the proposed method with existing algorithms, using the Princeton shape correspondence benchmark [33] and the evaluation procedure suggested therein. The experiments were conducted on both TOSCA [12] and SCAPE [5] datasets. For other methods, we used the information provided in Kim et al. and Ovsjanikov et al. [33, 44], and the results provided by Pokrass et al. [50]. The evaluation was performed as follows: in the protocol of Kim et al. [33], the ground-truth correspondences between small subset of feature points on the shapes are given. Then, given a predefined set of pairs of shapes, each pair belonging to the same group (cat, dog, etc.), we compute a mapping between each pair, and measure the geodesic distances between true locations of these feature points, and their mappings. These geodesic distances are then normalized by the shape's squared root of the area. Figure 13.2 presents the distortion curves of different algorithms, given by the percentage of the points falling within a certain geodesic distance from their true location, and averaged over all pairs of shapes. The proposed method produces accurate matching results, outperforming the existing methods, except for the recently proposed Iterative Closest Spectral Kernel Maps (ICSKM) by Shtern and Kimmel [60]. The ICSKM iterates between map estimation and descriptor computation, while the proposed approach uses a single iteration of the least squares solver, to estimate the matching. It is possible to apply the ICSKM as a refinement step for the proposed method, to further improve its performance. The results of the proposed method were obtained using a small set of seven known initial correspondences between pairs of shapes, formulated as functional constraints. Note that similar constraints, formulated as a initial pointiwse or region-wise correspondence, or as eigenvector pre-alignment, were also employed by [44, 50, 60], for obtaining meaningful correspondence results.

The effect of different combinations of descriptor and distance preservation terms Figure 13.3 presents the distortion curves, obtained, as detailed above, for TOSCA dataset, with different algorithm configurations. First, the method was applied using

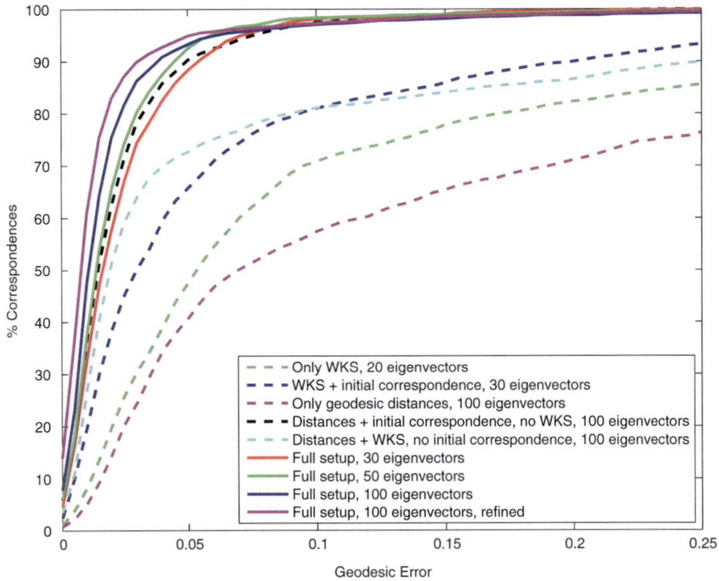

Fig. 13.3 Comparison of different algorithm setups, TOSCA dataset

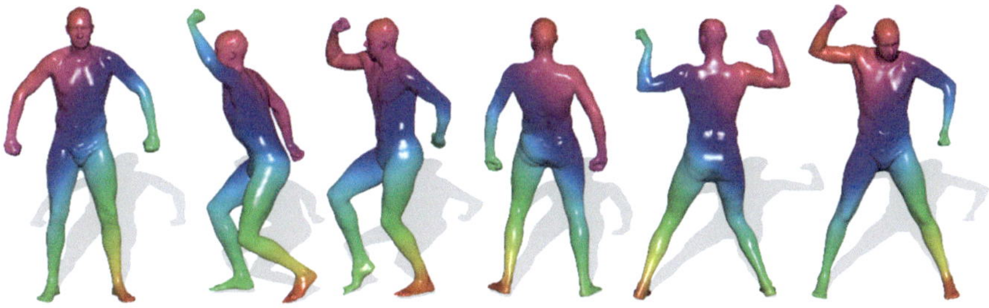

Fig. 13.4 Dense point-to-point correspondence between six almost isometric shapes from the SCAPE dataset

only the descriptor preservation term, with and without initial sparse correspondence set, which is equivalent to the functional maps method [44]. In this setting, the method is effectively limited to using only up to 30 eigenvectors; because of the low descriptor rank, when more eigenvectors are used, the least squares problem (13.32) becomes ill-posed.

When adding the spectral distance representation term, it is possible to extend the eigenbasis and include more eigenfunctions, to significantly improve accuracy of the results. Thus, the distance preservation term acts as a regularization for the least squares problem. In our test cases, we used 30–100 first eigenfunctions of the Laplace-Beltrami operator – much more than in the previous setup, with only the descriptor preservation term. Note that the matching accuracy gets higher as more eigenvectors are used. Having an initial sparse correspondence between the shapes further improves the algorithm results. Finally, when both spectral distance and descriptor representations are used, the proposed method achieves best results, outperforming all previous setups. The results can be slightly improved further, using the refinement procedure, suggested in Ovsjanikov et al. [44].

Additional correspondence examples Figure 13.4 visualizes point-to-point correspondences between several almost isometric shapes from the SCAPE dataset [5], obtained using the proposed method. Figures 13.5 and 13.6 visualize the mapping quality for shapes from the TOSCA dataset, by transferring the eigenvectors of the Laplace-Beltrami operator and smoothed delta functions, from one shape to another. In all the examples, the proposed method produces visibly plausible correspondence results. Note that there exists an inherent correspondence ambiguity problem when matching intrinsically symmetric shapes [21, 47, 51]. The propose method would produce one of the possible matches, which could be affected by the choice of the initial correspondence or surface descriptors.

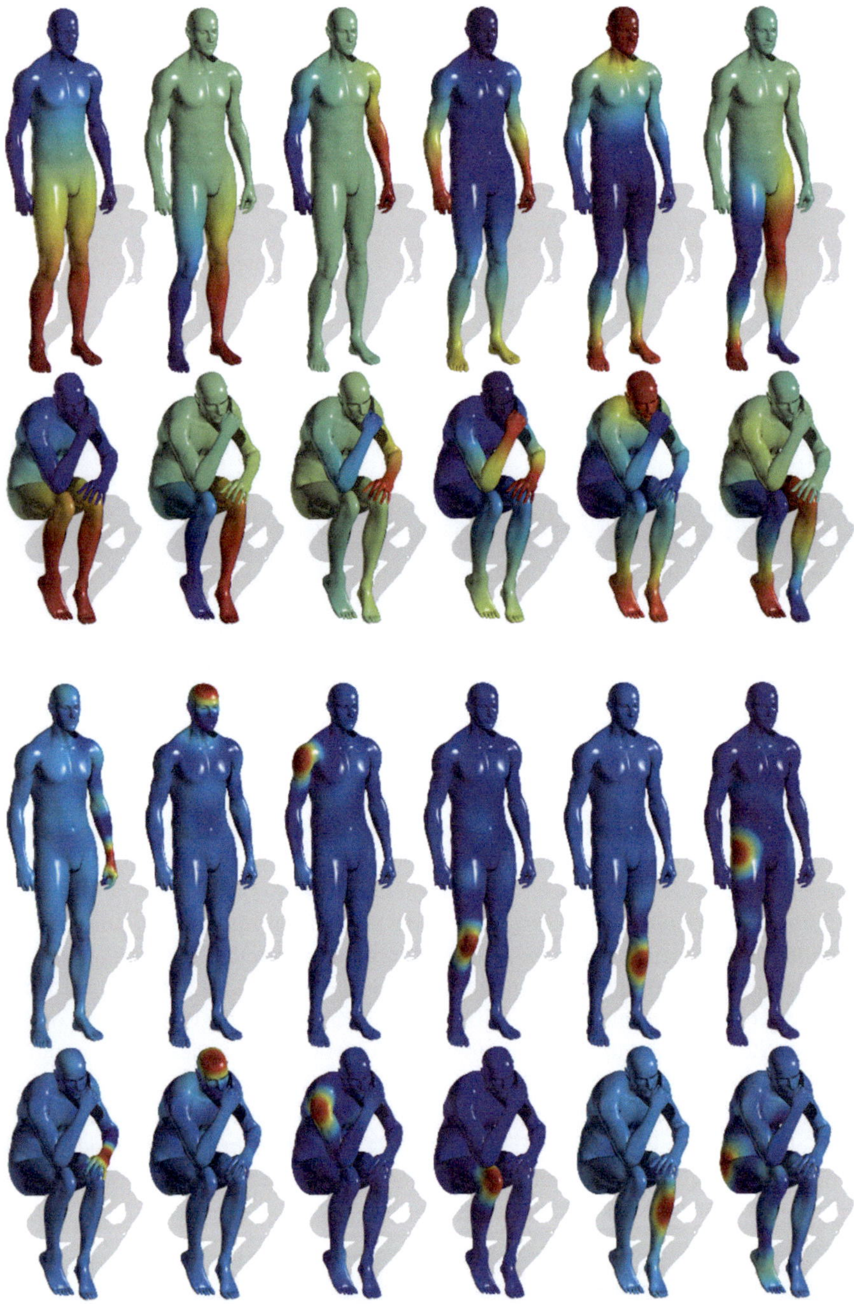

Fig. 13.5 Mapping functions between two almost isometric shapes via our spectral matching. *Top*: eigenfunctions of the Laplace-Beltrami operator. *Bottom*: smoothed point indicator functions

Fig. 13.6 Mapping functions between two almost isometric shapes via our spectral matching. *Top*: eigenfunctions of the Laplace-Beltrami operator. *Bottom*: smoothed point indicator functions

13.6 Conclusions

In this paper, we suggested extending the spectral generalized multidimensional scaling (SGMDS) method, by incorporating additional information from pointwise surface descriptors. The discrepancy measure, minimized by the algorithm, was defined as a sum of the metric distortion, and the surface descriptor distortion, introduced by the mapping. We showed that combination of these two distortion measures into a single optimization problem improves accuracy of the matching, compared to the case when each of them is used separately. By exploiting the smoothness of the inter-geodesic distances and surface descriptors, we were able to translate the problem into the spectral domain, where the matching computation is extremely efficient. In our future research, we plan to extend the proposed method to detect correspondence between non-isometric shapes, or shapes with local scale differences, but with similar global structures.

Acknowledgements We thank Alon Shtern for fruitful discussions. This research was supported by European Community's FP7- ERC program, grant agreement no. 267414.

References

1. Aflalo, Y., Dubrovina, A., Kimmel, R.: Spectral generalized multi-dimensional scaling. Int. J. Comput. Vis. **29**, 1–13 (2016)
2. Aflalo, Y., Kimmel, R.: Spectral multidimensional scaling. Proc. Natl. Acad. Sci. **110**(45), 18052–18057 (2013)
3. Aiger, D., Mitra, N.J., Cohen-Or, D.: 4-points congruent sets for robust pairwise surface registration. In: ACM Transactions on Graphics (TOG), vol. 27, pp. 85. ACM (2008)
4. Anguelov, D., Srinivasan, P., Koller, D., Thrun, S., Rodgers, J., Davis, J.: SCAPE: shape completion and animation of people. In: Proceedings of ACM Transactions on Graphics (SIGGRAPH), Los Angeles, vol. 24, pp. 408–416 (2005)
5. Anguelov, D., Srinivasan, P., Pang, H.-C., Koller, D., Thrun, S., Davis, J.: The correlated correspondence algorithm for unsupervised registration of nonrigid surfaces. Adv. Neural Inf. Process. Syst. **17**, 33–40 (2004)
6. Aubry, M., Schlickewei, U., Cremers, D.: Pose-consistent 3d shape segmentation based on a quantum mechanical feature descriptor. In: Pattern Recognition, pp. 122–131. Springer (2011)
7. Aubry, M., Schlickewei, U., Cremers, D.: The wave kernel signature: a quantum mechanical approach to shape analysis. In: 2011 IEEE International Conference on Computer Vision Workshops (ICCV Workshops), Barcelona, pp. 1626–1633. IEEE (2011)
8. Ben-Tal, A., Zibulevsky, M.: Penalty/barrier multiplier methods for convex programming problems. SIAM J. Optim. **7**(2), 347–366 (1997)
9. Bérard, P., Besson, G., Gallot, S.: Embedding Riemannian manifolds by their heat kernel. Geom. Funct. Anal. **4**(4), 373–398 (1994)
10. Besl, P.J., McKay, N.D.: Method for registration of 3-d shapes. In: Robotics-DL Tentative, pp. 586–606. International Society for Optics and Photonics (1992)
11. Borg, I., Groenen, P.: Modern Multidimensional Scaling: Theory and Applications. Springer, New York (1997)
12. Bronstein, A.M., Bronstein, M.M., Kimmel, R.: Numerical Geometry of Non-rigid Shapes. Springer, New York (2008)
13. Bronstein, A.M., Bronstein, M.M., Kimmel, R.: Efficient computation of isometry-invariant distances between surfaces. SIAM J. Sci. Comput. **28**(5), 1812–1836 (2006)
14. Bronstein, A.M., Bronstein, M.M., Kimmel, R.: Generalized multidimensional scaling: a framework for isometry-invariant partial surface matching. Proc. Natl. Acad. Sci. USA **103**(5), 1168–1172 (2006)
15. Bronstein, A.M., Bronstein, M.M., Kimmel, R., Mahmoudi, M., Sapiro, G.: A Gromov-Hausdorff framework with diffusion geometry for topologically-robust non-rigid shape matching. IJCV **89**(2–3), 266–286 (2010)
16. Bronstein, M., Bronstein, A.M.: Shape recognition with spectral distances with spectral distances. IEEE Trans. Pattern Anal. Mach. Intell. (PAMI) **33**(5), 1065–1071 (2011)
17. Burago, D., Burago, Y., Ivanov S.: A Course in Metric Geometry, vol. 33. American Mathematical Society Providence, Providence (2001)
18. Chen, Y., Medioni, G.: Object modeling by registration of multiple range images. In: Proceedings of the 1991 IEEE International Conference on Robotics and Automation, Sacramento, pp. 2724–2729. IEEE (1991)
19. Coifman, R.R., Lafon, S.: Diffusion maps. Appl. Comput. Harmon. Anal. **21**(1), 5–30 (2006). Special Issue: Diffusion Maps and Wavelets

20. Donoser, M., Bischof, H.: Efficient maximally stable extremal region (MSER) tracking. In: 2006 IEEE Computer Society Conference on Computer Vision and Pattern Recognition, New York, vol. 1, pp. 553–560. IEEE (2006)
21. Dubrovina, A., Kimmel, R.: Matching shapes by eigendecomposition of the laplace_belrami operator. In: Proceedings of the Symposium on 3D Data Processing Visualization and Transmission (3DPVT), Paris (2010)
22. Dubrovina, A., Kimmel, R.: Approximately isometric shape correspondence by matching pointwise spectral features and global geodesic structures. Adv. Adapt. Data Anal. **3**(1–2), 203–228 (2011)
23. Elad, A., Kimmel, R.: On bending invariant signatures for surfaces. Trans. Pattern Anal. Mach. Intell. (PAMI) **25**(10), 1285–1295 (2003)
24. Gębal, K., Bærentzen, J.A., Aanæs, H., Larsen, R.: Shape analysis using the auto diffusion function. In: Computer Graphics Forum, vol. 28, pp. 1405–1413. Wiley Online Library (2009)
25. Gonzalez, T.F.: Clustering to minimize the maximum intercluster distance. Theor. Comput. Sci. **38**, 293–306 (1985)
26. Gromov, M.: Structures metriques pour les varietes riemanniennes. Textes Mathematiques, vol. 1. CEDIC/Fernand Nathan, Paris (1981)
27. Hochbaum, D., Shmoys, D.: A best possible heuristic for the *k*-center problem. Math. Oper. Res. **10**(2), 180–184 (1985)
28. Hu, J., Hua, J.: Salient spectral geometric features for shape matching and retrieval. Vis. Comput. **25**(5–7), 667–675 (2009)
29. Huang, Q., Koltun, V., Guibas, L.: Joint shape segmentation with linear programming. In: ACM Transactions on Graphics (TOG), vol. 30, p. 125. ACM (2011)
30. Johnson, A.: Spin-images: a representation for 3-D surface matching. PhD thesis, Robotics Institute, Carnegie Mellon University, Pittsburgh (1997)
31. Karni, Z., Gotsman, C.: Spectral compression of mesh geometry. In: Proceedings of the 27th Annual Conference on Computer Graphics and Interactive Techniques, pp. 279–286. ACM Press/Addison-Wesley Publishing Co., New York (2000)
32. Kim, T.H., Lee, K.M., Lee, S.U.: Learning full pairwise affinities for spectral segmentation. IEEE Trans. Pattern Anal. Mach. Intell. **35**(7), 1690–1703 (2013)
33. Kim, V.G., Lipman, Y., Funkhouser, T.: Blended intrinsic maps. In: ACM SIGGRAPH 2011 papers (SIGGRAPH '11), New York, pp. 79:1–79:12. ACM (2011)
34. Kimmel, R., Sethian, J.A.: Computing geodesic paths on manifolds. Proc. Natl. Acad. Sci. (PNAS) **95**(15), 8431–8435 (1998)
35. Kovnatsky, A., Bronstein, M.M., Bronstein, A.M., Glashoff, K., Kimmel, R.: Coupled quasi-harmonic basis. Comput. Graph. Forum (EUROGRAPHICS) **32**, 439–448 (2013)
36. Lévy, B.: Laplace-Beltrami eigenfunctions towards an algorithm that "understands" geometry. In: IEEE International Conference on Shape Modeling and Applications (SMI 2006), Washington, DC, pp. 13–13. IEEE (2006)
37. Lipman, Y., Daubechies, I.: Conformal Wasserstein Distances: Comparing Surfaces in Polynomial Time Yaron Lipman, Ingrid Daubechies. Adv. Math. **227**(3) (2011)
38. Lipman, Y., Funkhouser, T.: Möbius voting for surface correspondence. ACM Trans. Graph. (Proc. SIGGRAPH) **28**(3), 72 (2009)
39. Mateus, D., Horaud, R., Knossow, D., Cuzzolin, F., Boyer, E.: Articulated shape matching using Laplacian eigenfunctions and unsupervised point registration. In: IEEE Conference on Computer Vision and Pattern Recognition (CVPR 2008), Anchorage, pp. 1–8. IEEE (2008)
40. Mellado, N., Aiger, D., Mitra, N.J.: Super 4 pcs fast global pointcloud registration via smart indexing. In: Computer Graphics Forum, vol. 33, pp. 205–215. Wiley Online Library (2014)
41. Memoli, F.: On the use of Gromov-Hausdorff distances for shape comparison. In: Botsch, M., Pajarola, R., Chen, B., Zwicker, M. (eds.) Symposium on Point Based Graphics, Prague. Eurographics Association, pp. 81–90
42. Memoli, F., Sapiro, G.: A theoretical and computational framework for isometry invariant recognition of point cloud data. Found. Comput. Math. **5**(3), 313–347 (2005)

43. Meyer, M., Desbrun, M., Schroder, P., Barr, A.H.: Discrete differential-geometry operators for triangulated 2-manifolds. Vis. Math. **III**, 35–57 (2003)
44. Ovsjanikov, M., Ben-Chen, M., Solomon, J., Butscher, A., Guibas, L.: Functional maps: a flexible representation of maps between shapes. ACM Trans. Graph. **31**(4), 30:1–30:11 (2012)
45. Ovsjanikov, M., Mérigot, Q., Mémoli, F., Guibas, L.: One point isometric matching with the heat kernel. In: Eurographics Symposium on Geometry Processing (SGP), Lyon (2010)
46. Ovsjanikov, M., Mérigot, Q., Pătrăucean, V., Guibas, L.: Shape matching via quotient spaces. In: Computer Graphics Forum, vol. 32, pp. 1–11. Wiley Online Library (2013)
47. Ovsjanikov, M., Sun, J., Guibas, L.J.: Global intrinsic symmetries of shapes. In: Computer Graphics Forum, vol. 27, pp. 1341–1348 (2008)
48. Pinkall, U., Polthier, K.: Computing discrete minimal surfaces and their conjugates. Exp. Math. **2**(1), 15–36 (1993)
49. Pokrass, J., Bronstein, A.M., Bronstein M.M.: A correspondence-less approach to matching of deformable shapes. In: Scale Space and Variational Methods in Computer Vision, pp. 592–603. Springer, Berlin/New York (2012)
50. Pokrass, J., Bronstein, A.M., Bronstein, M.M., Sprechmann, P., Sapiro, G.: Sparse modeling of intrinsic correspondences. Comput. Graph. Forum (EUROGRAPHICS) **32**, 459–268 (2013)
51. Raviv, D., Bronstein, A.M., Bronstein, M.M., Kimmel, R.: Full and partial symmetries of non-rigid shapes. Int. J. Comput. Vis. (IJCV) (2009)
52. Raviv, D., Dubrovina, A., Kimmel, R.: Hierarchical shape matching. In: Proceedings of the Scale Space and Variational Methods (SSVM), Ein-Gedi (2011)
53. Raviv, D., Dubrovina, A., Kimmel, R.: Hierarchical matching of non-rigid shapes. In: Scale Space and Variational Methods in Computer Vision, pp. 604–615. Springer, Berlin (2012)
54. Reuter, M., Wolter, F.-E., Peinecke, N.: Laplace-Beltrami spectra as "shape-DNA" of surfaces and solids. Comput. Aided Design **38**, 342–366 (2006)
55. Rodola, E., Bulo, S.R., Windheuser, T., Vestner, M., Cremers, D.: Dense non-rigid shape correspondence using random forests. In: IEEE Conference on Computer Vision and Pattern Recognition (CVPR), Columbus (2014)
56. Rustamov, R., Ovsjanikov, M., Azencot, O., Ben-Chen, M., Chazal, F., Guibas, L.: Map-based exploration of intrinsic shape differences and variability. In: SIGGRAPH, Hong Kong. ACM (2013)
57. Rustamov, R.M.: Laplace-Beltrami eigenfunctions for deformation invariant shape representation. In: Proceedings of the Symposium on Geometry Processing (SGP), Barcelona, pp. 225–233 (2007)
58. Sahillioğlu, Y., Yemez, Y.: Coarse-to-fine combinatorial matching for dense isometric shape correspondence. In: Computer Graphics Forum, vol. 30, pp. 1461–1470. Wiley Online Library (2011)
59. Shami, G., Aflalo, Y., Zibulevsky, M., Kimmel, R.: Classical scaling revisited. In: Proceedings of the IEEE international conference on computer vision (ICCV), pp. 2255–2263 (2015)
60. Shtern, A., Kimmel, R.: Iterative closest spectral kernel maps. In: International Conference on 3D Vision (3DV), Tokyo (2014)
61. Shtern, A., Kimmel, R.: Matching the LBO eigenspace of non-rigid shapes via high order statistics. Axioms **3**(3), 300–319 (2014)
62. Sun, J., Ovsjanikov, M., Guibas, L.: A concise and provably informative multi-scale signature based on heat diffusion. In: Proceedings of the Symposium on Geometry Processing (SGP '09), Aire-la-Ville, pp. 1383–1392. Eurographics Association (2009)
63. Tevs, A., Bokeloh, M., Wand, M., Schilling, A., Seidel, H.-P.: Isometric registration of ambiguous and partial data. In: IEEE Computer Society Conference on Computer Vision and Pattern Recognition (CVPR 2009), Miami Beach, pp. 1185–1192. IEEE Computer Society (2009)
64. Zaharescu, A., Boyer, E., Varanasi, K., Horaud, R.: Surface feature detection and description with applications to mesh matching. In: IEEE Conference on Computer Vision and Pattern Recognition (CVPR 2009), Miami, pp. 373–380. IEEE (2009)

65. Zeng, Y., Wang, C., Wang, Y., Gu, X., Samaras, D., Paragios, N.: Dense non-rigid surface registration using high-order graph matching. In: 2010 IEEE Conference on Computer Vision and Pattern Recognition (CVPR), San Francisco, pp. 382–389. IEEE (2010)
66. Zhang, H., Sheffer, A., Cohen-Or, D., Zhou, Q., van Kaick, O., Tagliasacchi, A.: Deformation-driven shape correspondence. Comput. Graph. Forum (Proc. SGP) **27**(5), 1431–1439 (2008)
67. Zigelman, G., Kimmel, R., Kiryati, N.: Texture mapping using surface flattening via multi-dimensional scaling. IEEE Trans. Vis. Comput. Graph. **8**(2), 198–207 (2002)

Chapter 14
The Perspective Face Shape Ambiguity

William A.P. Smith

Abstract When a face is viewed under perspective projection, its shape (i.e. the 2D position of features) changes dramatically as the distance between face and camera varies. This causes substantial variation in appearance which is significant enough to disrupt human recognition of unfamiliar faces. However, a face viewed at any distance is still perceived as natural and humans are bad at interpreting the subject-camera distance given only a face image. We show that perspective viewing of faces leads to an ambiguity. Namely, that observed configurational information (position of projected vertices) and shading can be explained by a continuous class of possible faces. To demonstrate the ambiguity, we propose a novel method for efficiently fitting a 3D morphable model to 2D vertex positions when the subject-camera distance is known. By varying this distance, we obtain a subspace of faces, all of which are consistent with the observed data. We additionally show that faces within this subspace can all produce approximately the same shading pattern via a spherical harmonic lighting model.

14.1 Introduction

When a human face is viewed under perspective projection, its projected shape varies with the distance between the camera and subject. The change in the relative distances between facial features can be quite dramatic. When a face is close to the camera, it appears taller and slimmer with the features closest to the camera (nose and mouth) appearing relatively larger and the ears appearing smaller and partially occluded. As distance increases and the shape converges towards the orthographic projection, faces appear broader and rounder with ears that protrude further and the internal features more concentrated towards the centre of the face. We show some examples of this effect in Fig. 14.1. Images from the Caltech Multi-Distance Portraits database [10] are shown in which subjects are viewed at a distance of 60 cm and 490 cm. Each face is cropped and rescaled such that the interocular distance is the same. The distortion caused by perspective transformation is clearly visible.

W.A.P. Smith (✉)
Department of Computer Science, University of York, York, UK
e-mail: william.smith@york.ac.uk

M. Breuß et al. (eds.), *Perspectives in Shape Analysis*, Mathematics and Visualization, DOI 10.1007/978-3-319-24726-7_14

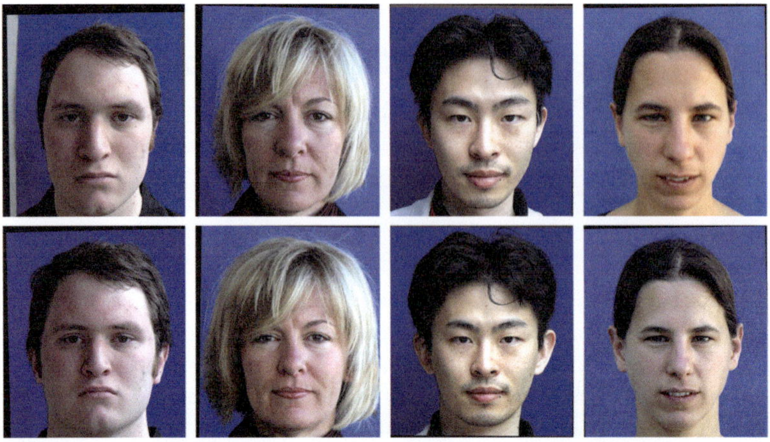

Fig. 14.1 Perspective transformation of real faces (From [10]). The subject is the same in each column but the change in viewing distance induces a significant change in projected shape

When the faces are unfamiliar, it is difficult to believe that the identity of the faces in the first row are the same as those in the second.

The change in face appearance under perspective projection has been widely noted before, for example in art history [20] and psychology [22, 23]. However, the vast majority of 2D face analysis methods that involve estimation of 3D face shape or fitting of a 3D face model neglect this effect and assume an affine camera (e.g. scaled orthographic or "weak perspective"). Such a camera does not introduce any perspective transformation. While this assumption is justified in applications where the subject-camera distance is likely to be large, any situation where a face may be viewed from a small distance must account for the effects of perspective.

While such close viewing conditions may appear contrived, there are many examples of scenarios where this occurs in both machine and human vision. In the former case, so-called "selfies" are an example of a widely popular picture format in which the subject-camera distance is small. Another example would be secure door entry systems where a subject presents themselves directly in front of the camera. The latter case includes security peepholes or even a mother nursing a child (where, presumably, crucial learning of the mother's face is occurring).

We do not believe that the perspective effect has previously been viewed as an *ambiguity*. Namely that, two different faces viewed at different distances could give rise to the same (or very similar) configuration and appearance. We call this the *perspective face shape ambiguity*. This ambiguity has implications for face recognition, 3D face shape estimation, forensic image analysis and establishing model-image dense correspondence.

Variation in face shape and appearance over a population is highly amenable to description using a linear statistical model. In particular, a 3D morphable model has been shown to accurately capture 3D face shape and generalise well to novel, unseen faces. We use such a model to represent prior knowledge about the space of face shapes. We address the face shape ambiguity by presenting a novel method for fitting a 3D morphable model to projected 2D vertex positions under perspective projection and with a specified subject-camera distance. Hence, observed 2D vertex positions provide a continuous class of solutions as the subject-camera distance is varied. We verify that, indeed, multiple explanations of observed 2D shape data is possible. We show that two faces with significantly different 3D shape can produce almost identical 2D projected shapes. We then go further by showing that a change in illumination (using a diffuse spherical harmonic model) can produce almost identical shading and hence appearance. This suggests that the ambiguity is not only geometric but also photometric.

14.2 Related Work

Faces under perspective projection The effect of perspective transformation on face appearance has been studied from both a computational and psychological perspective previously.

In art history, Latto and Harper [20] discuss how uncertainty regarding subject-artist distance when viewing a painting results in distorted perception. To investigate this further, they conducted a study which showed that perceptions of body weight from face images are influenced by subject-camera distance. Perona et al. [9, 27] investigated a different effect, noting that perspective distortion influences social judgements of faces. In psychology, Liu et al. [22, 23] show that human face recognition performance is degraded by perspective transformation.

There have been two recent attempts to address the problem of estimating subject-camera distance from monocular, perspective views of a face [10, 12]. The idea is that the configuration of projected 2D face features conveys something about the degree of perspective transformation. Flores et al. [12] approach the problem using exemplar 3D face models. They fit the models to 2D landmarks using the EPnP algorithm [21] and use the mean of the estimated distances as the estimated subject-camera distance. Burgos-Artizzu et al. [10] on the other hand work entirely in 2D. Their idea is to describe 2D landmarks in terms of their offset from mean positions, with the mean calculated either across views at different distances of the same face, or across multiple identities at the same distance. They can then perform regression to relate offsets to distance.

Our results highlight the difficulty that both of these approaches face. Namely that many interpretations of 2D facial landmarks are possible, all with varying subject-camera distance. We approach the problem in a different way by showing how to solve for shape parameters when the subject-camera distance is known. We can then show that multiple explanations are possible.

3D face shape from 2D geometric features Facial landmarks, i.e. points with well defined correspondence between identities, are used in a number of ways in face processing. Most commonly, they are used for registration and normalisation, as is done in training an Active Appearance Model [11]. For this reason, there has been sustained interest in building feature detectors capable of accurately labelling face landmarks in uncontrolled images [29].

The robustness and efficiency of 2D facial feature detectors has improved significantly in recent years. This has motivated the use of 2D facial landmarks as a cue for the recovery of 3D face shape. In particular, by fitting a 3D morphable model to these detected landmarks [2, 6, 19, 25]. All of these methods assume an affine camera and hence the problem reduces to a multilinear problem in the unknown shape and camera parameters.

Some work has considered other 2D shape features besides landmark points. Keller et al. [17] fit a 3D morphable model to contours (both silhouettes and inner contours due to texture, shape and shadowing). A related problem is to describe the remaining flexibility in a statistical shape model that is partially fixed. In other words, if the position of some points, curves or subset of the surface is known, the goal is to characterise the space of shapes that approximately fit these observations. Albrecht et al. [1] show how to compute the subspace of faces with the same profile. Lüthi et al. [24] extended this approach into a probabilistic setting.

We emphasise that the ambiguity occurs only in monocular, uncalibrated images. For example, Amberg et al. [3] describe an algorithm for fitting a 3D morphable model to stereo face images. In this case, the stereo disparity cue used in their objective function conveys depth information which helps to resolve the ambiguity. However, note that even here, their solution is unstable when camera parameters are unknown. They introduce an additional heuristic constraint on the focal length, namely they restrict it to be between 1 and 5 times the sensor size.

Other ambiguities There are other known ambiguities in the perception of 3D shape, some of which have been studied in the context of faces.

The *bas relief ambiguity* [5] shows that certain transformations of a surface can yield identical images when the lighting and albedo are also appropriately transformed. Specifically, a Generalised Bas Relief (GBR) transformation applied to a surface represented as an orthographic depth map yields ambiguous images (under the assumption of Lambertian reflectance). The GBR is a linear transformation and the bas relief ambiguity is exact (two different surfaces can produce identical appearance).

On the other hand, the perspective face ambiguity is nonlinear (perspective transformation has a nonlinear effect on projected shape) and approximate (we minimise error between observed and fitted vertex positions). It is also predominantly a geometric ambiguity – it is concerned with the projection of vertex positions to 2D, rather than appearance (although we show that shading can be approximately recreated). However, most importantly, the perspective face ambiguity is statistically constrained. The transformed faces stay within the span of a statistical model and, hence, remain plausible face shapes. A GBR transformation of a face surface will inevitably produce shapes that are not plausible faces.

Exploiting this fact, Georghiades et al. [13] resolve the bas-relief ambiguity by exploiting the symmetries and similarities in faces. Specifically they assume:

bilateral symmetry; that the forehead and chin should be at approximately the same depth; and that the range of facial depths is about twice the distance between the eyes. Such assumptions would not resolve the perspective face ambiguity that we describe as all fitted faces lie within the span of a statistical model and hence are plausible.

In the *hollow face illusion* [16], shaded images of concave faces are interpreted as convex faces with inverted illumination. The illusion even holds when hollow face is moving, with rotations being interpreted as reversed. This illusion is nothing other than the convex/concave ambiguity encountered in single image shape-from-shading. In human vision, this is always resolved for faces using a convex interpretation since experience of face shape makes the concave interpretation extremely unlikely. Again, the convex/concave ambiguity is not related to the perspective face ambiguity since a concave face would be impossible in the context of a statistical face model.

14.3 Preliminaries

Our approach is based on fitting a 3DMM to observations under perspective projection. Hence, we begin by describing the 3D morphable model and pinhole camera model.

14.3.1 3D Morphable Model

A 3D morphable model is a deformable mesh $\mathscr{M}(\boldsymbol{\alpha}) = (\mathscr{K}, \mathbf{s}(\boldsymbol{\alpha}))$, whose shape is determined by the shape parameters $\boldsymbol{\alpha} \in \mathbb{R}^S$. Shape is described by a linear model learnt from data using Principal Components Analysis (PCA) [7]. So, the shape of any object from the same class as the training data can be approximated as:

$$\mathbf{s}(\boldsymbol{\alpha}) = \mathbf{P}\boldsymbol{\alpha} + \bar{\mathbf{s}}, \tag{14.1}$$

where $\mathbf{P} \in \mathbb{R}^{3N \times S}$ contains the S principal components, $\bar{\mathbf{s}} \in \mathbb{R}^{3N}$ is the mean shape and the vector $\mathbf{s}(\boldsymbol{\alpha}) \in \mathbb{R}^{3N}$ contains the coordinates of the N vertices, stacked to form a long vector: $\mathbf{s} = [u_1\ v_1\ w_1\ \dots\ u_N\ v_N\ w_N]^{\mathrm{T}}$. Hence, the ith vertex is given by: $\mathbf{v}_i = [s_{3i-2}\ s_{3i-1}\ s_{3i}]^{\mathrm{T}}$.

The connectivity or topology of the deformable mesh is fixed and is given by the simplicial complex $\mathscr{K}$, which is a set whose elements can be vertices $\{i\}$, edges $\{i,j\}$ or triangles $\{i,j,k\}$ with the indices $i,j,k \in [1..N]$.

For convenience, we denote the sub-matrix corresponding to the ith vertex as $\mathbf{P}_i \in \mathbb{R}^{3 \times S}$ and the corresponding vertex in the mean face shape as $\bar{\mathbf{s}}_i \in \mathbb{R}^3$, such that the ith vertex is given by: $\mathbf{v}_i = \mathbf{P}_i\boldsymbol{\alpha} + \bar{\mathbf{s}}_i$. Similarly, we define the row corresponding to the u component of the ith vertex as $\mathbf{P}_{iu}$ (similarly for v and w) and define the u component of the ith mean shape vertex as $\bar{s}_{iu}$ (similarly for v and w).

14.3.2 Pinhole Camera Model

The perspective projection of the 3D point $\mathbf{v} = [u\ v\ w]^T$ onto the 2D point $\mathbf{x} = [x\ y]^T$ is given by the pinhole camera model $\mathbf{x} = \mathbf{pinhole}[\mathbf{v}, \boldsymbol{\Lambda}, \boldsymbol{\Omega}, \boldsymbol{\tau}]$ where

$$\mathbf{pinhole}[\mathbf{v}, \boldsymbol{\Lambda}, \boldsymbol{\Omega}, \boldsymbol{\tau}] = \begin{bmatrix} \frac{\phi(\omega_{11}u + \omega_{12}v + \omega_{13}w + \tau_x)}{\omega_{31}u + \omega_{32}v + \omega_{33}w + \tau_z} + \delta_x \\ \frac{\phi(\omega_{21}u + \omega_{22}v + \omega_{23}w + \tau_y)}{\omega_{31}u + \omega_{32}v + \omega_{33}w + \tau_z} + \delta_y \end{bmatrix} \tag{14.2}$$

where

$$\boldsymbol{\Omega} = \begin{bmatrix} \omega_{11} & \omega_{12} & \omega_{13} \\ \omega_{21} & \omega_{22} & \omega_{23} \\ \omega_{31} & \omega_{32} & \omega_{33} \end{bmatrix}$$

is a rotation matrix and $\boldsymbol{\tau} = \begin{bmatrix} \tau_x & \tau_y & \tau_z \end{bmatrix}^T$ is a translation vector which relate model and camera coordinates (the extrinsic parameters). The matrix:

$$\boldsymbol{\Lambda} = \begin{bmatrix} \phi & 0 & \delta_x \\ 0 & \phi & \delta_y \\ 0 & 0 & 1 \end{bmatrix}$$

contains the intrinsic parameters of the camera, namely the focal length ϕ and the principal point (δ_x, δ_y).

This nonlinear projection can be written in linear terms by using homogeneous representations $\tilde{\mathbf{v}} = [u\ v\ w\ 1]^T$ and $\tilde{\mathbf{x}} = [x\ y\ 1]^T$:

$$\lambda\tilde{\mathbf{x}} = \boldsymbol{\Lambda}\begin{bmatrix} \boldsymbol{\Omega} & \boldsymbol{\tau} \end{bmatrix}\tilde{\mathbf{v}}, \tag{14.3}$$

where λ is an arbitrary scaling factor. Without loss of generality, we work with a zero-centred image (i.e. $\delta_x = \delta_y = 0$).

14.4 Perspective Fitting to 2D Projections

In this section we present an algorithm for fitting a 3D morphable model to the 2D positions of the projected model vertices under perspective projection with an uncalibrated camera. As we will show, this process is ambiguous so we solve the problem for the case when the subject-camera distance is known. We do not consider the problem of computing correspondence between the model and observed data, since this is unnecessary for the demonstration of the ambiguity. Unknown correspondences could only increase the space of solutions consistent with the observations. Our approach is based on a linear approximation to the underlying objective function which we derive based on the direct linear transform method.

Our observations are the projected 2D positions $\mathbf{x}_i = [x_i\ y_i]^T$ $(i = 1 \dots L)$ of the L vertices that are visible (unoccluded). Without loss of generality, we assume that the ith 2D position corresponds to the ith vertex in the morphable model. The objective of fitting a morphable model to these observations is to obtain the shape parameters that minimise the reprojection error between observed and predicted 2D positions:

$$\boldsymbol{\alpha}^* = \arg\min_{\boldsymbol{\alpha}} \sum_{i=1}^{L} \|\mathbf{x}_i - \mathbf{pinhole}\left[\mathbf{P}_i\boldsymbol{\alpha} + \bar{\mathbf{s}}_i, \boldsymbol{\Lambda}, \boldsymbol{\Omega}, \boldsymbol{\tau}\right]\|^2. \tag{14.4}$$

This optimisation is non-convex due to the nonlinearity of perspective projection. Moreover, the intrinsic and extrinsic parameters may also be unknown. Nevertheless, a good approximate solution can be found using linear methods. This initial estimate provides a suitable initialisation for local nonlinear optimisation to further refine the shape parameters.

14.4.1 Direct Linear Transform

From Equations 14.1 and 14.3 we can relate each model vertex and observed 2D position via a linear similarity relation:

$$\begin{bmatrix} \mathbf{x}_i \\ 1 \end{bmatrix} \sim \boldsymbol{\Lambda} \begin{bmatrix} \boldsymbol{\Omega} & \boldsymbol{\tau} \end{bmatrix} \begin{bmatrix} \mathbf{P}_i\boldsymbol{\alpha} + \bar{\mathbf{s}}_i \\ 1 \end{bmatrix}, \tag{14.5}$$

where $\sim$ denotes equality up to a non-zero scalar multiplication. Such sets of relations can be solved using the direct linear transformation (DLT) algorithm [15]. Accordingly, we write

$$\begin{bmatrix} \mathbf{x}_i \\ 1 \end{bmatrix}_\times \boldsymbol{\Lambda} \begin{bmatrix} \boldsymbol{\Omega} & \boldsymbol{\tau} \end{bmatrix} \begin{bmatrix} \mathbf{P}_i\boldsymbol{\alpha} + \bar{\mathbf{s}}_i \\ 1 \end{bmatrix} = \mathbf{0} \tag{14.6}$$

where $\mathbf{0} = [0\ 0\ 0]^T$ and $[.]_\times$ is the cross product matrix:

$$[\mathbf{x}]_\times = \begin{bmatrix} 0 & -x_3 & x_2 \\ x_3 & 0 & -x_1 \\ -x_2 & x_1 & 0 \end{bmatrix}. \tag{14.7}$$

This means that each vertex yields three linear equations in the unknown shape parameters $\boldsymbol{\alpha}$ (although only two are linearly independent). However, the intrinsic and extrinsic parameters are, in general, also unknown.

To simplify our consideration, we ignore the effects of rotation (i.e. $\boldsymbol{\Omega} = \mathbf{I}_3$). Note that introducing rotations would only increase the ambiguity since it would allow the model to explain a broader set of observations.

Since we are interested in the effect of varying subject-camera distance, we limit translations to the z direction, hence $\boldsymbol{\tau} = [0\ 0\ \tau_z]^T$. It has been shown previously that translating a face away from the centre of projection (i.e. in the x and y directions) does not affect human recognition performance [23]. We believe that this is because the relatively small field of view in a typical camera means that the change in perspective appearance has only a small dependence on such translations. For this reason, we do not study its effect here and concentrate only on subject-camera distance.

Substituting these simplifications yields:

$$\begin{bmatrix} 0 & -\phi & y_i & \tau_z y_i \\ \phi & 0 & -x_i & -\tau_z x_i \\ -\phi y_i & \phi x_i & 0 & 0 \end{bmatrix} \begin{bmatrix} \mathbf{P}_i\boldsymbol{\alpha} + \bar{\mathbf{s}}_i \\ 1 \end{bmatrix} = \mathbf{0}. \tag{14.8}$$

The only remaining unknown besides the shape parameters $\boldsymbol{\alpha}$ is the focal length of the camera ϕ. Recall that changing the focal length amounts only to a uniform scaling of the projected points in 2D. Note that this corresponds exactly to the scenario in Fig. 14.1. There, subject-camera distance was varied before each image was rescaled such that the interocular distance was constant. We now seek to explore the ambiguity by varying subject-camera distance, solving for the best shape parameters whilst choosing the 2D scaling that minimises distance between observed and predicted 2D vertex positions.

14.4.2 Alternating Least Squares Solution

This problem is bilinear in the unknown shape and focal length parameters. We solve this problem using alternating linear least squares . Hence, we begin by writing the linear equations for each vertex in terms of the shape parameters, leading to a system of linear equations for all visible vertices:

$$\underbrace{\begin{bmatrix} y_1\mathbf{P}_{1w} - \phi\mathbf{P}_{1v} \\ \phi\mathbf{P}_{1u} - x_1\mathbf{P}_{1w} \\ x_1\mathbf{P}_{1v} - y_1\mathbf{P}_{1u} \\ \vdots \\ y_L\mathbf{P}_{Lw} - \phi\mathbf{P}_{Lv} \\ \phi\mathbf{P}_{Lu} - x_n\mathbf{P}_{Lw} \\ x_L\mathbf{P}_{Lv} - y_n\mathbf{P}_{Lu} \end{bmatrix}}_{\mathbf{C}} \boldsymbol{\alpha} = \underbrace{\begin{bmatrix} \phi\bar{s}_{1v} - y_1\bar{s}_{1w} - \tau_z y_1 \\ x_1\bar{s}_{1w} - \phi\bar{s}_{1u} + \tau_z x_1 \\ y_1\bar{s}_{1u} - x_1\bar{s}_{1v} \\ \vdots \\ \phi\bar{s}_{Lv} - y_n\bar{s}_{Lw} - \tau_z y_L \\ x_L\bar{s}_{Lw} - \phi\bar{s}_{Lu} + \tau_z x_L \\ y_L\bar{s}_{Lu} - x_L\bar{s}_{Lv} \end{bmatrix}}_{\mathbf{d}}. \tag{14.9}$$

Hence, we have a linear system of the form $\mathbf{C}\boldsymbol{\alpha} = \mathbf{d}$. Since the number of vertices is much larger than the number of model dimensions, the problem is over constrained. Hence, we solve in a least squares sense subject to an additional constraint to ensure plausibility of the solution. We follow Brunton et al. [8] and use a hyperbox constraint on the shape parameters. This ensures that each parameter lies within k standard deviations of the mean by introducing a linear inequality constraint on the shape parameters. We use a hard hyperbox constraint in preference to a soft elliptical prior as it avoids mean-shape bias and having to choose a regularisation weight.

To solve for focal length, we again form a linear system of equations which leads to a simple linear regression problem with a straightforward closed form solution:

$$\phi^* = \frac{\sum_{i=1}^{L} \left[(x_i(\mathbf{P}_{iu}\boldsymbol{\alpha} + s_{iu}) + y_i(\mathbf{P}_{iv}\boldsymbol{\alpha} + \bar{s}_{iv}) \right] \left[\tau_z + \mathbf{P}_{iw}\boldsymbol{\alpha} + \bar{s}_{iw} \right]}{\sum_{i=1}^{L} (\mathbf{P}_{iv}\boldsymbol{\alpha} + \bar{s}_{iv})^2 + (\mathbf{P}_{iu}\boldsymbol{\alpha} + \bar{s}_{iu})^2} \tag{14.10}$$

We alternate between solving Equations 14.9 and 14.10, alternately fixing $\boldsymbol{\alpha}$ and ϕ. This process converges rapidly and usually 5 iterations are sufficient. We initialise by using the mean shape from the morphable model to solve for focal length first, i.e. we substitute the zero vector $\boldsymbol{\alpha} = \mathbf{0}$ into Equation 14.10. The overall approach can be viewed as solving the following minimisation problem:

$$\boldsymbol{\alpha}(\tau_z) = \arg\min_{\boldsymbol{\alpha}} \min_{\phi} \|\mathbf{C}\boldsymbol{\alpha} - \mathbf{d}\|^2, \text{ s.t. } \begin{bmatrix} \mathbf{I}_m \\ -\mathbf{I}_m \end{bmatrix} \boldsymbol{\alpha} \leq \begin{bmatrix} k\sigma_1 \\ \vdots \\ k\sigma_m \\ k\sigma_1 \\ \vdots \\ k\sigma_m \end{bmatrix} \tag{14.11}$$

where σ_i is the standard deviation of the ith shape parameter. Note that solving this minimisation is not equivalent to solving the original objective in Equation 14.4. Hence, we can further refine the solution by applying nonlinear optimisation over $\boldsymbol{\alpha}$ and ϕ, using the original objective function. In practice, the improvement obtained by doing this is very small – typically the fitting energy is reduced by less than 1 % with no visible difference in the fitted model. So in our experimental results we simply use the alternating least squares solution with 5 iterations.

14.4.3 The Perspective Face Shape Ambiguity

Given 2D observations $\mathbf{x}_i$, we therefore have a continuous space of solutions $\boldsymbol{\alpha}(\tau_z)$ as a function of subject-camera distance. This is the perspective face shape ambiguity.

Note that this can be viewed as a transformation within the shape parameter space of the morphable model. If the target observations $\mathbf{x}_i$ are provided by projecting a 3D face obtained from Equation 14.1 with shape parameters $\boldsymbol{\alpha}_1$ and distance $\tau_z = d_1$, then solutions $\boldsymbol{\alpha}(d_2)$ can be seen as a nonlinear transformation within parameter space, yielding a new set of shape parameters $\boldsymbol{\alpha}_2$, as a function of the fitted distance $\tau_z = d_2$. When $d_1 = d_2$ the fitted face will be approximately equal to the target face.

14.5 Fitting Lighting to Diffuse Shading

The fitting process described in the previous section aims to minimise the distance between the 2D projected positions of target and fitted vertices. In other words, it recreates the 2D configuration of features present in the target face. However, this does not mean that the two faces will have the same appearance. Since the 3D shape of the faces is different (as will be shown in the experimental results), the surface normals at corresponding points will be different. Under the same illumination, this will lead to different shading and hence appearance.

We now show how the shape obtained using the method in the previous section can be shaded so as to minimise the difference in appearance between the target and fitted face. We do not consider the effect of surface texture (i.e. diffuse albedo). The effect of albedo on appearance is to simply scale the diffuse shading. Hence, it plays no role in the perspective shape ambiguity. In fact, if albedo is also allowed to vary between target and fitted face, it may be able to improve the approximation of the observed appearance and hence enhance the ambiguity. We show here simply how to make the diffuse shading pattern approximately equal.

If $\mathbf{n}_i \in \mathbb{R}^3$ is the surface normal at vertex i, with $\|\mathbf{n}_i\| = 1$, the order 2 spherical harmonic lighting basis vector for that vertex (ignoring constant factors) is given by [4]:

$$\mathbf{b}_i = \begin{bmatrix} 1 & n_{i,x} & n_{i,y} & n_{i,z} & 2n_{i,z}^2 - n_{i,x}^2 - n_{i,y}^2 & n_{i,x}n_{i,y} & n_{i,x}n_{i,z} & n_{i,y}n_{i,z} & n_{i,x}^2 - n_{i,y}^2 \end{bmatrix}. \quad (14.12)$$

Hence, the matrix of basis vectors for the L observed vertices is given by:

$$\mathbf{B} = \begin{bmatrix} \mathbf{b}_1 \\ \vdots \\ \mathbf{b}_L \end{bmatrix}. \tag{14.13}$$

We compute basis matrices $\mathbf{B}_T$ and $\mathbf{B}_F$ for the target and fitted faces respectively. If the target face is illuminated by known spherical harmonic lighting vector $\mathbf{l}_T$ then the diffuse shading for the mesh is given by: $\mathbf{B}_T\mathbf{l}_T$. The lighting vector that minimises the difference in appearance of the fitted face to the target is given by solving the linear system of equations:

$$\mathbf{l}_F^* = \arg\min_{\mathbf{l}_F} \|\mathbf{B}_T\mathbf{l}_T - \mathbf{B}_F\mathbf{l}_F\|^2. \tag{14.14}$$

This provides the optimal transformation of lighting for the fitted face.

14.6 Experimental Results

We use the Basel Face Model [26] (BFM) which is a 3D morphable model comprising 53,490 vertices and which is trained on 200 faces. We use the shape component of the model only. The model is supplied with 10 out-of-sample faces which are scans of real faces that are in correspondence with the model. Unusually, the model does not factor out scale, i.e. faces are only aligned via translation and rotation. This means that the vertex positions are in absolute units of distance. This allows us to specify camera-subject distance in physically meaningful units.

We begin with a target face. For this purpose, we either use the BFM out-of-sample faces or we randomly generate a face. We do this by sampling randomly from the multivariate normal distribution with zero mean and covariance matrix $\boldsymbol{\Sigma} = \text{diag}(\sigma_1^2, \ldots, \sigma_n^2)$ to yield a shape parameter vector and hence shape. We arbitrarily set the focal length $\phi = 1$ and choose the subject-camera distance. We then project every vertex of the target face to provide 2D observations. We use all $S = 199$ model dimensions and constrain parameters to be within $k = 3$ standard deviations of the mean.

14.6.1 Subspace of Ambiguity

We begin by visualising the subspace associated with the perspective face shape ambiguity for a single target face. We randomly generate shape parameters, yielding the target face shown in column 1 of Fig. 14.2. Note that in the figure the face is

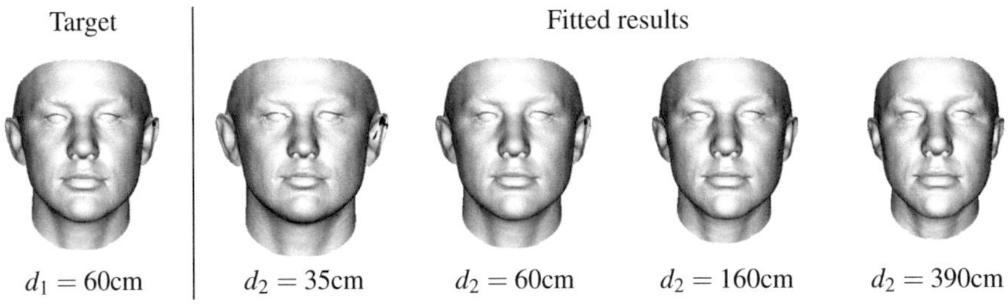

Fig. 14.2 Target (column 1) and fitted results (columns 2–5) shown under orthographic projection. When the target is viewed under perspective projection at distance d_1 and the fitted faces at distances d_2, they give almost identical 2D projections

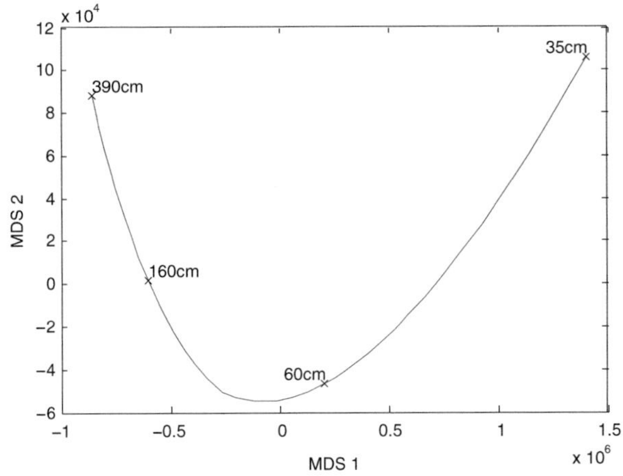

Fig. 14.3 An illustration of the nonlinearity of the perspective face ambiguity. We plot the fitted parameter vectors in a 2D MDS space as the subject-camera distance is varied. The target face is the same as in Fig. 14.2, again with $d_1 = 60$ cm

shown in orthographic projection. For our observations, we project the face under perspective projection at a distance of $d_1 = 60$ cm. We then solve for the optimal fit at distances ranging from $d_2 = 35$ to 390 cm. We show a sample of these fitted results, again under orthographic projection, in columns 2–5 of Fig. 14.2. There is significant variation in the shape of the face, yet all produce the same projected 2D positions when viewed at different distances.

To verify that the transformation is indeed nonlinear, in Fig. 14.3 we perform multidimensional scaling (MDS) on the fitted parameter vectors. We then plot the trajectory of the fitted faces through the space formed by the first two MDS dimensions. We highlight the positions in MDS space associated with the fitting results from Fig. 14.2. It is clear that the trajectory, and hence the perspective ambiguity, is highly nonlinear.

14.6.2 Shape Fitting

In Figs. 14.4 and 14.5 we show the result of fitting to 4 of the BFM scans (i.e. the targets are real, out-of-sample faces). We experiment with two subject-camera distances for either extreme ($\tau_z = 30$ cm) or moderate ($\tau_z = 90$ cm) perspective distortion. In Fig. 14.4, the target face is close to the camera ($\tau_z = 30$ cm) and we fit the model at a far distance ($\tau_z = 90$ cm). This configuration is reversed in Fig. 14.5. For visualisation we show the results both as shaded surfaces and with the texture of the real target face.

The target face is shown under perspective and orthographic projection in the first and third columns respectively. The fitted face is similarly shown in the second and fourth columns. Hence, the observations are provided by column 1 and the fitted result in column 2. The orthographic views in columns 3 and 4 enable comparison between the target and fitted shape under the same projection. This demonstrates clearly that two faces with significantly different 3D shape can give rise to almost identical 2D landmark positions under perspective projection.

Quantitatively, d_S is the mean Euclidian distance between the target and fitted surface. In all cases, d_S is significant, sometimes as much as 1 cm. On the other hand, in all cases, the mean distance between fitted and target landmarks is less than a pixel (and less than 1 % of the interocular distance). Note that Burgos-Artizzu et al. [10] found that the difference between landmarks on the same face placed by two different humans was typically 3 % of the interocular distance. Similarly, the 300 faces in the wild challenge [29] found that even the best methods did not obtain better than 5 % accuracy for more than 50 % of the landmarks. Hence, the vertex fitting error is substantially smaller than the accuracy of either human or machine placed landmarks.

There are clear differences in shading with the fittings in Fig. 14.4 exhibiting sharper features and hence more dramatic shading and in Fig. 14.5, flatter features and hence flatter shading. This is seen more clearly in Fig. 14.6 where we show rotated views of the target and two fitted surfaces.

14.6.3 Illumination Fitting

We now show how a change in illumination can enable the fitted face to produce almost identical shading to the target, despite the large difference in 3D shape. For this experiment, we render the target face under perspective projection with frontal illumination and Lambertian shading. We then solve for the spherical harmonic lighting parameters that minimise the error between this target shading and that of the fitted face. In Figs. 14.7 and 14.8 we show the results of this experiment, again for two scenarios of near and distant target.

In the top row we show the fitted face rendered with the same illumination as the target. In the middle row we show the target face. It is clear that there is a significant

Fig. 14.4 Fitting results (near target): target at 30 cm, fitted result at 90 cm

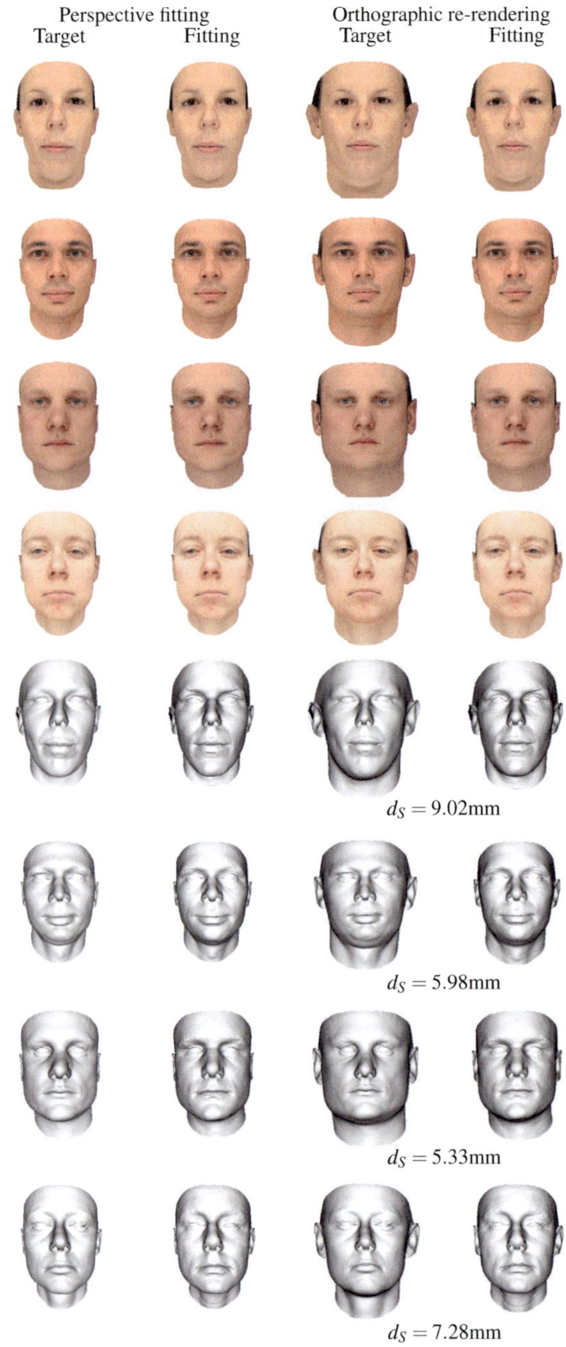

Fig. 14.5 Fitting results (distant target): target at 90 cm, fitted result at 30 cm

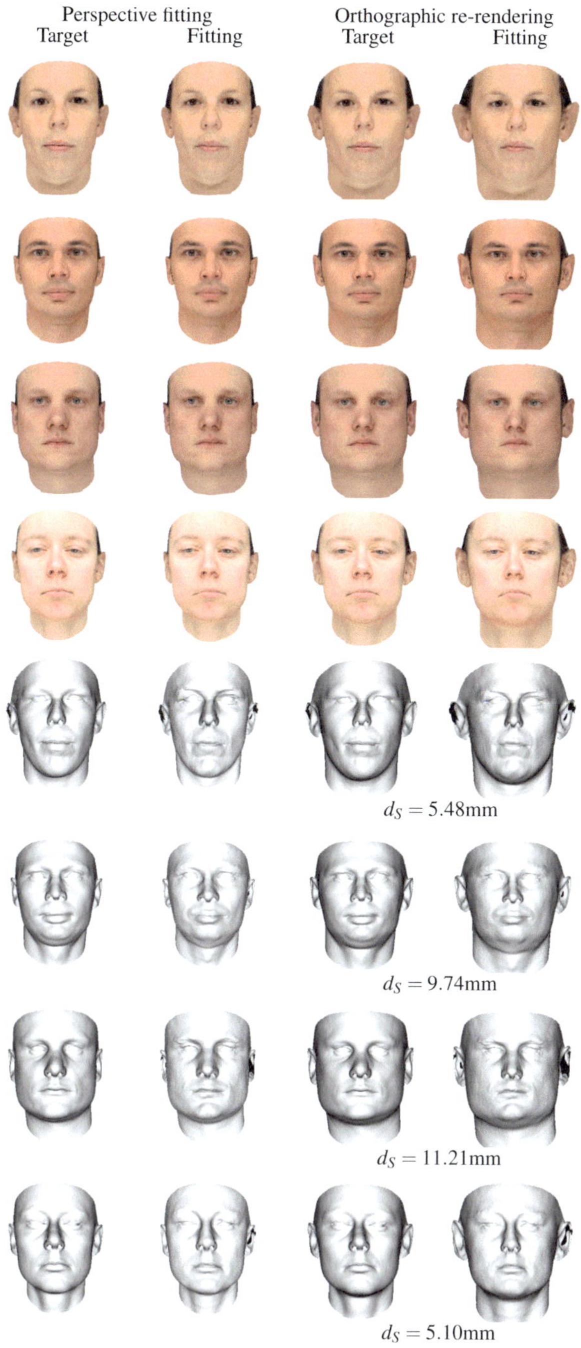

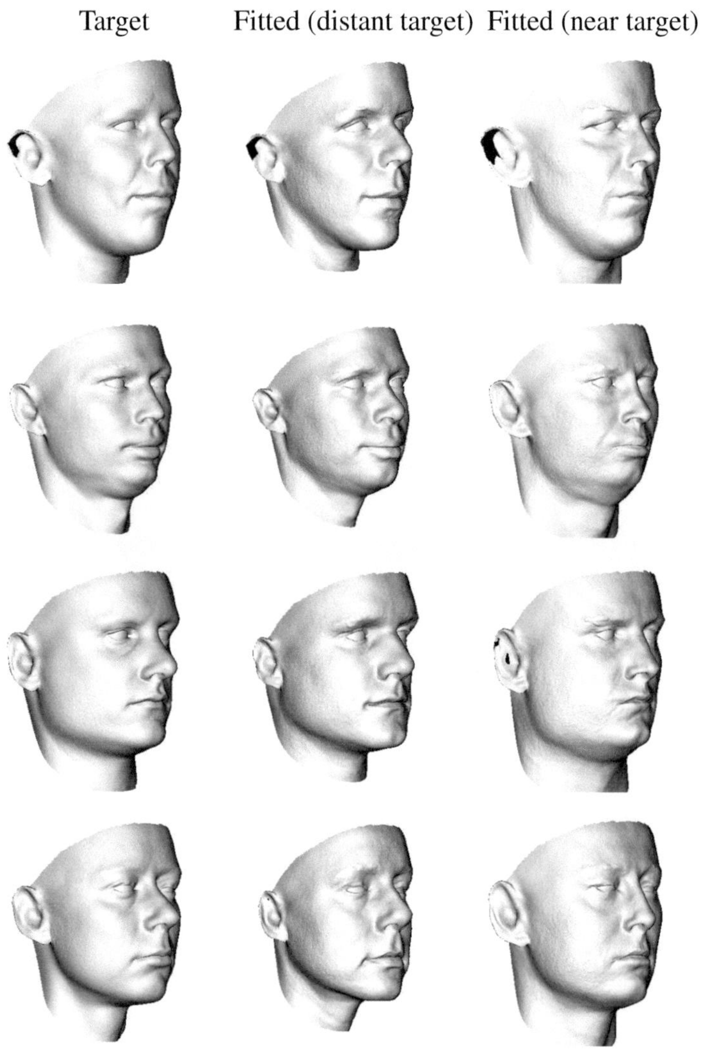

Fig. 14.6 Rotated views of target (*left*), fitting to distant target (*middle*) and fitting to near target (*right*). The faces in each row can produce almost identical projected 2D shapes

difference in shading. In the bottom row, we show the target face rendered with fitted spherical harmonic lighting. Notice that the shading is now much closer to that of the target face. This perceptual improvement is corroborated quantitatively where it can be seen that the RMS error in the image intensity reduces in all cases.

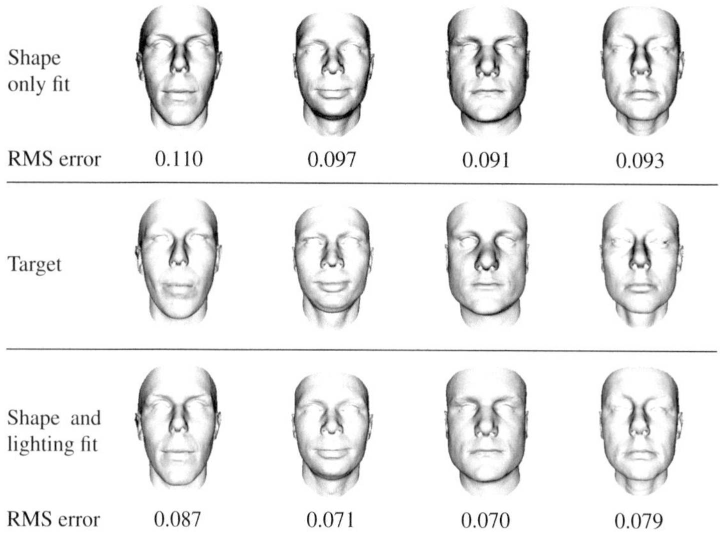

Fig. 14.7 Illumination fitting results (close target): target at 30 cm, fitted result at 90 cm. RMS errors are computed for intensity of foreground pixels

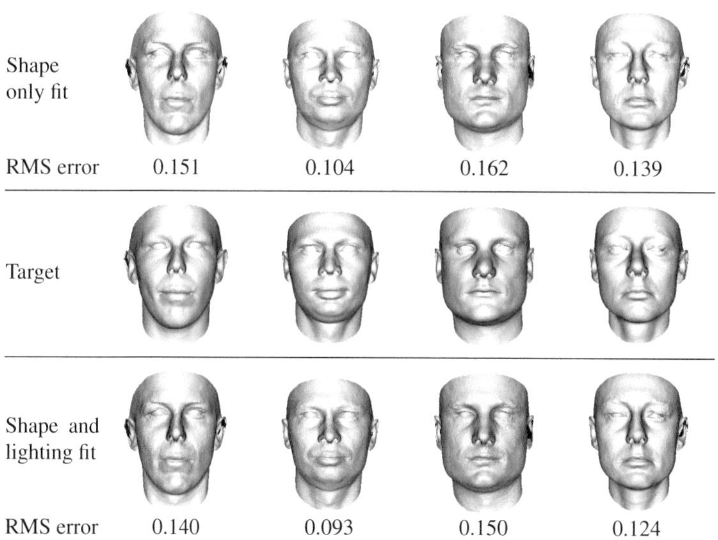

Fig. 14.8 Illumination fitting results (distant target): target at 90 cm, fitted result at 30 cm. RMS errors are computed for intensity of foreground pixels

14.7 Discussion

In this paper we have introduced a new ambiguity which arises when faces are viewed under perspective projection. We have shown that 2D shape and shading can be explained by a space of possible faces which vary significantly in 3D shape. There are a number of interesting implications of this ambiguity. First, any attempt to recover 3D facial shape from 2D shape or shading observations is ill-posed under perspective projection, even with a statistical constraint. Second, metric distances between landmark points in 2D images are not unique. We have shown that faces with very different shapes can give rise to almost identical projected 2D shapes (with mean differences less than 1 % of interocular distance in all cases). This casts doubt on the use of metric distances between features as a way of comparing the identity of two face photographs. This has previously been used in forensic imaging [28]. The perspective face shape ambiguity perhaps partially explains the studies that have demonstrated the weakness of these approaches [18].

We consider it surprising that the natural variability in face shape (at least as far as is captured by a morphable model) should include variations consistent with perspective transformation. An intuitive interpretation of this is that there are faces which look like they have been subjected to perspective transformation when they have not. There must be a limit to this. For example, to fit to a target face that is distant requires the close fitted face to have large protruding ears (see Fig. 14.5). If this fitted face was then used as a distant target, the ears would need to increase in size again for a close fitting. Clearly, repeating this process would quickly take the fitted result outside the span of the model (or the hyper box constraint would simply limit the ability of the model to explain the observations).

14.7.1 Generality of Assumptions

The perspective face ambiguity applies in an uncalibrated scenario, i.e. when camera focal length or pixel size is unknown and therefore the subject-camera distance cannot be estimated from the size of the face in the image. Images taken by digital cameras usually contain meta data including the focal length and camera model. The pixel size is fixed for a particular camera model and so could, in principle, be stored in a database. Hence, it appears that in practice some calibration information is likely to be available and the ambiguity resolved. In fact, there are two reasons why this is not the case:

1. In a fully calibrated situation (i.e. when camera focal length and pixel size is known) then the size of a face in the image does give some indication as to the subject-camera distance. However, head size varies significantly across the population: e.g. the bitragion breadth (i.e. face width) ranges from 12.51 cm to 15.87 cm for males and females [14] – a variation of over 25 %. With an uncertainty in the distance estimate of $\sim$25 %, the perspective ambiguity remains

significant, particularly when the face is close to camera. Moreover, in statistical shape modelling, the scale of each sample is often factored out when generalised Procrustes analysis is used to register the training data. This means that the statistical shape model has no explicit scale, rendering the size cue even less accurate for distance estimation.

2. Images that have been modified in any way, e.g. cropped, resized or compressed, will often no longer contain meta data or the meta data will incorrectly describe the effective camera geometry. This is likely to be the case for a large proportion of the images on the web (and the images in Fig. 14.1 are perfect examples: these files contain no metadata). In this case, no calibration information is available and the ambiguity is exactly as described in this paper.

14.7.2 Future Work

There are many ways in which the work can be extended. First, there are a number of simplifications that we made which could be relaxed and their effect investigated. This includes allowing rotations and hence considering the ambiguity in non-frontal poses. There appears to be very little work investigating the effect of perspective transformation on non-frontal faces. Intuitively, the effects may be less dramatic since it is the large (relative) depth variation between nose tip and ears that makes the effect so noticeable. We also ignored the effect of the skew parameter and translations in x and y away from the centre of projection. A more complex camera model could even be used, for example considering radial distortion.

Next, our shape estimation approach could be cast in probabilistic terms. We take a rather simplistic approach, simply seeking to minimise the 2D vertex error in a least squares sense. As has been shown previously [1, 24], partially fixing a statistical shape model still leaves flexibility. Hence, our fitting algorithm could return the subspace of faces that is approximately consistent with the observed vertices. Shape fitting could also be extended to edge features such as silhouettes. These are interesting because there is no longer a one-to-one correspondence between 2D shape features and model vertices. This suggests that the ambiguity would be even more significant in this case. An interesting follow-up to the work of Amberg et al. [3] would be to investigate whether there is an ambiguity in uncalibrated *stereo* face images.

Our consideration of appearance was limited to diffuse shading under a spherical harmonic illumination model. It is known that light source attenuation is a useful cue for the interpretation of shading under perspective projection so this may be an interesting avenue for future work. Similarly, cast shadows and specular reflections may also help resolve the ambiguity.

Acknowledgements I would like to thank the reviewers for their thoughtful comments which helped improve the chapter significantly.

References

1. Albrecht, T., Knothe, R., Vetter, T.: Modeling the remaining flexibility of partially fixed statistical shape models. In: Proceedings of the Workshop on the Mathematical Foundations of Computational Anatomy (2008)
2. Aldrian, O., Smith, W.A.P.: Inverse rendering of faces with a 3D morphable model. IEEE Trans. Pattern Anal. Mach. Intell. **35**(5), 1080–1093 (2013)
3. Amberg, B., Blake, A., Fitzgibbon, A., Romdhani, S., Vetter, T.: Reconstructing high quality face-surfaces using model based stereo. In: Proceedings of the ICCV, Rio de Janeiro (2007)
4. Basri, R., Jacobs, D.W.: Lambertian reflectance and linear subspaces. IEEE Trans. Pattern Anal. Mach. Intell. **25**(2), 218–233 (2003)
5. Belhumeur, P.N., Kriegman, D.J., Yuille, A.L.: The bas–relief ambiguity. Int. J. Comput. Vis. **35**(1), 33–44 (1999)
6. Blanz, V., Mehl, A., Vetter, T., Seidel, H.P.: A statistical method for robust 3D surface reconstruction from sparse data. In: Proceedings of the 3DPVT, Thessaloniki, pp. 293–300 (2004)
7. Blanz, V., Vetter, T.: Face recognition based on fitting a 3D morphable model. IEEE Trans. Pattern Anal. Mach. Intell. **25**(9), 1063–1074 (2003)
8. Brunton, A., Salazar, A., Bolkart, T., Wuhrer, S.: Review of statistical shape spaces for 3D data with comparative analysis for human faces. Comput. Vis. Image Underst. **128**, 1–17 (2014)
9. Bryan, R., Perona, P., Adolphs, R.: Perspective distortion from interpersonal distance is an implicit visual cue for social judgments of faces. PloS one **7**(9), e45,301 (2012)
10. Burgos-Artizzu, X.P., Ronchi, M.R., Perona, P.: Distance estimation of an unknown person from a portrait. In: Proceedings of the ECCV, Zurich, pp. 313–327 (2014)
11. Cootes, T.F., Edwards, G.J., Taylor, C.J.: Active appearance models. In: Proceedings of the ECCV, Freiburg, pp. 484–498 (1998)
12. Flores, A., Christiansen, E., Kriegman, D., Belongie, S.: Camera distance from face images. In: Proceedings of the ISVC, Rethymnon, pp. 513–522 (2013)
13. Georghiades, A., Belhumeur, P., Kriegman, D.: From few to many: illumination cone models for face recognition under variable lighting and pose. IEEE Trans. Pattern Anal. Mach. Intell. **23**(6), 643–660 (2001)
14. Gordon, C.C., Churchill, T., Clauser, C.E., Bradtmiller, B., McConville, J.T.: Anthropometric survey of US army personnel: methods and summary statistics 1988. Technical report NATICK/TR-89/044, DTIC Document (1989)
15. Hartley, R., Zisserman, A.: Multiple View Geometry in Computer Vision. Cambridge university press, Cambridge/New York (2003)
16. Hill, H., Bruce, V.: A comparison between the hollow–face and 'hollow–potato' illusions. Perception **23**, 1335–1337 (1994)
17. Keller, M., Knothe, R., Vetter, T.: 3D reconstruction of human faces from occluding contours. In: Proceedings of the Mirage, Rocquencourt, pp. 261–273 (2007)
18. Kleinberg, K.F., Vanezis, P., Burton, A.M.: Failure of anthropometry as a facial identification technique using high-quality photographs. J. Forensic Sci. **52**(4), 779–783 (2007)
19. Knothe, R., Romdhani, S., Vetter, T.: Combining PCA and LFA for surface reconstruction from a sparse set of control points. In: Proceedings of the International Conference on Automatic Face and Gesture Recognition, Southampton, pp. 637–644 (2006)
20. Latto, R., Harper, B.: The non-realistic nature of photography: further reasons why Turner was wrong. Leonardo **40**(3), 243–247 (2007)
21. Lepetit, V., Moreno-Noguer, F., Fua, P.: EPnP: an accurate $O(n)$ solution to the PnP problem. Int. J. Comput. Vis. **81**(2), 155–166 (2009)
22. Liu, C.H., Chaudhuri, A.: Face recognition with perspective transformation. Vis. Res. **43**(23), 2393—2402 (2003)
23. Liu, C.H., Ward, J.: Face recognition in pictures is affected by perspective transformation but not by the centre of projection. Perception **35**(12), 1637–1650 (2006)

24. Lüthi, M., Albrecht, T., Vetter, T.: Probabilistic modeling and visualization of the flexibility in morphable models. In: Proceedings of the Thirteenth IMA Conference on Mathematics of Surfaces, York, pp. 251–264 (2009)
25. Patel, A., Smith, W.A.P.: 3D morphable face models revisited. In: Proceedings of the CVPR, Miami, pp. 1327–1334 (2009)
26. Paysan, P., Knothe, R., Amberg, B., Romdhani, S., Vetter, T.: A 3D face model for pose and illumination invariant face recognition. In: Proceedings of the IEEE International Conference on Advanced Video and Signal Based Surveillance (2009)
27. Perona, P.: A new perspective on portraiture. J. Vis. **7**(9), 992–992 (2007)
28. Porter, G., Doran, G.: An anatomical and photographic technique for forensic facial identification. Forensic Sci. Int. **114**(2), 97–105 (2000)
29. Sagonas, C., Tzimiropoulos, G., Zafeiriou, S., Pantic, M.: 300 faces in-the-wild challenge: the first facial landmark localization challenge. In: Proceedings of the ICCV Workshop on Automatic Facial Landmark Detection in-the-Wild Challenge, Sydney, pp. 397–403 (2013)

Chapter 15
On Shape Recognition and Language

Petros Maragos, Vassilis Pitsikalis, Athanasios Katsamanis, George Pavlakos, and Stavros Theodorakis

Abstract Shapes convey meaning. Language is efficient in expressing and structuring meaning. The main thesis of this chapter is that by integrating shape with linguistic information shape recognition can be improved in performance. It broadens the concept of shape to visual shapes that include both geometric and optical information and explores ways that additional linguistic information may help with shape recognition. Towards this goal, it briefly describes some shape categories which have the potential of better recognition via language, with emphasis on gestures and moving shapes of sign language, as well as on cross-modal relations between vision and language in videos. It also draws inspiration from psychological studies that explore connections between gestures and human languages. Afterwards, it focuses on the broad class of multimodal gestures that combine spatio-temporal visual shapes with audio information. In this area, an approach is reviewed that significantly improves multimodal gesture recognition by fusing 3D shape information from motion-position of gesturing hands/arms and spatio-temporal handshapes in color and depth visual channels with audio information in the form of acoustically recognized sequences of gesture words.

15.1 Introduction

This chapter explores the fusion of shape and linguistic information for improving shape recognition. While its main objective is to address the computer vision problem of shape recognition for shape categories where linguistic information is available by using statistical pattern classification methodologies, it also draws inspiration from psychological studies that explore connections between shapes and human languages. Towards this goal, we broaden the meaning of "shape" to

P. Maragos (✉) • V. Pitsikalis • A. Katsamanis • S. Theodorakis
School of Electrical and Computer Engineering, National Technical University of Athens, Athens 15773, Greece
e-mail: maragos@cs.ntua.gr; vpitsik@cs.ntua.gr; nkatsam@cs.ntua.gr; sth@cs.ntua.gr

G. Pavlakos
Computer and Information Science, University of Pennsylvania, Philadelphia, PA 19104, USA
e-mail: pavlakos@seas.upenn.edu

M. Breuß et al. (eds.), *Perspectives in Shape Analysis*, Mathematics and Visualization, DOI 10.1007/978-3-319-24726-7_15

include not only geometric but additional optical attributes; this augmented shape information is referred to as "visual shape". Thus, as explained in Sect. 15.2.1, shape is meant here in a broader sense of visual information that may encompass brightness, color, depth and time dynamics, even if the main channel is the 2D geometrical shape (or the projection-silhouette of a 3D shape) as time evolves. It focuses on gesture shapes, inspired by the long-term studies of the importance of gestures for the origins of human language and their synergy with speech [3, 25, 31, 47].

We begin with Sect. 15.2 that clarifies how we mean the information conveyed by a shape and in which ways it can be supplemented by linguistic information. We also use statistical inference to intuitively explain how shape recognition may benefit from additional linguistic information. Next, Sect. 15.3 provides a brief survey of shape categories which have the potential of better recognition by combining visual with linguistic information, with emphasis on gestures and moving shapes of sign language, as well as on cross-modal relations between vision and language in videos. This is followed by a motivating Sect. 15.4 on the importance of gestures for human communication. Afterwards, we focus in Sect. 15.5 on the main paradigm of the chapter, which is the broad class of multimodal gestures combining spatio-temporal shapes and other visual cues with audio information in the form of sequences of spoken commands accompanying the gestures; in this section we review an approach [37, 38] that fuses shapes with linguistic information, which is audio-visually expressed, for significantly improving the automated recognition of multimodal gestures. While discussing the examples of both Sects. 15.3 and 15.5 we draw analogies with the main ideas of this chapter.

15.2 Visual Shapes and Linguistic Information

15.2.1 Visual Shapes

Shapes are traditionally perceived and understood as objects of geometry, two-dimensional (2D) or three-dimensional (3D). For a better understanding, perception attributes may be added to them, e.g. as in Gestalt psychology. For automated shape recognition, the computer vision community further explores broader appearance characteristics of shapes by viewing them (whenever possible) as gray intensity images that have both shape and texture. Thus, if 2D shapes are perceived from images, they obtain a third dimension of brightness texture. Instead of brightness, we may also add color to a 2D shape. Temporal dynamics are also important in recognizing moving shapes. Another way of adding a third dimension to a 2D shape to be recognized as a projected silhouette of a 3D object is by using depth. For 3D shapes, including brightness or time evolution will add a fourth dimension.

Thus, in addition to their 2D main projection or silhouette, shapes of world objects can have some additional geometric attributes such as depth and region summary as exemplified by their skeleton axis and its branch points, or even *optical attributes*, e.g. intensity, color, as well as motion (in case of moving shapes) possibly represented by dynamics of the above attributes as time evolves. We shall call this augmented shape information a **visual shape**, meaning that it contains attributes both from geometry (2D or 3D) and optics (photometry and motion). A rich category of such visual shapes that include all the above attributes and will be the main paradigm of this chapter are *gestures*. Both from a human perception and a computer representation viewpoint, gestures comprise several information streams which include a main 2D shape information such as the projection of the handshapes and possibly the moving arms on the image plane, color information, 3D shape, 3D motion, and by using appropriate sensors or computer vision algorithms they can be supplemented with depth and skeleton information. This is illustrated in Fig. 15.1 through an example showing a user performing the Italian gesture "basta" ("that's enough!"). We sampled the video of a user performing this gesture and selected non-uniformly five frames to depict the most important states of the "basta" gesture as time evolves. We supplement the RGB frames with skeleton and depth information, as well as images of the right or both handshapes. In this example the RGB, depth

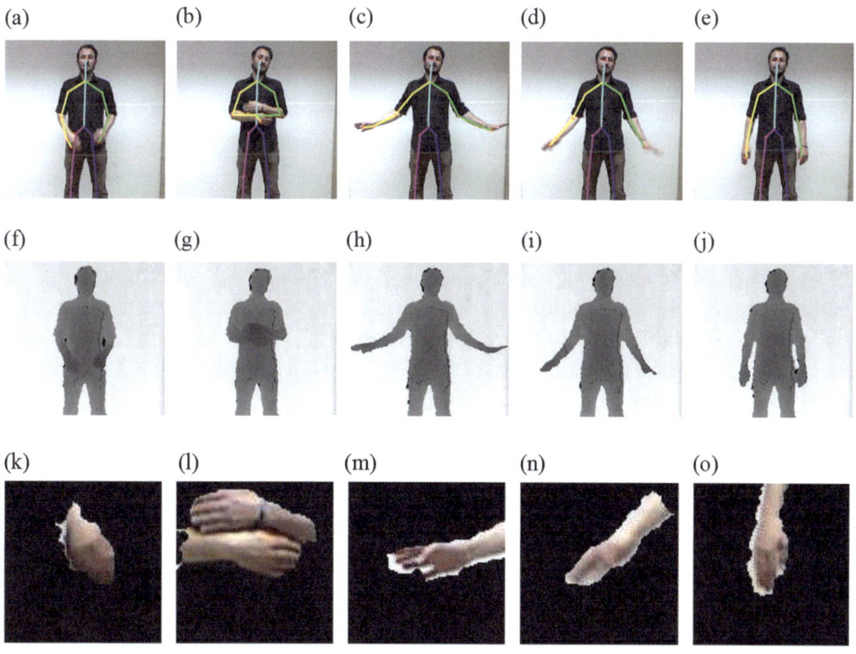

Fig. 15.1 Sequence of frames sampled for a video of a user performing the Italian gesture "basta" ("that's enough!"), obtained with a Kinect sensor. Each column corresponds to a different temporal section of the gesture performance, covering the overall range of motion. Given the start and the end frames here, the duration of the gesture is 36 frames, i.e. 1.8 s (with the frame rate at 20 fps). *First row*: RGB frames accompanied with the skeleton of the user that is superimposed on them. *Second row*: The respective depth frames. *Third row*: Images of the segmented handshapes

and skeleton data were provided by a Kinect sensor. The skeleton information of this sensor includes the human skeleton axis and its branch points, such as hands' centers, elbows, shoulders, face, knees and other critical points.

As exemplified in Fig.15.1, shape information can become richer, and hence its recognition easier, if we augment the geometry of a shape with optical attributes. We offer an intuitive explanation from the domain of statistical pattern classification, by using Bayesian inference. Let S_i represent the i-th class from a collection of shape classes. Suppose we are given measurable data $\mathscr{D}$, which may contain either only geometric information G or geometric and optical information O, and the goal is to infer the shape class given the data via the maximum-a-posteriori principle. If we have information only from geometry, then $\mathscr{D} = \{G\}$, and

$$P(S_i/\mathscr{D}) = \frac{P(S_i)P(G/S_i)}{P(\mathscr{D})} \tag{15.1}$$

where $P(\cdot)$ denotes probability or likelihood. In the case of geometry plus optics, $\mathscr{D} = (G, O)$ and hence

$$P(S_i/\mathscr{D}) = \frac{P(S_i)P(G/S_i)P(O/S_i, G)}{P(\mathscr{D})} \tag{15.2}$$

In the above combined case, the deciding numerator of the right hand side, excluding the prior class probability $P(S_i)$ which is common in both cases (15.1) and (15.2), is a product of two terms, the probability of optical data given the shape class and geometry times the probability of the geometric data given the shape class. By exploiting these two terms we may be able to increase the discriminatory potential of their product. Thus, we may improve the classification of the shape by using statistical knowledge about both its corresponding geometric and optical data, whenever such information is available.

From the domain of philosophy, an extreme such example of the richness of visual shapes versus geometric shapes is Plato's allegory of the cave (presented in his work "The Republic") where silhouettes of real world objects, whose fire-produced shadows are cast on a cave wall while the real objects are being moved behind human spectators, cannot be recognized. In contrast, if the same real objects are seen with direct eye contact and under the sunlight, they reveal their true identity. In Fig. 15.2 we attempted to create an example that illustrates only the visual aspects of the cave allegory. Namely, Fig. 15.2 shows time snapshots from a video of people running and contrasts the complete visual perception provided by the video RGB frames (shape geometry plus color) versus the obviously poorer insufficient information of the 2D silhouettes (shape geometry only) of the moving objects. As *motion* played an important role in the previous gesture sequence of Fig.15.1, we also see in Fig. 15.2 that motion is an important visual cue for understanding of moving shapes.

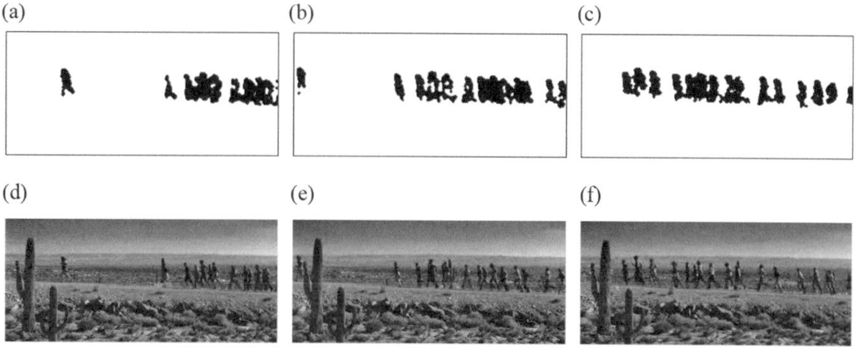

Fig. 15.2 Moving shapes: time sequence of sample frames from a video showing people running (*bottom row*) and their silhouettes (*top row*). The frame rate for this video is at 30 fps. Second frame column is apart from the first by 34 frames (1.13 s), while third frame column is apart from the second by 48 frames (1.6 s)

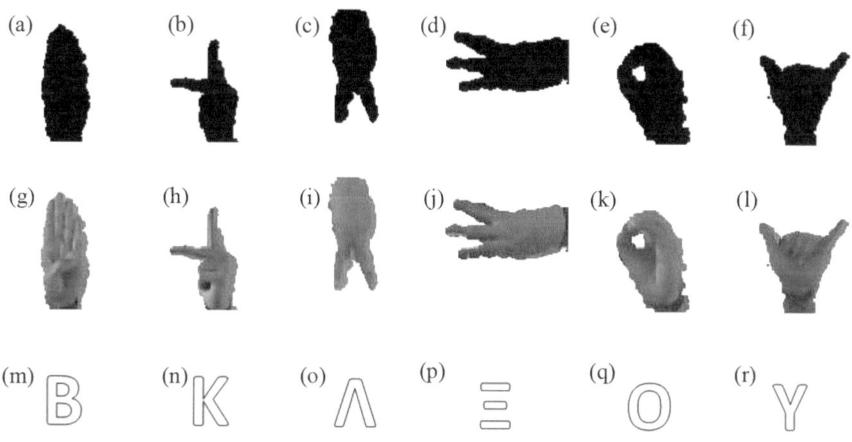

Fig. 15.3 Greek sign language alphabet: shapes, images, letters

So far, one main conclusion is that a *geometric shape*, defined as a 2D or 3D set of points representing an object in the Euclidean space, is a minimalistic form of a *visual shape*, where "visual" means augmenting geometry with optical attributes.

15.2.2 Adding Linguistic Information

The main message from the previous discussion, i.e. that shape inference is enriched if we couple geometry with optics, is further illustrated in the first two rows of Fig. 15.3. The top row shows only silhouettes of handshapes from a sign language. The silhouette only information has some ambiguities, one of which is the question whether the front or back side of the handshapes is visible. In contrast, the middle row shows their corresponding gray images (shape plus brightness texture), i.e. an example of what we call visual shapes, which disambiguate both the visible side of the handshape and add texture details on the visible surface. If a viewer did not know sign language, the first two rows of Fig. 15.3 would just be some handshapes with shape differences among them. However, if we add the information of the third row which corresponds these handshapes with distinct letters of the Greek sign language alphabet, then we have augmented information of a visual shape plus language. This addition of linguistic information can improve the recognition of such handshapes, both from an intuitive viewpoint and from a Bayesian inference viewpoint. To detail the latter (as inspired from statistical speech recognition [23, 40]), assume for example that we are given a time sequence $S = (s_1, s_2, \ldots, s_T)$ of shapes s_i from a visual language, in the form of spatio-temporal visual data, and each shape corresponds to a word w_i, then we can recognize the unknown sequence of visual words $W = (w_1, w_2, \ldots, w_T)$ by estimating it via the maximum-a-posteriori principle:

$$W^* = \arg\max_W P(W/S) = \arg\max_W \frac{P(S/W)P(W)}{P(S)} \tag{15.3}$$

Thus, the likelihood $P(S/W)$ of the visual shape sequence given its linguistic structure is combined with the prior probability $P(W)$ of the linguistic sequence; this can potentially improve the recognition by exploiting statistical knowledge of the language, e.g. if the n-gram probabilities $P(w_i/w_{i-1} \cdots w_{i-n+1})$ are known.

One way of creating a correspondence between visual shapes and words of some language is via *clustering*. As further elaborated in Sect. 15.4 on the importance of gestures for human communication, imagine given a sequence of visual shape data that span a domain of visual realizations of concepts or objects common to some human community and are represented by visual feature vectors. Then, by some clustering method such as for example the K-means algorithm we can partition the data over this domain into cells (which are regions of the feature space), each representing a concept or object. The mapping of visual shapes in each cell to the cell centroid is some form of feature encoding known as *vector quantization*. Then, these centroids can play the role of words or subword units in some language. In addition to its general usefulness in pattern recognition and machine learning [5, 14, 46], clustering via vector quantization has also been used in signal processing for data compression [20], in speech recognition for converting continuous feature vectors into discrete patterns [40], and in computer vision for action or object recognition based on the *bag of visual words* approach [19, 27, 43].

In the following sections we shall briefly describe some paradigms where visual shape information is supplemented by additional linguistic information. We distinguish three cases:

(1) Relationships between visual shapes and linguistic information. These include (i) direct correspondences as for example in Fig. 15.3 and the pictograms

mentioned in the beginning of Sect. 15.3; (ii) cross-modal relationships in Sect. 15.3.2 between visual objects, represented by their shape information, and linguistic information as corresponding words in text or related audio sounds, employed in a multimedia analysis framework.

(2) In the second case, we employ linguistic information from sign language at the level of visual phonetics. For example, in sign language recognition (Sect. 15.3.1) the video segment corresponding to the visual word of an isolated sign is decomposed into a time sequence of subunits that have a phonetic meaning.

(3) In the third case, which focuses on multimodal gesture recognition (Sect. 15.5), linguistic information is expressed in parallel audio and visual modalities: in the visual stream, gestures occur in a time sequence; in parallel, in the audio stream a sequence of corresponding keywords (or spoken commands) accompanies the visual gestures and provides additional linguistic information.

In all the above paradigms the linguistic information we employ stays only in the specific examples as case studies, and at the level of words or word-subunits; for instance, we do not discuss linguistic structure at the level of sentences.

15.3 Shape and Language Paradigms

Among the earliest paradigms of correspondences between shape and language are the ideographic and logographic writing systems. In the ideographic system the graphemes are the ideograms which are graphic symbols expressing pictorially some concept, independently of any specific language but often assuming some prior convention. A special case are the *pictograms* which further provide a pictorial resemblance with a physical object. Thus, in pictograms there is a direct connection between shape and language. The logographic system is based on logograms which are graphemes that represent words or morphemes and may also contain phonetic elements. Examples of logograms include numerous Egyptian hieroglyphs and Chinese characters. A famous example that may fit in one of the above cases are the shapes on the Phaistos Disk, which was discovered in 1908 at the Minoan palace of Phaistos on the Greek island of Crete, possibly dating from the 2nd millennium B.C.; see Fig. 15.4. Although the ancient Egyptian hieroglyphs have been deciphered after the discovery of the Rosetta Stone in 1799, the glyphs on the Phaistos Disk still remain an archeological mystery.

In the two following subsections we highlight some ideas relevant to this chapter from two broad categories of moving shapes where we encounter numerous correspondences between shapes and language: (i) sign gestures and facial expressions encountered in sign language and (ii) multimodal relationships between vision plus language (audio or text) that are abundant in movie videos.

Fig. 15.4 Phaistos disk (At the archaeological museum of Heraklion, Crete)

15.3.1 Sign Language

Human languages include both spoken and sign languages. Sign languages are natural languages communicable purely by vision via sequences of time-varying 3D shapes. They serve for communication in the Deaf communities, as well as among deaf and hearing people if the latter learn to sign. They convey information and meaning via spatio-temporal visual patterns, which are formed by manual (handshapes) and non-manual cues (facial expressions and upper body motion). A coarse correspondence of a word in spoken language is a sign in sign language. See [15, 28] for surveys of linguistic and cognitive aspects of sign language. The area of computer-based processing and recognition of sign videos is also broadly related to vision-based human-computer interaction using gesture recognition [22].

While significant progress exists in the field of automatic sign language recognition from the computer vision and pattern recognition fields, e.g. see [1, 8, 32, 44, 45, 49] and the references therein, it still remains a quite challenging task especially for continuous sign language. In addition to signs having a complex multi-cue 4D space-time structure, the difficulty in their automatic recognition is also due to the large variability with respect to inter-signer or intra-signer variations of signing while expressing the same concept-word. An example exhibiting such variations is shown in Fig. 15.5. This variation is due to various sources: (i) the physiology of each signer and the manner of his/her signing, (ii) the coarticulation – continuous variability that causes multiple pronunciations, and (iii) the existence of multiple pronunciations per se (e.g. from different dialects). Due to the above variability, instead of recognizing each sign as a whole word, a more efficient approach (inspired by speech recognition) is to decompose signs into *subunits*, resembling the phonemes of speech, and recognize them as a specific sequence of subunits by using some statistical model, e.g. via Hidden Markov Models (HMMs). Clearly, the subunits approach performs much better on large vocabularies and continuous

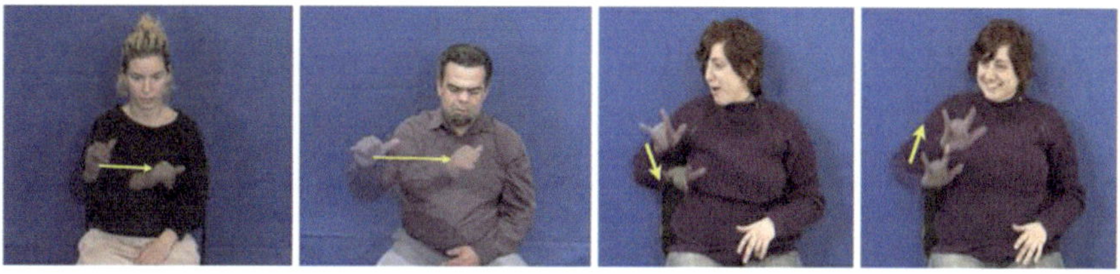

(a) AIRPLANE-1 (b) AIRPLANE-1 (c) AIRPLANE-2 (d) AIRPLANE-3

Fig. 15.5 Multiple realizations for sign /airplane/. (**a**) and (**b**) are due to inter-signer variability. (**c**) and (**d**) are due to intra-signer variability. On each image we superimpose the beginning and end frames of the sign with an *arrow*

language; further, the subunits are reusable and help with signer adaptation. In lack of a lexicon, a computational technique to find such subunits is *data-driven*, i.e. perform unsupervised clustering on a large database and use the cluster centroids as subunits. This performs well in several instances, especially when the subunits are pre-classified and statistically modeled based on visual features into dynamic vs. static, as done by Theodorakis et al. [45], where the dynamic or static refers to the type of the signer's hands and arms motion. However, a superior performance accompanied with phonetic interpretability may be obtained if the chosen subunits are also based on the phonetic structure of a sign, as for example by incorporating the Posture-Detention-Transition-Steady Shift (PDTS)[1] system [24] of phonetic labels. A sequence of PDTS phonetic subunits is shown in Fig. 15.6. Pitsikalis et al. [39] combined the phonetic information provided by the PDTS transcriptions of sign videos with the automatically extracted visual features to create (1) statistically trained *phonetic subunits* and a corresponding lexicon, which were then used for (2) optimally aligning (via Viterbi decoding) the data with the phonetic labels and hence providing the missing temporal segmentation, as well as (3) better sign recognition. Thus, we have a clear paradigm of improved shape recognition when the visual information is coupled with linguistic information.

While information and meaning in sign languages are mainly conveyed by moving handshapes, they are also conveyed in part by non-manual cues such as facial expressions. These expressions can be visually modeled by deformable models that encode both geometric shape and brightness texture information. Such a class of models often used in computer vision are the active appearance models (AAMs) [11]. Examples of the deformable geometric masks of such facial AAMs are illustrated in Fig. 15.7, which shows a few frames from a sign sequence that involves eye blinking. The transient phenomenon of eye blinking, where the eyes may take one of the open/closed states, conveys low-level linguistic information such as sentence – and possibly sometimes sign – boundaries, as described in Anton-

[1] In the PDTS system, D is a "hold" but for shorter duration than P. S is a "movement" without acceleration. T is more abrupt motion.

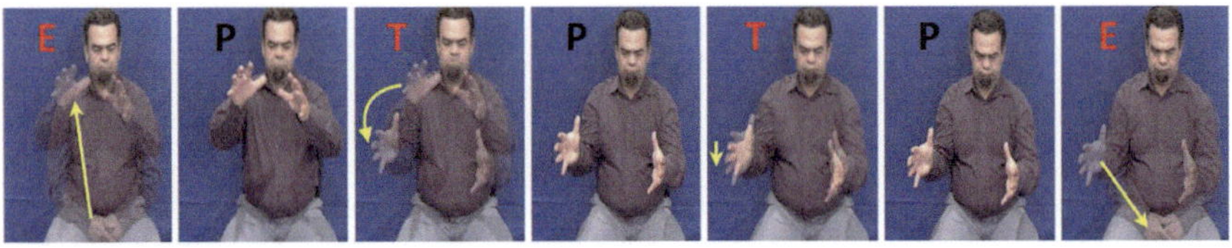

Fig. 15.6 Sample frames from the sign /pile/ from the Greek sign language. Images marked with "T" and "E" represent dynamic segments with the phonetic labels "Transition (T)" and "Epenthesis (E)", visualized by superimposing on the same image the beginning and end frames with an *arrow*. Images marked with "P" represent static segments with the phonetic label "Posture (P)", visualized by a single frame (Figure courtesy of Pitsikalis, Theodorakis, Vogler and Maragos [39])

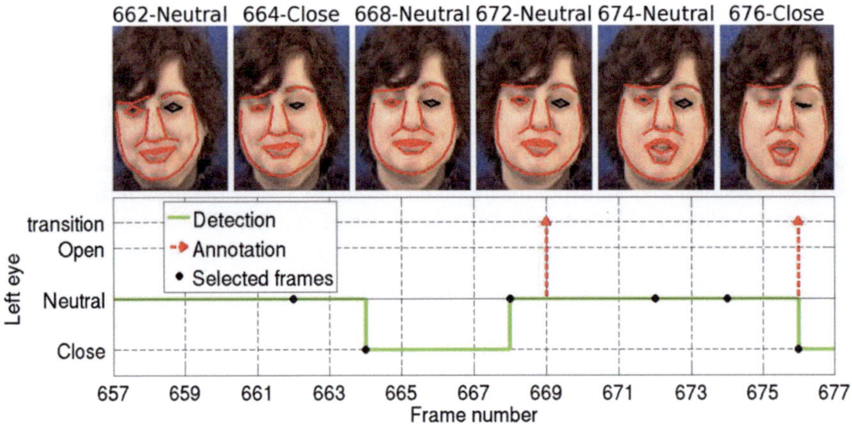

Fig. 15.7 Sign boundary detection based on eye blinking detection on a Greek Sign Language database. Indicative frames (*up*) are marked with a *black dot* in the detection diagram (*down*) (Figure courtesy of Antonakos, Pitsikalis and Maragos [2])

akos et al. [2] and the references therein. The detection of the eye opening/closing transitions can be detected from the changes in the corresponding AAM parameters. Figure 15.7 presents an example of such a detection between neutral-close-neutral (neutral is considered as intermediate) and its correspondence with the annotated sign boundaries. This is another paradigm of synergy between visual shape and language.

15.3.2 Multimodal Relations Between Shapes and Language

Every day communication between people is a blend of different modalities. Humans often combine different pieces of information, e.g. visual and linguistic, in order to communicate and interact. In multimedia data such as multimodal videos, visual, auditory and linguistic information coexist as well. In multimodal videos we encounter a variety of visual objects that we can recognize more easily when there exists a concurrent linguistic reference either in the text domain or as an acoustic event. (Note that linguistic information can also exist in a video without text or audio, e.g. in sign language videos as described in Sect. 15.3.1.) This is one aspect of a broader class of phenomena with audio-visual modality integration, which is an active research area in behavioral psychophysics, e.g. see [48], and in neuroscience where, for instance, brain activity during watching TV programs as measured by fMRI reveals correlations between audio and visual stimuli [7]. From a computational viewpoint, this audio-visual synergy can improve recognition performance in multimedia systems via cross-modal integration, as surveyed in [29] and the references therein. In general, there has been significant evidence that human perception is multimodal and hence perception of visual objects can be improved when different modalities are synergetically employed.

A corpus-based framework for analyzing and modeling multimedia dialectics is the COSMOROE framework [36] which describes the semantic interplay between verbal and non-verbal communication; specifically, the cross-media semantic interrelation between images, language (in the form of either spoken language transcription, graphic/scene text shown on the video, or acoustic stimuli, e.g. human/animal or environmental sounds) and body movement. In Fig. 15.8 we provide two such examples from cross-modal relations between visual shapes and linguistic information. For instance, in a quite complex scene as presented in Fig. 15.8a, where there is interaction between people (with clothing that attracts human attention), the image of the *dog* could go unnoticed; however, the fact that the dog is *barking* guides our look towards it. Same observation applies to Fig. 15.8b as well; the lamp could easily get overlooked if the acoustic stimulus as in the phrase *"Take the lamp out on the porch"* did not take place. This association of a visual object with the linguistic information may render the recognition procedure easier for humans and more robust for computers.

In short, the COSMOROE framework [36] aims at finding and analyzing relations from linguistics to other modalities, especially visual shapes, in multimodal corpora. In parallel, there is a recent trend in computer vision in the opposite

(a) (b)

Fig. 15.8 Correspondence between shapes and linguistic information (aural or textual) in movie videos. (**a**) Acoustic event: dog barking. (**b**) Utterance: "Take the lamp out on the porch"

direction, i.e. associating visual objects with linguistic attributes, which can benefit recognitions problems such as action recognition [27] and person recognition [12] in movie videos, as well as general object recognition [18, 35].

15.4 Gestures in Human Communication

In Sect. 15.3.1 on sign language we summarized that certain types of moving bodily shapes can convey linguistic messages that represent complete languages. Here we further extend this idea by providing a brief survey on how gestures have been of great significance in human communication. In particular, according to specific theories [3, 31, 47], they have supported the beginnings of language formation, after which gesture shapes and language can reinforce each other.

By gestures we mean visible actions involving shapes of manual and non-manual bodily motions and postures; most of them are dynamic (i.e. time-varying for part of their duration) and use the hands. Kendon [25] classifies human gestures in (1) Gesticulations, (2) Speech-framed, (3) Pantomimes, (4) Emblems (quotable gestures), and (5) Sign language. The above sequence has been called *Kendon's continuum* [30]. As the numerical index of the gesture class increases, the degree to which speech should accompany a gesture decreases whereas the degree to which a gesture shows language-like properties increases.

The theory that gesture-based human communication evolved first whereas conventional languages evolved later has had many supporters from the antiquity until it became more definite in the eighteenth century; afterwards gesture and sign languages started being studied as natural languages. Wittgenstein in his work [52] on the philosophy of language argued that "What we call meaning must be connected with the primitive language of gestures". In search of the origins of human communication, Tomasello [47] has provided ample evidence about the critical importance of gestures, in particular of the pointing and pantomiming types, for humans to develop (i) social cognitive skills that create a common conceptual ground, including joint attention, shared experience and common cultural knowledge, and (ii) social motives such as requesting, informing, helping and sharing with others. These developments of social cognition and motivation create a *shared intentionality*, as is called by some modern philosophers of action, e.g. [42]. Quoting from [47], "pointing (deictic gestures) direct the attention of a recipient to something in the immediate perceptual environment, whereas pantomiming (iconic gestures) direct the imagination of a recipient to something that typically is not in the immediate perceptual environment by simulating an action, relation, or object". Interestingly apes have also developed pointing (attention-getters) and pantomiming (intention-movements) gestures for their communication. One big difference between the gesture-based ape versus human communication is that for apes it serves individual intentionality, whereas for humans it serves shared intentionality. This shared intentionality is at the heart of the *cooperative model for human communication* [47].

Thus, according to the theory and evidences in [47], the human social cognitive skills and social motivation create a cooperative psychological infrastructure of human communication based on gestures, which laid the foundations for the later development of conventional languages. By "conventional language" we mean a symbolic communicative code, which assumes some preexisting codified form of communication like the gesture modality. Such a linguistic code is based on a non-linguistic infrastructure of intentional understanding and common conceptual ground [47, 51]. From a computational viewpoint, we may conjecture that nowadays, if we are given a collection of gestures referring to a common perceptual ground of objects, then by clustering and feature encoding we could in theory map gestures to some abstract language words which could be the cluster centroids. Of course, after their early development, human conventional languages, mainly spoken languages, evolved into a very creative and versatile form of communication which, despite its complexity, has fundamentally supported and propelled human civilization. In contrast to gestures, the vocal modality in nonhuman mammals remained inflexible and has not created a language. Quoting from [47], "for all mammals, including nonhuman primates, *vocal displays* are mostly unlearned, genetically fixed, emotionally urgent, involuntary, inflexible responses to evolutionarily important events that benefit the vocalizer. In stark contrast, a significant number of nonhuman primate *gestures*, especially those of great apes, are individually learned and flexibly produced communicative acts, involving an understanding of important aspects of individual intentionality."

Another supporter of the "gesture-first" conjecture is Arbib [3] who supports a theory that human language evolved as a result of biological and cultural evolution starting from simple manual gestures we share with apes, progressing to the imitation of manual skills and pantomime, and culminating to the development of sign language and speech.

In addition to the gesture-first theory which advocates that human language started as non-spoken gestures and signs, there are also combined theories that advocate a fusion of the gesture and speech modality. For example, based on evidence from neurological and psychological data, McNeill [31] argues for a two-phase development of language acquisition in children: The first phase is based only on gestures without speech. Later, when the required brain structures have matured at age about 3–4, the second phase begins and involves both speech and gestures. This *gesture-speech unity* continues in adult life and uniquely characterizes the human language that we have actually evolved as a species.

It is this *multimodal* view of the language, containing both imagery via gestures and linguistic codes via speech, that we further pursue in this chapter by discussing computational approaches to automate its recognition, as explained next in the paradigm of audio-visual gesture recognition.

15.5 Multimodal Gesture Recognition

Multimodal gestures, i.e. time sequences of isolated gestures with simultaneous utterance of the corresponding keyword (or spoken command), is a primary domain where the fusion of visual shapes (gestures) with linguistic information (spoken commands) leads to significantly improved recognition over visual only recognition. They are becoming increasingly useful for human-computer interaction [6, 22, 26, 34]. In this section we highlight the main ideas and method of the chapter authors' recent works in [37] and [38] for the effective recognition of multimodally expressed gestures as performed freely by multiple users. The experiments were performed on a demanding dataset [17] which was acquired via Kinect for the purpose of the ChaLearn multimodal gesture recognition challenge (in conjunction with ACM ICMI 2013) [16]. It comprises multimodal cultural-anthropological gestures of everyday life, in multi-user spontaneous realizations of both spoken and hand-gesture articulations, intermixed with other random and irrelevant hand or body movements and spoken phrases. The use of Kinect enables multimodal capturing and provides four information streams, three visual (RGB color video, depth video, and skeleton with tracking of its branch points) and one aural (audio stream), all essential to multimodal processing. In the next subsections, we briefly review the approach in [37, 38] for multimodal gesture recognition, where the additional employment of speech significantly improves the performance of recognition over using only visual shape information (handshape and skeleton).

15.5.1 Methodology

The multimodal gesture recognition system exploits the color, depth, skeleton and audio signals captured by the Kinect sensor. See Fig. 15.9 for an overall view of the proposed fusion scheme. It extracts features for the handshape configuration, the movement of the hands and the speech signal, and it essentially implements a two-level[2] fusion approach:

1st Pass (P1): To independently account for the specificities of each of the modalities involved, we first train separate gesture-word models for each modality. These unimodal models are then used to generate a set of possible gesture-word sequence hypotheses for a given recording. Then, this original set of hypotheses is multimodally rescored and resorted.

[2]In the work of [38] the P1/P2 terms are not employed any more compared to [37], since [38] includes several other contributions, the discussion of which is beyond the scope of this chapter. Herein we keep the P1/P2 terms only for descriptive reasons.

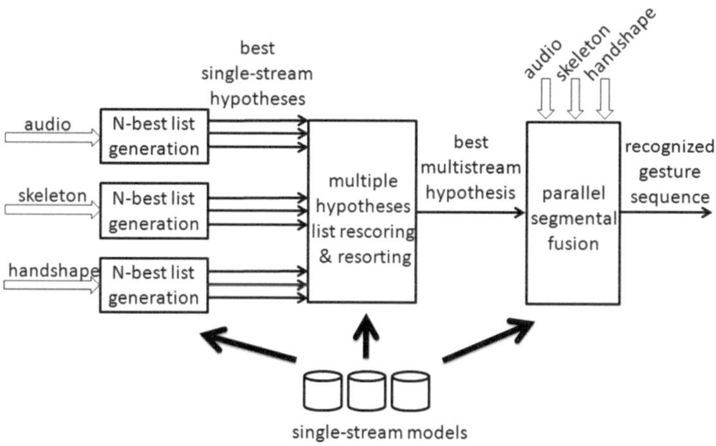

Fig. 15.9 Overview of the multimodal fusion scheme for gesture recognition based on multimodal hypotheses rescoring. Single-stream models are first used to generate possible hypotheses for the observed gesture sequence. The hypotheses are then rescored by all streams and the best one is selected. Finally, the observed sequence is segmented at the temporal boundaries suggested by the selected hypothesis and parallel fusion is applied to classify the resulting segments. Details are given in Sect. 15.5.1.2 (Figure courtesy of Pitsikalis, Katsamanis, Theodorakis and Maragos [38])

2nd Pass (P2): Based on the temporal boundaries of the gestures in the best fused hypothesis, a parallel segmental fusion step as in [49] exploiting all three modalities further improves recognition.

Gestures in our case occur in parallel with their semantically corresponding speech words, without implying however strictly synchronous realizations in all modalities. Given a vocabulary $V = \{g_i\}$, $i = 1, \ldots, |V|$, of multimodal gestures g_i that are to be detected and recognized in a recording and a set $C = \{\mathbf{O}_m\}$, $m = 1, \ldots, |C|$, of measurements from multiple information channels/streams that are concurrently observed, our goal is to generate the best multimodal hypothesis $\mathbf{h}$ for the sequence of gesture appearances, based on these observations. In our experiments, the latter set comprises three streams, namely handshape features, skeleton features and audio spectral features. In essence, any set of information streams can be employed in this framework, although the combination of visual and audio cues significantly enhances recognition results.

15.5.1.1 Single Information Stream Modeling

The modeling methodology essentially follows the keyword-filler paradigm for speech [41, 50] and is based on hidden Markov models (HMMs). For a tutorial on HMMs and their application to speech recognition, the reader is referred to [23, 40]. The problem of recognizing a limited number of gesture-words in a video possibly comprising other heterogeneous events as well, is seen as a keyword detection problem. The gesture-words to be recognized are the keywords and all the rest is ignored. Each gesture-word is modeled by a left-to-right HMM with a common number of states and with Gaussian mixture models (GMMs) representing the state-dependent observation probability distributions. There are also two separate filler HMMs to represent either silence/inactivity, or all other possible events (called "background model" – BM) appearing in that stream.

15.5.1.2 Multimodal Fusion

N-Best Rescoring and Resorting: Using the single stream gesture models and a gesture grammar G, which defines the set of alternative hypotheses allowed, a list of N-best possible hypotheses is initially generated for the unknown sequence for each stream. Specifically, by applying Viterbi decoding [40] we can estimate the best hypothesis $\hat{\mathbf{h}}_m$ per stream:

$$\hat{\mathbf{h}}_m = \arg\max_{\mathbf{h}\in G} \log P(\mathbf{O}_m|\mathbf{h}, \lambda_m), \quad m = 1, \ldots, |C|, \tag{15.4}$$

where $\mathbf{O}_m$ is the observation sequence for modality m, λ_m is the corresponding set of HMM models, and G is the set of alternative hypotheses allowed by the gesture grammar.

Similarly, in the more general case, we can generate a complete list of the N-best gesture-word sequences per stream, and form a set $H = \{\mathbf{h}_1, \ldots, \mathbf{h}_L\}$ of all the hypotheses (L in total) for the available modalities. Given this set, we sort the hypotheses [10, 21, 33] and identify the most likely hypothesis exploiting all modalities. In this direction, we estimate a combined score for each possible gesture sequence as a weighted sum of standardized modality based scores:

$$v_i = \sum_{m=1}^{|C|} w_m v^s_{m,i}, \quad i = 1, \ldots, L \tag{15.5}$$

where the weights w_m for each modality m can be determined experimentally (by maximizing the recognition score on the validation set). The modality-based scores $v^s_{m,i}$ are standardized versions of $v_{m,i}$ which are estimated by means of Viterbi decoding:

$$v_{m,i} = \max_{\mathbf{h}\in G_{h_i}} \log P(\mathbf{O}_m|\mathbf{h}, \lambda_m), \quad i = 1, \ldots, L, \quad m = 1, \ldots, |C|, \tag{15.6}$$

This maximization searches over acceptable gesture sequences that follow a specific hypothesis-dependent finite-state grammar G_{h_i}. Thus, this is a constrained recognition problem where the search space of possible state sequences includes only sequences corresponding to the hypothesis $\mathbf{h}_i$ plus possible variations by keeping the appearances of target gestures unaltered and only allow SIL (silence) and BM (background model) labels to be inserted, deleted and substituted with each other. The most probable gesture-word sequence hypothesis $\mathbf{h}^* = \mathbf{h}_{i^*}$, where $i^* = \arg\max_i v_i$, after this step is the one with the maximum combined score.

Segmental Parallel Step: Herein we exploit the modality-specific time boundaries (found via forced alignment) for the most likely gesture sequence and segment each observation stream, to reduce the recognition problem to a segmental classification one. For every segment and each stream, we compute the log probability:

$$LL^t_{m,j} = \max_{\mathbf{q}\in Q} \log P(\mathbf{O}^t_m, \mathbf{q}|\lambda_{m,j}), \quad j = 1, \ldots, |V|, \tag{15.7}$$

where t is the time index of the segment, $\lambda_{m,j}$ are the parameters of the HMM model for the gesture g_j and the stream m; $\mathbf{q}$ is a possible state sequence. These segmental scores are linearly combined across modalities to get a multimodal score:

$$LL^t_j = \sum_{m=1}^{|C|} w'_m LL^t_{m,j} \tag{15.8}$$

where w'_m is the stream-weight for modality m set to optimize recognition performance of this step. Finally, the recognized gesture for each segment t is the one with the highest multimodal score. This final stage is expected to give additional improvements, allowing local refinements by exploiting possible benefits of a segmental classification process.

15.5.2 Experimental Results

15.5.2.1 Multimodal Gesture Dataset

For the experimental work we employed the ChaLearn multimodal gesture challenge dataset [17], which focuses on multiple-instance, user-independent learning of gestures from multimodal data. It provides via Kinect RGB and depth images of face and body, user masks, skeleton information, as well as concurrently recorded audio including the speech utterance accompanying the gesture. See top row of Fig. 15.10 for an example of the data. The vocabulary contains 20 Italian cultural-anthropological gestures, performed by 39 users in 13,858 gesture-word instances in total. Gesture recognition over this dataset presents several challenges: presence of distracting gestures, large number of categories, length of gesture sequences, user variety and corresponding variability in gestures and spoken dialects, variations in background and lighting; see Fig. 15.11.

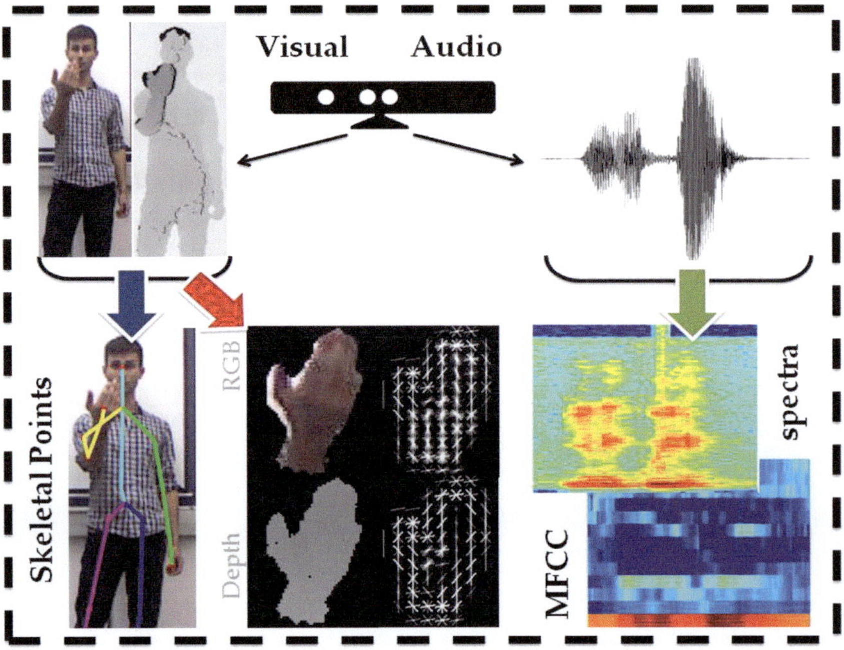

Fig. 15.10 A collection of sample cues as well as extracted features for each modality. *Top row*: visual data (RGB and depth) and audio data. *Bottom row*: visual features (skeletal points, HOGs in the RGB and depth channels) and audio features (MFCCs)

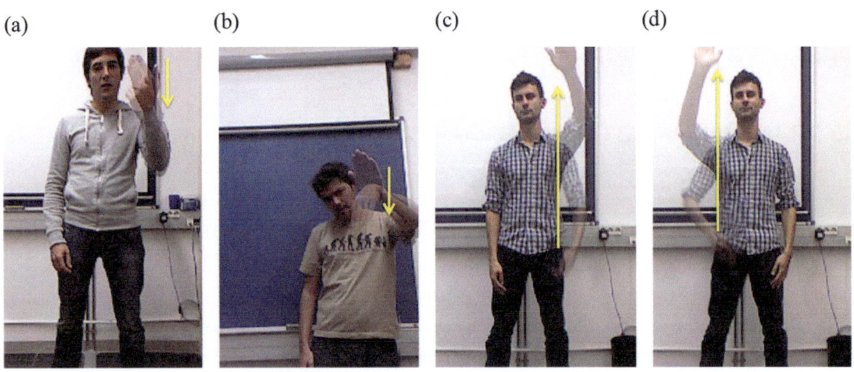

Fig. 15.11 (**a,b**) Arm position variation (low, high) for gesture "vieni qui" ("come here"); (**c,d**) Left- and right-handed instances of gesture "vattene" ("go away"). Gesture motion is visualized by superimposing on the same image the beginning and end frames with an *arrow*

15.5.2.2 Multimodal Features

We statistically train separate HMMs at the level of word-gestures per each modality, i.e. handshape, skeleton and audio.

Table 15.1 Single modalities evaluation expressed as accuracy (in %)

Audio	*Skeleton*	*HandShape*
87.2	49.1	20.2

Table 15.2 Our approach in comparison with the first three places of the ACM 2013 Gesture Challenge

Approach	Accuracy %
Ours [38]	93.3
iva.mm [53]	87.2
wweight [17]	84.6
E.T. [4]	82.9

Handshape Cue: The features employed are Histograms of Oriented Gradients (HOGs) [13] as extracted in both hands' segmented images for both RGB and depth modality. We segment the hands by employing the hand's tracking and by performing threshold depth segmentation. Essentially, any visual descriptor could be computed on the handshape information; HOGs are just an example that is used widely in the literature (e.g. in [9]).

Skeleton Cue: The features employed for the skeleton cue include: the hands' and elbows' 3D position, the hands' 3D position with respect to the corresponding elbow, the 3D direction of the hands' movement, and the 3D distance of hands' centroids.

Audio Cue: To efficiently capture the spectral properties of speech signals, our frontend generates 39 acoustic features every 10 ms. Each feature vector comprises 13 Mel Frequency Cepstral Coefficients (MFCCs) along with their first and second derivatives.

A visualization of the extracted features for all the available modalities is presented in bottom row of Fig. 15.10.

15.5.2.3 Recognition Results

We summarize the most recent[3] experimental results from [38].

In Table 15.1 we show the recognition results for each modality. The results are expressed in *accuracy* (%), which is computed as 100 − WER where WER is the percent word error rate that includes insertions, deletions and substitutions. As observed, the audio modality is the strongest one.

Table 15.2 shows the performance of the proposed multimodal two-pass fusion scheme [38] in comparison with other approaches who participated in the Gesture Challenge [17]. Our scheme begins with a first-pass fusion step (*P1*) leading to

[3]The multimodal gesture recognition system in [38] is an extension of [37], where additional components are included such as voice and gesture activity detection and a gesture-loop grammar, which improve the recognition results.

the best fused hypothesis as a result of the N-best rescoring. Then follows the *P2* component as the second-pass fusion step; in this we employ the gesture-word level segmentation of the above best fused hypothesis, leading to the second-pass fused result and the final recognized words. This multimodal fusion yields a recognition accuracy of 93.3 %, which outperforms the other approaches and reduces the smallest previous error by a relative 47 %.

A gesture sequence decoding example is shown in Fig. 15.12. Herein we illustrate both audio and visual modalities for a word sequence accompanied with the ground truth word-level transcriptions (row: "REF"). In addition we show the decoding output employing the single-audio modality (AUDIO) and the three presented fusion cases ($P1$, $P2$ and $P1+P2$). As we observe there are several cases where the subject pronounces an out-of-vocabulary (OOV) word and either performs a gesture or not. This indicates the difficulty of the task as these cases should be ignored. By focusing on the recognized word sequence that employs the single-audio modality we notice two insertions (words "PREDERE" and "FAME"). By employing either the $P1$ or $P2$ the above word insertions are corrected as the visual modality is integrated and helps identifying that these segments correspond to OOV words. Further, the single pass fusion components lead to errors which the proposed approach manages to deal with: $P1$ causes insertion of "OK", $P2$ of a word deletion "BM". These are in contrast to $P1 + P2$ which recognizes correctly the whole sentence.

Note that for the above audio-visual fusion on the Gesture Challenge dataset, we implicitly address inter-stream differences, since (a) our modeling deals with not perfectly aligned audio and visual information (we enforce different boundaries for each stream), and (b) with fusion we can handle cases where one stream is less informative than the others. In fact, Fig. 15.12 presents cases (third and sixth frame) where the audio modality is ambiguous (and estimates the wrong word), whereas

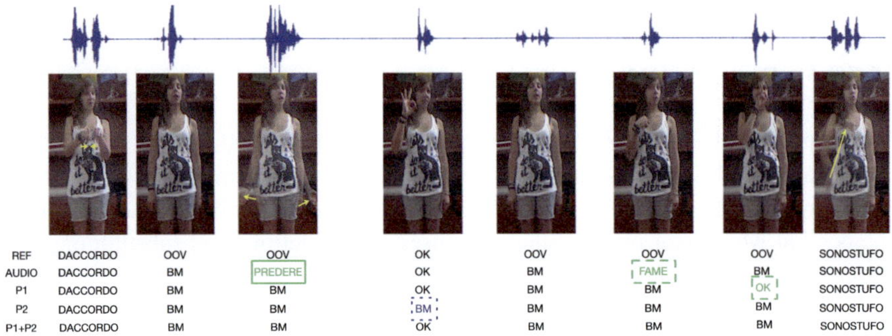

Fig. 15.12 An example of recognizing a gesture-word sequence. Audio (*top*) and visual modalities (second) via a sequence of images for a word sequence. Ground truth transcriptions ("REF"). Decoding results for the single-audio modality (AUDIO) and the three different fusion schemes (P1, P2 and P1+P2). Errors are highlighted: deletions (*blue* color) and insertions (*green* color). A background model (BM) models the out-of-vocabulary (OOV) words (Figure courtesy of Pitsikalis, Katsamanis, Theodorakis and Maragos [38])

for the visual streams we are more confident about the gesture, so with fusion of the results we get the correct gesture-word for these segments.

15.6 Conclusions

In this chapter we have proposed a broader view of shapes and their temporal sequences as communicative devices. In particular, we have emphasized the connections between shape and language and have argued for improving shape recognition by adjoining linguistic information. To illustrate this idea we have provided several paradigms including examples from sign recognition and shape-language relations in multimodal videos. Then, we have focused on the class of multimodal gesture sequences and showed the great improvement in gesture recognition achievable by fusing visual gesture shapes with spoken commands in multimodal videos. These paradigms employed some specific methodologies from pattern recognition, i.e. HMMs, motivated by the relative success they have had in speech recognition on integrating acoustic with linguistic information, but there are also alternative machine learning approaches that could be applied. However, despite the possibility of employing more efficient methodologies, the main thesis of this chapter remains the capability of improving shape recognition by adding linguistic information. This is possible and meaningful for those categories of shapes whose modeling can be considered in a linguistic context.

Acknowledgements We wish to thank Niki Efthymiou and Nancy Zlatintsi at NTUA CVSP Lab for Fig. 15.8 and discussions related to Sect. 15.3.2. This research work was supported by the project COGNIMUSE which is implemented under the ARISTEIA Action of the Operational Program Education and Lifelong Learning and is co-funded by the European Social Fund and Greek National Resources. It was also partially supported by the European Union under the project MOBOT with grant FP7-ICT-2011-9 2.1 – 600796.

References

1. Agris, U., Zieren, J., Canzler, U., Bauer, B., Kraiss, K.F.: Recent developments in visual sign language recognition. Univ. Access Inf. Soc. **6**, 323–362 (2008)
2. Antonakos, E., Pitsikalis, V., Maragos, P.: Classification of extreme facial events in sign language videos. EURASIP J. Image Video Process. **2014**, 14 (2014)
3. Arbib, M.A.: How the Brain Got Language: The Mirror System Hypothesis. Oxford University Press, New York (2012)
4. Bayer, I., Silbermann, T.: A multi modal approach to gesture recognition from audio and video data. In: Proceedings of the ACM International Conference on Multimodal Interaction, Sydney, pp. 461–466 (2013)
5. Bishop, C.M.: Pattern Recognition and Machine Learning. Springer, New York (2006)
6. Bolt, R.A.: Put-that-there: voice and gesture at the graphics interface. ACM Comput. Graph. **14**(3), 262–270 (1980)

7. Bordier, C., Puja, F., Macaluso, E.: Sensory processing during viewing of cinematographic material: computational modeling and functional neuroimaging. NeuroImage **67**, 213–226 (2013)
8. Bowden, R., Windridge, D., Kadir, T., Zisserman, A., Brady, M.: A linguistic feature vector for the visual interpretation of sign language. In: Proceedings of the European Conference on Computer Vision (ECCV), Prague (2004)
9. Buehler, P., Everingham, M., Zisserman, A.: Learning sign language by watching TV (using weakly aligned subtitles). In: Proceedings of the IEEE International Conference on Computer Vision & Pattern Recognition (CVPR), Miami, pp. 2961–2968 (2009)
10. Chow, Y.-L., Schwartz, R.: The N-best algorithm: an efficient procedure for finding top N sentence hypotheses. In: HLT'89 Proceedings of the Workshop on Speech and Natural Language, Morristown, pp. 199–202 (1989)
11. Cootes, T.F., Edwards, G.J., Taylor, C.J.: Active appearance models. IEEE Trans. Pattern Anal. Mach. Intell. **23**(6), 681–685 (2001)
12. Cour, T., Sapp, B., Nagle, A., Taskar, B.: Talking pictures: temporal grouping and dialog-supervised person recognition. In: Proceedings of the IEEE International Conference on Computer Vision & Pattern Recognition (CVPR), San Francisco (2010)
13. Dalal, N., Triggs, B.: Histograms of oriented gradients for human detection. In: Proceedings of the IEEE International Conference on Computer Vision & Pattern Recognition (CVPR), San Diego, pp. 886–893 (2005)
14. Duda, R.O., Hart, P.E., Stork, D.G.: Pattern Classification, 2nd edn. Wiley, New York (2001)
15. Emmorey, K.: Language, Cognition, and the Brain: Insights from Sign Language Research. Lawrence Erlbaum Associates, Mahwah (2002)
16. Escalera, S., Gonzàlez, J., Baró, X., Reyes, M., Guyon, I., Athitsos, V., Escalante, H., Sigal, L., Argyros, A., Sminchisescu, C., Bowden, R., Sclaroff, S.: ChaLearn multi-modal gesture recognition 2013: grand challenge and workshop summary. In: Proceedings of the ACM International Conference on Multimodal Interaction, Sydney, pp. 365–368 (2013)
17. Escalera, S., Gonzàlez, J., Baró, X., Reyes, M., Lopes, O., Guyon, I., Athistos, V., Escalante, H.J.: Multi-modal gesture recognition challenge 2013: dataset and results. In: Proceedings of the ACM International Conference on Multimodal Interaction, pp. 445–452 (2013)
18. Farhadi, A., Endres, I., Hoiem, D., Forsyth, D.A.: Describing objects by their attributes. In: Proceedings of the IEEE International Conference on Computer Vision & Pattern Recognition (CVPR), Miami (2009)
19. Fei-Fei, L., Perona, P.: A Bayesian hierarchical model for learning natural scene categories. In: Proceedings of the IEEE International Conference on Computer Vision & Pattern Recognition (CVPR), San Diego (2005)
20. Gersho, A., Gray, R.M.: Vector Quantization and Signal Compression. Springer Science & Business Media, Boston (1992)
21. Glotin, H., Vergyr, D., Neti, C., Potamianos, G., Luettin, J.: Weighting schemes for audio-visual fusion in speech recognition. In: Proceedings of the IEEE International Conference on Acoustics, Speech and Signal Processing (ICASSP), Salt Lake City, pp. 173–176 (2001)
22. Jaimes, A., Sebe, N.: Multimodal human–computer interaction: a survey. Comput. Vis. Image Underst. **108**(1), 116–134 (2007)
23. Jelinek, F.: Statistical Methods for Speech Recognition. MIT Press, Cambridge (1997)
24. Johnson, R.E., Liddell, S.K.: A segmental framework for representing signs phonetically. Sign Lang. Stud. **11**(3), 408–463 (2011)
25. Kendon, A.: Gesture: Visible Action as Utterance. Cambridge University Press, Cambridge/New York (2004)
26. Kopp, S., Bergmann, K.: Automatic and strategic alignment of co-verbal gestures in dialogue. In: Wachsmuth, I., de Ruiter, J., Kopp, S., Jaecks, P. (eds.) Alignment in Communication: Towards a New Theory of Communication, pp. 87–107. John Benjamins Publ. Co., Amsterdam (2013)

27. Laptev, I., Marszalek, M., Schmid, C., Rozenfeld, B.: Learning realistic human actions from movies. In: Proceedings of the IEEE International Conference on Computer Vision & Pattern Recognition (CVPR), Anchorage (2008)
28. Liddell, S.K.: Grammar, Gesture and Meaning in American Sign Language. Cambridge University Press, Cambridge (2003)
29. Maragos, P., Gros, P., Katsamanis, A., Papandreou, G.: Cross-modal integration for performance improving in multimedia: a review. In: Maragos, P., Potamianos, A., Gros, P. (eds.) Multimodal Processing and Interaction: Audio, Video, Text, pp. 3–48. Springer, New York (2008)
30. McNeill, D.: Gesture: a psycholinguistic approach. In: The Encyclopedia of Language and Linguistics, pp. 1–15. Elsevier, Boston (2006)
31. McNeill, D.: Gesture-speech unity: phylogenesis, ontogenesis microgenesis. Lang. Interact. Acquis. **5**(2), 137–184 (2014)
32. Ong, S., Ranganath, S.: Automatic sign language analysis: a survey and the future beyond lexical meaning. IEEE Trans. Pattern Anal. Mach. Intell. **27**, 873–891 (2005)
33. Ostendorf, M., Kannan, A., Austin, S., Kimball, O., Schwartz, R., Rohlicek, J.R.: Integration of diverse recognition methodologies through reevaluation of N-best sentence hypotheses. In: HLT'91 Proceedings of the Workshop on Speech and Natural Language, pp. 83–87 (1991)
34. Oviatt, S., Cohen, P.: Perceptual user interfaces: multimodal interfaces that process what comes naturally. Commun. ACM **43**(3), 45–53 (2000)
35. Parikh, D., Grauman, K.: Relative attributes. In: Proceedings of the International Conference on Computer Vision (ICCV), Barcelona (2011)
36. Pastra, K.: COSMOROE: a cross-media relations framework for modelling multimedia dialectics. Multimed. Syst. **14**, 299–323 (2008)
37. Pavlakos, G., Theodorakis, S., Pitsikalis, V., Katsamanis, A., Maragos, P.: Kinect-based multimodal gesture recognition using a two-pass fusion scheme. In: Proceeding of the IEEE International Conference on Image Processing (ICIP), Paris, pp. 1495–1499 (2014)
38. Pitsikalis, V., Katsamanis, A., Theodorakis, S., Maragos, P.: Multimodal gesture recognition via multiple hypotheses rescoring. J. Mach. Learn. Res. **16**, 255–284 (2015)
39. Pitsikalis, V., Theodorakis, S., Vogler, C., Maragos, P.: Advances in phonetics-based sub-unit modeling for transcription alignment and sign language recognition. In: Proceedings of the IEEE Conference on Computer Vision & Pattern Recognition Workshops, Colorado Springs (2011)
40. Rabiner, L.R., Juang, B.H.: Fundamentals of Speech Recognition. Prentice Hall, Englewood Cliffs (1993)
41. Rose, R.C., Paul, D.B.: A hidden Markov model based keyword recognition system. In: Proceedings of the IEEE International Conference on Acoustics, Speech and Signal Processing (ICASSP), Albuquerque, pp. 129–132 (1990)
42. Searle, J.R.: Mind, Language, and Society: Philosophy in the Real World. Basic Books, New York (1999)
43. Sivic, J., Russell, B.C., Efros, A.A., Zisserman, A., Freeman, W.T.: Discovering objects and their location in images. In: Proceedings of the International Conference on Computer Vision (ICCV), Beijing, (2005)
44. Starner, T., Weaver, J., Pentland, A.: Real-time American sign language recognition using desk and wearable computer based video. IEEE Trans. Pattern Anal. Mach. Intell. **20**(12), 1371–1375 (1998)
45. Theodorakis, S., Pitsikalis, V., Maragos, P.: Dynamic–static unsupervised sequentiality, statistical subunits and lexicon for sign language recognition. Image Vis. Comput. **32**, 533–549 (2014)
46. Theodoridis, S., Koutroumbas, K.: Pattern Recognition, 4th edn. Academic Press (2008)
47. Tomasello, M.: Origins of Human Communication. MIT Press, Cambridge (2008)
48. Vatakis, A., Spence, C.: Audiovisual synchrony perception for music, speech, and object actions. Brain Res. **1111**, 134–142 (2006)

49. Vogler, C., Metaxas, D.: A framework for recognizing the simultaneous aspects of American sign language. Comput. Vis. Image Underst. **81**(3), 358–384 (2001)
50. Wilpon, J., Rabiner, L.R., Lee, C.H., Goldman, E.R.: Automatic recognition of keywords in unconstrained speech using hidden Markov models. IEEE Trans. Acoust. Speech Signal Process. **38**(11), 1870–1878 (1990)
51. Wittgenstein, L.: Philosophical Investigations. (Translated by Anscombe, G.E.M., and Editors Hacker, P.M.S., Schulte, J., 4th edn.). Wiley-Blackwell Publ. (2009) (1953)
52. Wittgenstein, L.: The Big Typescript: TS 213 (Edited and translated by Luckhardt, C.G., Aue, M.E.). Blackwell Publication (2005)
53. Wu, J., Cheng, J., Zhao, C., Lu, H.: Fusing multi-modal features for gesture recognition. In: Proceedings of the ACM International Conference on Multimodal Interaction, Sydney, pp. 453–460 (2013)

Chapter 16
Tongue Mesh Extraction from 3D MRI Data of the Human Vocal Tract

Alexander Hewer, Stefanie Wuhrer, Ingmar Steiner, and Korin Richmond

Abstract In speech science, analyzing the shape of the tongue during human speech production is of great importance. In this field, magnetic resonance imaging (MRI) is currently regarded as the preferred modality for acquiring dense 3D information about the human vocal tract. However, the desired shape information is not directly available from the acquired MRI data. In this chapter, we present a minimally supervised framework for extracting the tongue shape from a 3D MRI scan. It combines an image segmentation approach with a template fitting technique and produces a polygon mesh representation of the identified tongue shape. In our evaluation, we focus on two aspects: First, we investigate whether the approach can be regarded as independent of changes in tongue shape caused by different speakers and phones. Moreover, we check whether an average user who is not necessarily an anatomical expert may obtain acceptable results. In both cases, our framework shows promising results.

A. Hewer (✉)
Saarbrücken Graduate School of Computer Science, Saarbrücken, Germany

DFKI Language Technology Lab, Saarbrücken, Germany

Cluster of Excellence Multimodal Computing and Interaction, Saarland University, Saarbrücken, Germany
e-mail: hewer@coli.uni-saarland.de

S. Wuhrer
INRIA Grenoble Rhône-Alpes, Saint Ismier, France
e-mail: stefanie.wuhrer@inria.fr

I. Steiner
DFKI Language Technology Lab, Saarbrücken, Germany Cluster of Excellence Multimodal Computing and Interaction, Saarland University, Saarbrücken, Germany
e-mail: steiner@coli.uni-saarland.de

K. Richmond
Centre for Speech Technology Research, University of Edinburgh, Edinburgh, UK
e-mail: korin@cstr.ed.ac.uk

M. Breuß et al. (eds.), *Perspectives in Shape Analysis*, Mathematics and Visualization, DOI 10.1007/978-3-319-24726-7_16

16.1 Introduction

Shape analysis is of great importance in speech science. In this research area, analyzing and understanding the shape and the motions of the human tongue during the production of speech is of great interest. For example, a tongue model may be integrated into virtual avatars for multimodal spoken interaction or computer-aided pronunciation training. In the latter case, the user can be shown how to move the tongue to produce a specific sound [9]. Furthermore, such a tongue model could be used in articulatory speech synthesis to approximate the vocal tract area function.

Observing and imaging the tongue during speech is a challenging task, since it is inside the mouth and therefore almost completely hidden from view. Thus, traditional imaging modalities based on light, such as photography, are of limited use for acquiring information about the tongue. Currently, magnetic resonance imaging (MRI) can be regarded as the state-of-the-art technique for imaging the human vocal tract. This method is capable of providing 3D information about the inside of the mouth of a speaker without being hazardous or invasive.

The data acquired by MRI has to be further processed to extract the desired shape information, and manually extracting shape information from MRI scans can be a tedious and time-consuming task. This motivates an extended version of our framework [13] that combines image segmentation and template fitting to extract the tongue surface from a 3D MRI scan in a minimally supervised fashion. The only user input required by our method is a sparse set of annotated landmarks. Optionally, the user may additionally crop the MRI scan to the region containing the tongue for improved performance. We demonstrate experimentally that our method is stable with respect to inaccurate landmarks, which implies that a user who is not necessarily an anatomical expert is able to get acceptable results with only minimal input.

It is desirable to represent the extracted tongue surface using a high level representation. In this work, we choose as representation a polygon mesh. This representation has the advantage that it can be directly used in various fields of application, as meshes can be used to produce piecewise linear approximations of scenes of arbitrarily complex geometry and topology. The meshes can be textured and subsequently rendered in real-time to produce photo-realistic images. This even holds for large models, as polygon meshes can be easily represented in a hierarchy of resolutions using subdivision [5, Chapter 1]. Furthermore, polygon meshes are often employed in computer graphics to generate animations of complex objects [5, Chapter 9], and in computer vision to conduct a statistical analysis of a class of shapes, as for example faces [7]. By using polygon models, such deformations and statistical summaries can easily be computed for the extracted tongue surfaces. In speech processing, polygon models of tongues have been used to generate acoustical simulations [4], and using polygon models for our meshes allows us to use the extracted surfaces in existing simulation tools.

Our method uses a single generic template represented by a polygon mesh that was constructed based on an MRI scan by a non-expert. Experiments indicate that our approach has a success rate of 75 % for the dataset of Adam Baker [3] and the Ultrax project [1]. Furthermore, we show that our method is independent of shape changes caused by different speakers and phones.

This chapter is organized as follows. Section 16.2 gives an overview of related work and Sect. 16.3 describes our framework and elaborates on the motivation behind the design. Section 16.4 provides background information on the datasets used as the source of the 3D MRI scans in our experiments. It is worth noting that compared to our previous work, we had data from more speakers available. In Sect. 16.5, we focus on investigating whether our approach is speaker- and phone-independent, and whether a non-expert user can achieve acceptable results. Finally, Sect. 16.6 gives conclusions and discusses open problems.

16.2 Related Work

As it is tedious to manually extract information from MRI scans, a number of methods have been proposed to facilitate this process. Here, we provide a brief overview of recent methods.

The method of Peng et al. [22] aims at identifying the tongue's contour in a 2D mid-sagittal scan. It is based on an active contours approach [17] where a previously trained shape model is used to control the evolution of the contour. This technique was later extended by Eryildirim et al. [10] to align the contour's end points to the corresponding extremities of the tongue. More recently, Raessy et al. [23] showed that it is possible to train oriented active shape models [20] in such a way that they can be used to reliably identify the boundary of the tongue in 2D scans. These methods depend on manually preparing a training set and are restricted to the 2D case.

Lee et al. [16] proposed a framework for extracting the tongue from 3D dynamic MRI in a minimally supervised fashion. The random walker technique [11], which requires a user to manually place some seeds, was used as the base segmentation method. This framework produces a low-level volume segmentation.

Harandi et al. [12] used a template-matching technique to extract a mesh representation of the tongue from 3D MRI scans. A template is extracted from a source scan by an anatomical expert. This template is then fitted to a target scan using color information. Specifically, the mesh points are moved in such a way that the color at the original point in the source scan is similar to the deformed point in the target scan. This approach is limited by requiring an expert to provide the templates.

16.3 Framework

Our framework consists of three main steps. First, we apply an image segmentation technique to the MRI data to identify the spatial support of the tongue and related tissue.

Second, we extract the surface points of the tissue, thereby reducing the data to a purely geometric representation. This is motivated by the fact that it is relatively easy to combine geometric information from different sources. For example, the surface point cloud obtained from one scan might be incomplete. In this case, the information obtained from a second scan of the same speaker could be used to reconstruct certain missing data by simply adding the corresponding points to the point cloud of the first scan.

Third, we apply a template fitting technique to obtain a polygon mesh representation of the tongue surface from the point cloud. Using such a method has the advantage that we can exploit prior knowledge about the shape of the tongue in the form of a provided template. This is especially useful in situations where the point cloud is noisy, incomplete, or contains additional information other than the tongue.

16.3.1 Interpretation of a Scan as a 3D Image

Before discussing our proposed method, we describe how an MRI scan can be turned into a 3D image.

Formally, a scan is given by $g : S \to \mathbb{R}$ where $S \subset \mathbb{R}^3$ is a discrete domain in the form of a rectangular box. The scan domain S contains the positions $\mathbf{x}$ at which the scanner took the measurements. Thus, $g(\mathbf{x})$ represents the density of hydrogen molecules measured by the scanner at coordinate $\mathbf{x}$. Each sample position represents a point on a regular grid with grid spacings h_x, h_y, and h_z.

A 3D image, on the other hand, is given by $f : \Omega \to [0, 255]$ where $\Omega \subset \mathbb{R}^3$ is again a discrete domain in the form of a rectangular box. Here, $f(\mathbf{y})$ is the gray-value at voxel coordinate $\mathbf{y}$. In contrast to the sample positions, however, these voxel coordinates are arranged on a Cartesian grid with $h_x = h_y = h_z = 1$.

This means we first have to find a mapping $s : \Omega \to S$ from the voxel coordinates in our image representation to the sample positions of the scan. Here, we can use $\mathbf{y} = (x, y, z)^\top \in \Omega$ as an index to access the vertices of the regular grid in S, as

$$s(\mathbf{y}) := \left(xh_x, \, yh_y, \, zh_z\right)^\top . \tag{16.1}$$

To visualize the measured hydrogen density, we define a quantization operator $q : \mathbb{R} \to [0, 255]$ that maps the observed densities to 256 values. This allows us to

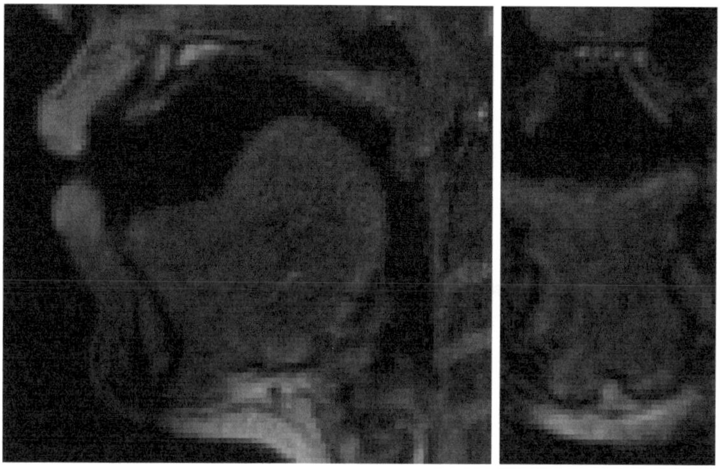

Fig. 16.1 Two different slice types of a 3D image showing the human vocal tract. *Left*: Sagittal slice. *Right*: Coronal slice

interpret the scan as 3D image $f : \Omega \to [0, 255]$ where

$$f(\mathbf{y}) = q\big(g(s(\mathbf{y}))\big) \tag{16.2}$$

can be seen as the quantized gray-value representation of the hydrogen density at sample position $s(\mathbf{y})$.

In the following, we assume that the data was recorded in a standard sagittal manner, and refer to an (x, y)-plane of an MRI scan as a sagittal slice and to a (y, z)-plane of an MRI scan as a coronal slice. Both types of slices are shown in Fig. 16.1.

16.3.2 Image Segmentation

The first step of our method aims to identify the spatial support of the tongue. That is, we wish to divide Ω into an object region Ω_O and a background region Ω_B. The object region Ω_O should contain points that are related to the tongue. However, it is also allowed to contain regions that belong to other organic tissue. This relaxation is necessary as in some images no boundary may be detectable between the tongue and other tissues with which it is in contact, such as the palate. The background region Ω_B consists of parts of the scan we have no interest in. These are, for example, bones, air, or other tissue not related to the tongue.

Figure 16.1 demonstrates that an object can be distinguished from the background by using color information. This motivates the use of image segmentation techniques that make use of color information to extract Ω_O.

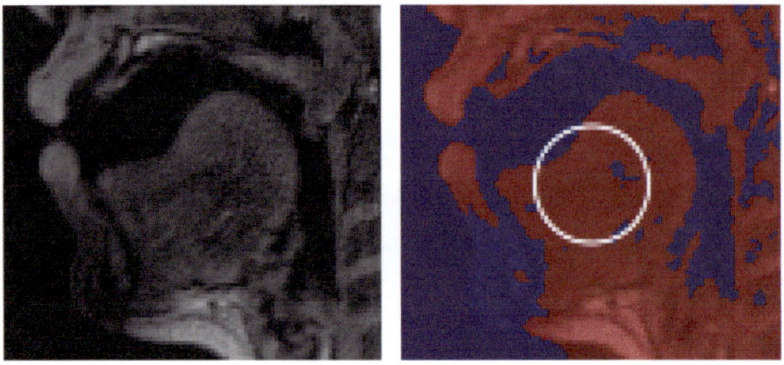

Fig. 16.2 Example result of Chan-Vese in 2D. *Left*: Sagittal slice. *Right*: Resulting segmentation. Ω_O is colored in *red*, Ω_B in *blue*. The initial contour used is shown as a *white circle*

As we aim to apply our method to large datasets, the segmentation method must satisfy two requirements. First, the required manual input from the user should be minimal. Second, the segmentation method should be robust. To satisfy both requirements, we compute segmentations using the method by Chan and Vese [8]. This method is robust and generates smooth boundaries between Ω_O and Ω_B, which can later be used to derive clean surface normals.

The method by Chan and Vese requires as initialization a closed contour C that separates Ω into Ω_O and Ω_B. In our approach, this initial contour can be computed automatically: Given a sparse set of manually annotated landmarks L as described in Sect. 16.5.1, a sphere can be placed at the centroid of these landmarks in Ω. Alternatively, it can be positioned at the center of Ω if the image mainly shows the tongue, as for example in Fig. 16.2.

The approach evolves the initial contour C such that the gray-value variance inside the regions is minimized, i.e.

$$E_{\mathrm{CV}}(C) = \sum_{\mathbf{x}\in\Omega_O} \left(f(\mathbf{x}) - \mu_{\Omega_O}\right)^2 + \sum_{\mathbf{x}\in\Omega_B} \left(f(\mathbf{x}) - \mu_{\Omega_B}\right)^2 + \lambda\,\mathrm{length}(C), \qquad (16.3)$$

where Ω_O and Ω_B are the regions induced by C and μ_X represents the average gray-value in region X. The method has a regularizer weighted by $\lambda > 0$ that tries to minimize the length of the contour. To minimize the energy, we apply the standard scheme of Chan and Vese. That is, we start with a continuous version of the energy that uses a level set representation [21] of the contour, and subsequently derive the Euler-Lagrange equation of this energy to set up a gradient descent approach that is discretized using a finite differences implicit scheme. Figure 16.2 shows an example result in 2D that used a circle as the initial contour.

Note that the remainder of our method is independent of the selected segmentation method, and any segmentation method can be freely selected if this is advantageous for a specific dataset. In our preliminary experiments [13], we also explored a graph cut method [6] for segmentation. However, we did not explore this option further as approaches of the graph cut family require a significant amount of manual input, rendering them impractical when processing large datasets.

16.3.3 Surface Point Extraction

Given a partition $\Omega = \Omega_O \cup \Omega_B$, we compute the surface information by extracting surface points $P^* := \{\mathbf{p}_i\}$ of Ω_O and normals $N := \{\mathbf{n}_i\}$ for P^*, such that $\mathbf{n}_i$ is the normal at $\mathbf{p}_i$. Surface points $\mathbf{p}_i$ are points of Ω_O that have at least one neighboring point $\mathbf{q}$ in Ω_B. Surface normals are chosen to point towards the outside of Ω_O. Note that due to the relaxation we formulated earlier for Ω_O, P^* may contain surface points belonging to other articulators than the tongue. Furthermore, P^* is a subset of Ω, i.e., the surface information was computed in the image domain. The template fitting, however, should operate on the domain of the observed vocal tract to be anatomically correct. Thus, we apply the mapping from Equation (16.1) to obtain the correct surface information P as

$$P := \{s(\mathbf{p}) \mid \mathbf{p} \in P^*\}. \tag{16.4}$$

The surface P consists of a loose collection of points, as shown in Fig. 16.3. Furthermore, the point cloud may be missing information and may contain data other than the tongue. Therefore, surface reconstruction approaches like the Poisson reconstruction [14] may produce undesirable results. To avoid this problem, in the following, we utilize the information that a subset of P forms part of the surface of a tongue.

16.3.4 Template Fitting

We use a template fitting technique [25] to jointly find the subset of P representing the tongue and a polygon mesh representation of the tongue surface. That is, we deform a template mesh $M := (V, F)$ to match the point cloud data P. We use a vertex-face representation of meshes, i.e., $V := \{\mathbf{v}_i\}$ denotes the vertex set of the mesh with $\mathbf{v}_i \in \mathbb{R}^3$ and F its face set. To obtain a deformation, the approach computes a set $A := \{A_i\}$ where $A_i : \mathbb{R}^3 \to \mathbb{R}^3$ is a rigid body motion for the vertex

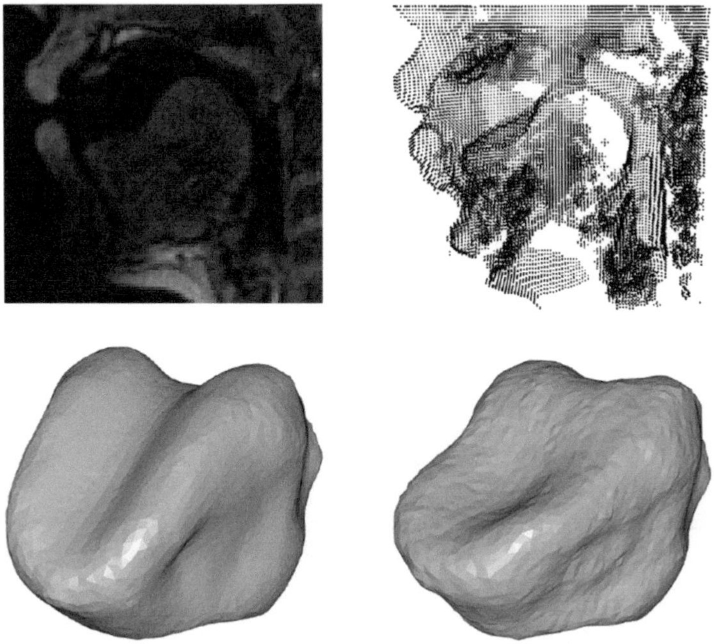

Fig. 16.3 Example result of the template fitting method. *Top row*: Sagittal slice of the used MRI scan (*left*) and obtained point cloud (*right*). *Bottom row:* Template used in our approach (*left*) and result of the template fitting (*right*)

$\mathbf{v}_i$ by minimizing the energy

$$E_{\mathrm{Def}}(A) = \alpha \frac{1}{|V^*|} \sum_{v_i \in V^*} \left(\mathrm{dist}_D \left(A_i(\mathbf{v}_i), \arg\min_{\mathbf{p}_j \in P} \|A_i(\mathbf{v}_i) - \mathbf{p}_j\| \right) \right)$$
$$+ \beta \frac{1}{|V|} \sum_{v_i \in V} \left(\sum_{v_j \in \mathcal{N}(v_i)} \mathrm{dist}_S \left(A_i, A_j \right) \right)$$
$$+ \gamma \frac{1}{|L|} \sum_{(\mathbf{v}_i, \mathbf{q}_i) \in L} \left(\mathrm{dist}_L \left(A_i(\mathbf{v}_i), \mathbf{q}_i \right) \right). \tag{16.5}$$

This energy consists of three terms. Each term is weighted by a non-negative value, α, β, or γ, that is normalized according to the number of participating vertices in the respective term. This normalization makes it easier to compare the influences of the different terms.

The data term $\mathrm{dist}_D(\cdot)$ measures the distance between the transformed vertex $A_i(\mathbf{v}_i)$ and the normal plane at its nearest neighbor. This term is minimized when the template is close to the point cloud P. In our implementation, this term is only evaluated at $V^* \subseteq V$ to increase robustness to noise. In particular, a vertex $\mathbf{v}_i$ is ignored if the Euclidean distance between $\mathbf{v}_i$ and its nearest neighbor is too large or if the angle between the outer normals of $\mathbf{v}_i$ and its nearest neighbor is too large. This commonly used heuristic [2, 18] is meant to distinguish valid data observations from invalid ones. Additionally, we do not consider vertices that are part of the landmark set L to avoid distorting any manually provided correspondences.

The deformation smoothness term $\text{dist}_S(\cdot)$ measures the difference in rigid body motion A_i between $\mathbf{v}_i$ and the vertices of the neighborhood $\mathcal{N}(\mathbf{v}_i)$ that consists of the one-ring neighbors of $\mathbf{v}_i$ and vertices of the mesh within distance of $2{\cdot}\text{res}(M)$ from $\mathbf{v}_i$ where $\text{res}(M)$ is the average edge length of the template mesh M. The minimization of $\text{dist}_S(\cdot)$ encourages the template to preserve its overall shape during deformation, which helps to keep the mesh away from data points that do not belong to the surface of the tongue and allows missing parts to be filled in smoothly. This term is active at all vertices.

Finally, the landmark term $\text{dist}_L(\cdot)$ is optional. This term computes the squared Euclidean distance between pairs of manually annotated vertices $\mathbf{v}_i \in V$ and corresponding coordinates $\mathbf{q}_i \in \mathbb{R}^3$ that are contained in a set of landmarks $L := \{(\mathbf{v}_i, \mathbf{q}_i)\}$. Note that the coordinates $\mathbf{q}_i$ do not have to be contained in P. By minimizing this term, the approach will move the selected vertices to the user-provided coordinates.

We discover that minimizing both the data and the smoothness terms will move the mesh to a subset of P that resembles a tongue-like surface.

We follow a similar strategy as Wuhrer et al. [25] to obtain a minimizer A of the energy. Before performing the optimization, we perform a rigid alignment of the template. This step uses the user-provided landmarks and the point cloud to find a good scale and position for the template.

The energy given in Equation (16.5) is not differentiable with respect to A, which prevents us from minimizing it directly. Therefore, we perform the optimization by minimizing a series of differentiable energies $E^t_{\text{Def}}(A^t)$ where $t \in [1, t_{\max}]$. The energy E^t_{Def} differs from the original energy E_{Def} in the following way: In E^t_{Def}, we use the minimizer of the previous energy in the series to transform the vertex in $\text{dist}_D(\cdot)$: $A_i^{t-1}(\mathbf{v}_i)$. This means that $\arg\min_{\mathbf{p}_j \in P}(\cdot)$ no longer depends on A^t. Thus, the energy becomes differentiable and we can use a quasi-Newton technique [19] to compute the minimizer. Moreover, for $t_{\max} > 1$, the weight β of the smoothness term changes in each iteration. Given a base value β, the weights β^t used in iteration t are computed as

$$\beta^t = 2\beta - (t-1)\frac{\beta}{t_{\max} - 1}. \tag{16.6}$$

This means that we start the optimization by promoting smooth transformations. The weight is then gradually reduced until we arrive at the base weight β in the last iteration.

After the minimization of the last energy, we obtain the sought transformations A as $A^{t_{\max}}$. Note that we use the identity $A_i^0(\mathbf{v}_i) = \mathbf{v}_i$ as A^0 that is needed in the first energy E^1_{Def}. Furthermore, we apply a coarse-to-fine strategy to cope with large deformations.

Figure 16.3 illustrates an example of the template fitting.

16.4 Datasets

This study is evaluated on a large dataset of 12 speakers, and extends our previous work [13], which only considered data from a single speaker. We use two MRI datasets to validate our method, that of Adam Baker [3], and the full dataset from the Ultrax project [1].

The Baker dataset contains static 3D MRI scans of a male speaker. Twenty-five of these scans are speech related and show vocal tract configurations for different phones. This data was acquired as part of the Ultrax project, but released separately.

The remainder of the Ultrax dataset consists of static 3D MRI scans of 11 adult speakers. Seven of these speakers are female and four are male. While scanning, the subjects, who were all phonetically trained, were asked to sustain the articulatory configurations for a given phone for around 20 s. Prompts were displayed to the subject using a laptop connected to video-goggles. Each subject recorded scans for the following phone set [i, e, ɛ, a, ɑ, ʌ, ɔ, o, u, ʉ, ə, s, ʃ], with an additional scan for the pose at rest. Simultaneous audio recordings were made using a FOMRI-III fiber optic microphone. This microphone is specially designed for use in MRI scanners, using both a pair of microphones and adaptive noise cancellation algorithms to reduce the level of MRI scanner noise. Though it is not possible to remove the scanner noise entirely, the use of this microphone does make it possible to monitor and verify the subject's phone production acoustically. The Ultrax dataset also contains other types of MRI scans for all subjects, but those were not used in this work.

The scans were acquired using a Siemens Verio 3T scanner at the Clinical Research Imaging Centre in Edinburgh. Each scan comprises 44 sagittal slices with a thickness of 1.2 mm and an image size (whole head) of 320 × 320 pixels in the image domain. In the scan domain, we have distances of $h_x = h_y = 1.1875$ mm and $h_z = 1.2$ mm, corresponding to a voxel size of 1.1875 mm × 1.1875 mm × 1.2 mm. The scans were acquired with an echo time of 0.93 ms and a repetition time of 2.36 ms.

16.5 Evaluation

The focus of this section is on investigating whether our approach can be regarded as independent of shape changes caused by different speakers and phones. To show this independence, we demonstrate that is possible to obtain satisfying results across different speakers and phones by always applying the same procedure. To this end, all parameters except for the landmarks are fixed for all scans.

In the following, we first outline how the template is created and how the scans are prepared. We then describe experiments to evaluate the stability of the weights in the template fitting, investigate whether our approach is applicable to different speakers and phones, and analyze the robustness of our approach to erroneously placed landmarks.

16.5.1 Template Creation

The template is manually extracted from a scan of the Baker dataset. After the extraction, we adjust the mesh to be symmetric to remove this particular bias towards the original speaker. Note that the template only models the upper part of the tongue surface and does not include its sublingual part. The template consists of 5864 vertices and 11,724 faces, and is shown in Fig. 16.4.

We select seven vertices as landmarks. These vertices and an example of the corresponding user-provided coordinates on an MRI scan are shown in Fig. 16.4. Five landmarks are distributed on a sagittal slice that is located roughly at the center of the tongue. Three of these landmarks are located at feature points that are relatively easy to locate for an average user, namely the tongue root near the epiglottis and the pharynx (green landmark), the tongue tip (red landmark), and the position where the tongue surface connects to its sublingual part (pink landmark). The remaining two landmarks in the mid-sagittal slice are placed at approximately $\frac{1}{3}$ and $\frac{2}{3}$ of the distance from the tongue tip to the root, corresponding to the tongue blade (yellow landmark) and back (orange landmark), respectively. We believe that using this feature-free approach to select the tongue blade and back facilitates the landmark placement. The tongue blade landmark serves as anchor for two additional lateral landmarks that may be positioned using a coronal slice. These are located

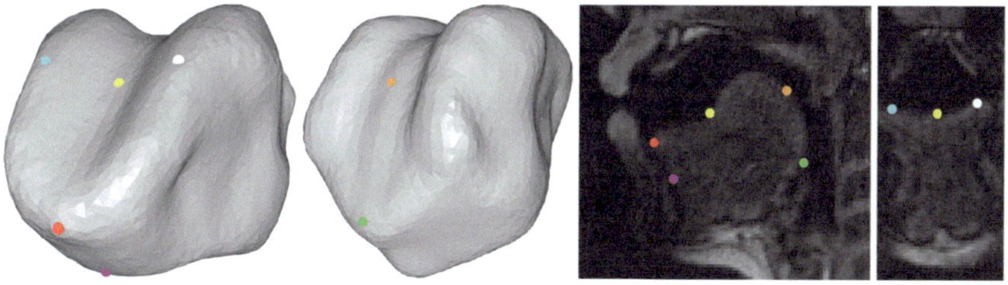

Fig. 16.4 Placement of the landmarks. *Left:* Selected vertices for the landmarks on the template. The left image shows a front view of the template, the right one a view from the back. *Right:* Sagittal and coronal slice showing an example of the corresponding user-provided landmarks on an MRI scan

near the left (blue landmark) and right (white landmark) boundaries of the tongue's upper surface and serve to add lateral information to the landmark set.

Note that not all landmarks are required for our approach. If the user does not provide coordinates for a subset of the landmarks, these landmarks will simply be ignored in the optimization process.

16.5.2 Scan Selection and Preparation

We consider the data of all available speakers to ensure high variance with respect to speaker-specific anatomy. To obtain a high variance of intra-speaker tongue shape, scans corresponding to the three corner vowels [ɑ, i, u] are considered for each speaker. These vowels show the tongue in different extreme positions, e.g. as far back and low in the mouth as possible for [ɑ] [15]. We discovered that one speaker showed a high activity of the soft palate leading to contacts with the tongue. Therefore, we removed scans of this specific speaker from further processing. Furthermore, we removed one scan from another speaker because a part of the tongue was not visible.

After this selection process, the data is pre-processed using three steps. First, each scan is cropped to a region of interest containing the vocal tract.

Second, each scan is segmented automatically using the Chan-Vese method. Here, we use $\lambda = 140$ and initialize C to a sphere of radius 15 located at the center of the image representation of the cropped scan. We found that this approach failed to properly segment the scans of one speaker, and all scans of this speaker were removed from further processing. After these steps, 29 point clouds derived from the scans were available for further experiments.

Third, we manually select the landmark coordinates in each scan. To facilitate this task, we developed a graphical user interface that allows landmarks to be placed on the image representation of the scan. Subsequently, the landmark positions are mapped to the scan domain. In our experiments, we encountered scans where the placement of the two lateral landmarks posed a problem. Due to contact with other tissue, the left and right boundaries of the tongue's upper surface were difficult to identify. We found that these landmarks are not always needed to obtain acceptable results. For our experiments, we use the 2 lateral landmarks for only 13 of the 29 scans.

Note that this workflow may be modified. In particular, it is possible to omit the cropping step, thereby reducing the amount of manual pre-processing required of the user. Working with the full scans produces the same results as working with the cropped scans if the Chan-Vese method is initialized after pre-aligning the scans based on the provided landmarks. However, working with the cropped scans decreases the processing time of the segmentation method and the memory requirements for computing the point cloud.

16.5.3 Experiments

As no ground truth is available, we evaluate the results by computing the Euclidean distances between vertices on the deformed template and their nearest neighbors in the point cloud. Since our template is incomplete, we ignore vertices at the bottom of the mesh, as they are not part of the tongue's boundary. To quantitatively summarize the results, we compute cumulative error functions. For a qualitative evaluation, we show the visual quality of some of the results.

In all following experiments, the parameters $\alpha = 1$ and $t_{max} = 20$ are fixed. In the data term, we use the same heuristic as [25] to identify valid data observations: We consider only vertices of the template mesh M whose nearest neighbor in the point cloud is at distance at most $5 \cdot \text{res}(M)$ and whose normal deviates at most 60°.

16.5.3.1 Influence of Parameters

We first evaluate the stability of the weights β and γ used in the optimization. This evaluation consists of two parts. First, we check if there is a weight β that produces acceptable results for all scans by setting $\gamma = 0$ and testing the ten weights $\beta = 1, 2, \ldots, 10$. In this experiment, the landmarks are used only for rigid alignment.

The parameter value $\beta_{\text{optimal}} = 4$ represents a good compromise between closeness to the data and smoothness of the resulting mesh. This can be seen in Fig. 16.5, which shows the results for an example scan from the Baker dataset. On the one hand, low weights for β lead to overfitting, which produces a very noisy mesh. On the other hand, high weights for β reduce the amount of alignment because the smoothness term has too much influence. Note that the very large distances visible in the cumulative error function are due to holes in the corresponding point cloud and can therefore be disregarded. We encountered 13 scans where this choice for β produced suboptimal results. The poor performance in 4 of those scans was related to palate contacts of the tongue or segmentation issues. In the remaining 9 scans, the poor performance stems from template fitting related problems. For example, the tongue tip of the template was aligned to the front palate region in some results. Additionally, in some scans, parts of the template were not aligned to the data, as shown in Fig. 16.6c.

Second, we analyze whether activating the landmark energy in equation (16.5) can improve the results for fixed β_{optimal}. Specifically, we consider weights $\gamma = 0.1, 0.2, \ldots, 1$. Hence, for this experiment, the landmarks were used in the template fitting.

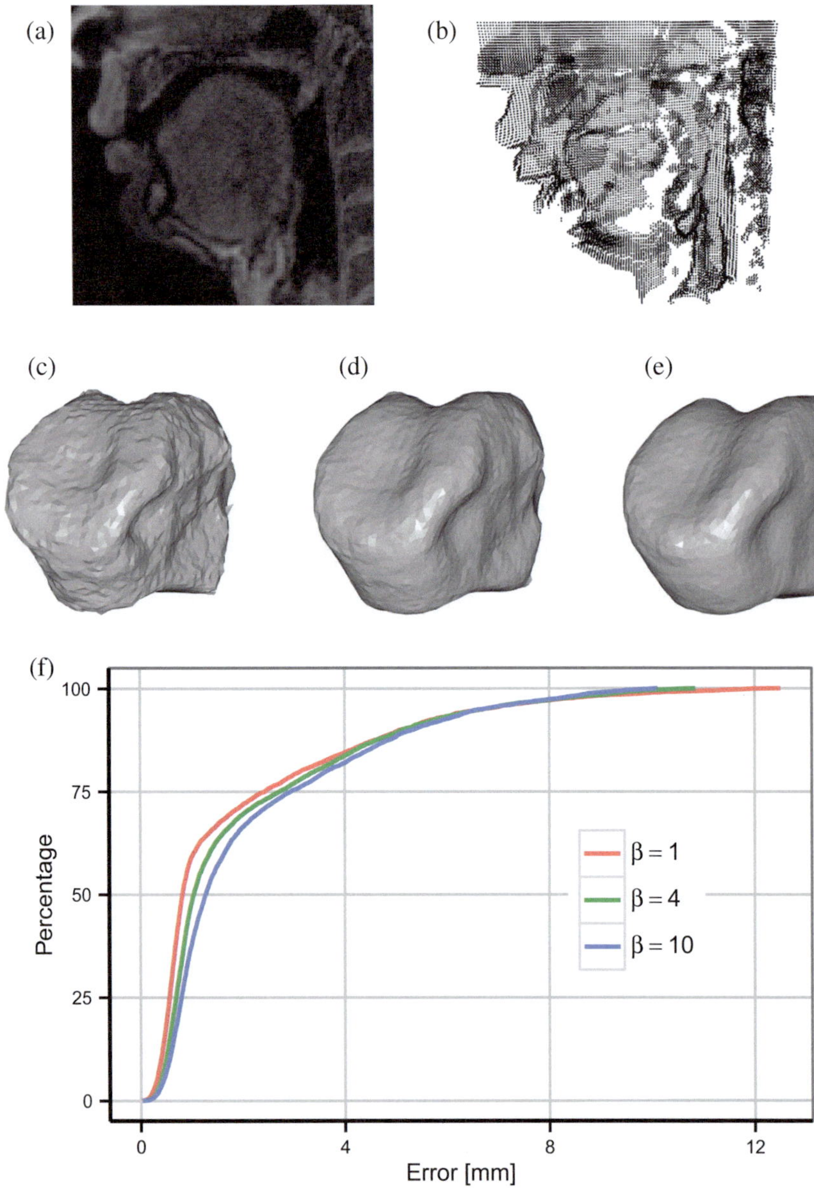

Fig. 16.5 Example showing how the weight β of the smoothness term affects the result. (**a**) Sagittal slice of the used MRI scan. (**b**) Generated point cloud. (**c**) Result for $\beta = 1$. (**d**) Result for $\beta = 4$. (**e**) Result for $\beta = 10$. (**f**) Cumulative error functions for the different results

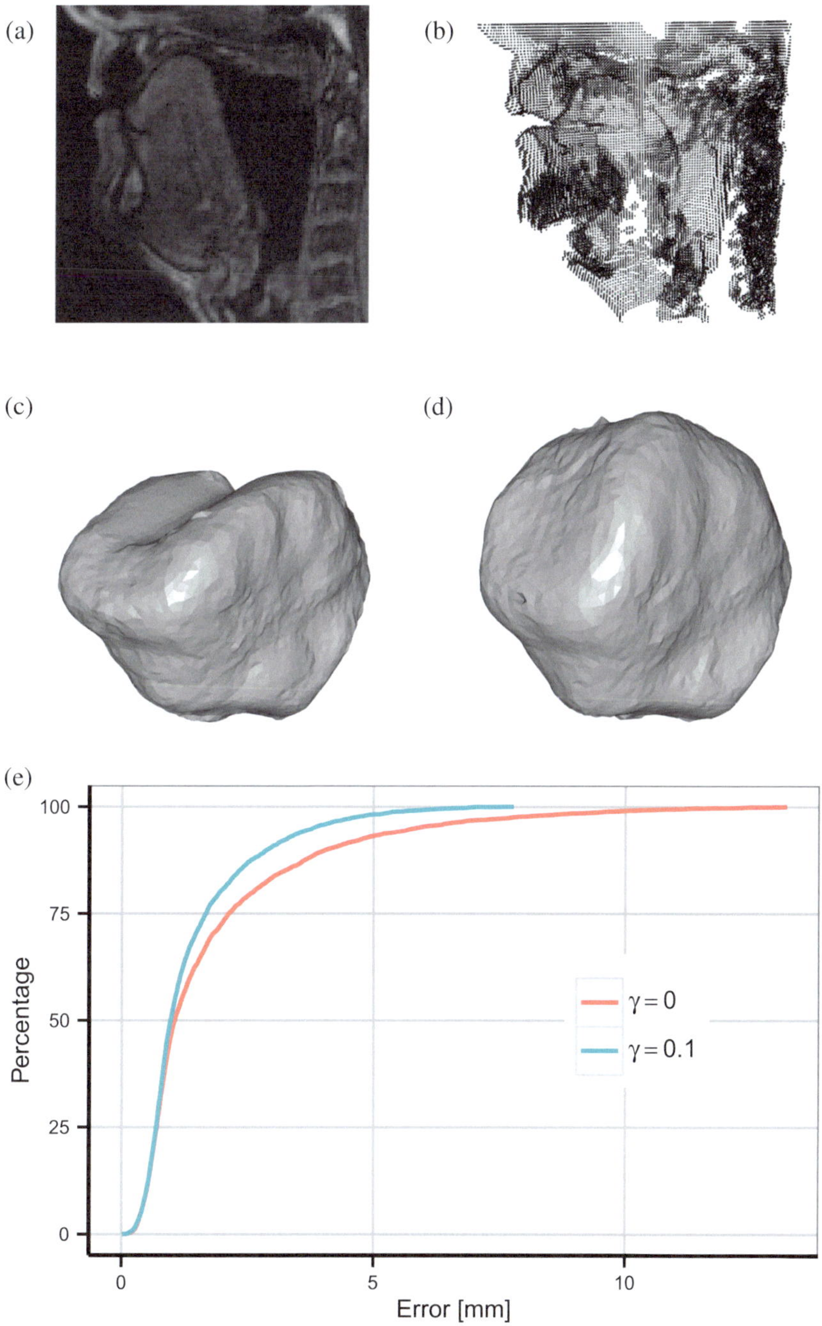

Fig. 16.6 Example showing how the landmark energy can help to improve the result. (**a**) Sagittal slice of the used MRI scan. (**b**) Generated point cloud. (**c**) Result for deactivated landmark energy ($\gamma = 0$). (**d**) Result for active landmark energy ($\gamma = 0.1$). (**e**) Cumulative error functions for the two results

Figure 16.6 shows that even using small values of γ can improve the results significantly. The figure shows a particular scan from the Ultrax dataset where activating the landmark energy drastically improves the mesh alignment. On our dataset, the value $\gamma_{\text{optimal}} = 0.1$ led to the best results. For this parameter setting, 6 of the 9 scans that had template fitting problems for $\gamma = 0$ are aligned correctly.

16.5.3.2 Evaluation of Independence of Speakers and Phones

We now evaluate the template fitting results obtained for parameters $\beta_{\text{optimal}} = 4$ and $\gamma_{\text{optimal}} = 0.1$ across different speakers and phones. For these parameter settings, our approach was successful for 22 of the 29 considered scans. These 22 scans include scans from all 10 speakers for which scan preparation was successful and scans from all three considered phones. To evaluate whether the method is biased towards specific speakers or phones, we consider the set of cumulative error plots across different phones and speakers. To avoid large distances originating from potential holes in the point cloud, we only consider distances below 5 mm in the error computation. Figure 16.7a shows the distribution of cumulative error plots for different phones, and Fig. 16.7b shows the distribution of cumulative error plots for different speakers. Note that all cumulative error plots are similar, and hence the variance between the plots is low. This shows that for our dataset, there is no significant bias towards any specific speaker or phone, and leads us to conclude that our approach is speaker- and phone-independent.

16.5.3.3 Evaluation of Noisy Landmark Placement

In the final experiment, we analyze the robustness of our approach against errors in the coordinates of the landmarks provided by the user. To this end, we add Gaussian noise with mean 0 mm and standard deviation 5 mm to each component of the original coordinates to simulate the input of an inexperienced user. We only consider the scans where our framework succeeded and used the optimal weights $\beta = \beta_{\text{optimal}}$ and $\gamma = \gamma_{\text{optimal}}$.

Errors in the landmarks do not have a significant effect on the results. In all but one of the tested scans, our approach obtains acceptable results even when noisy landmarks are used. Figure 16.8 shows a representative example of a deformed template computed using noisy landmarks. Note that the shape of the deformed templates obtained with clean and noisy landmarks is globally quite similar and only leads to localized differences. However, we encountered one scan where the noisy landmarks lead to a suboptimal result.

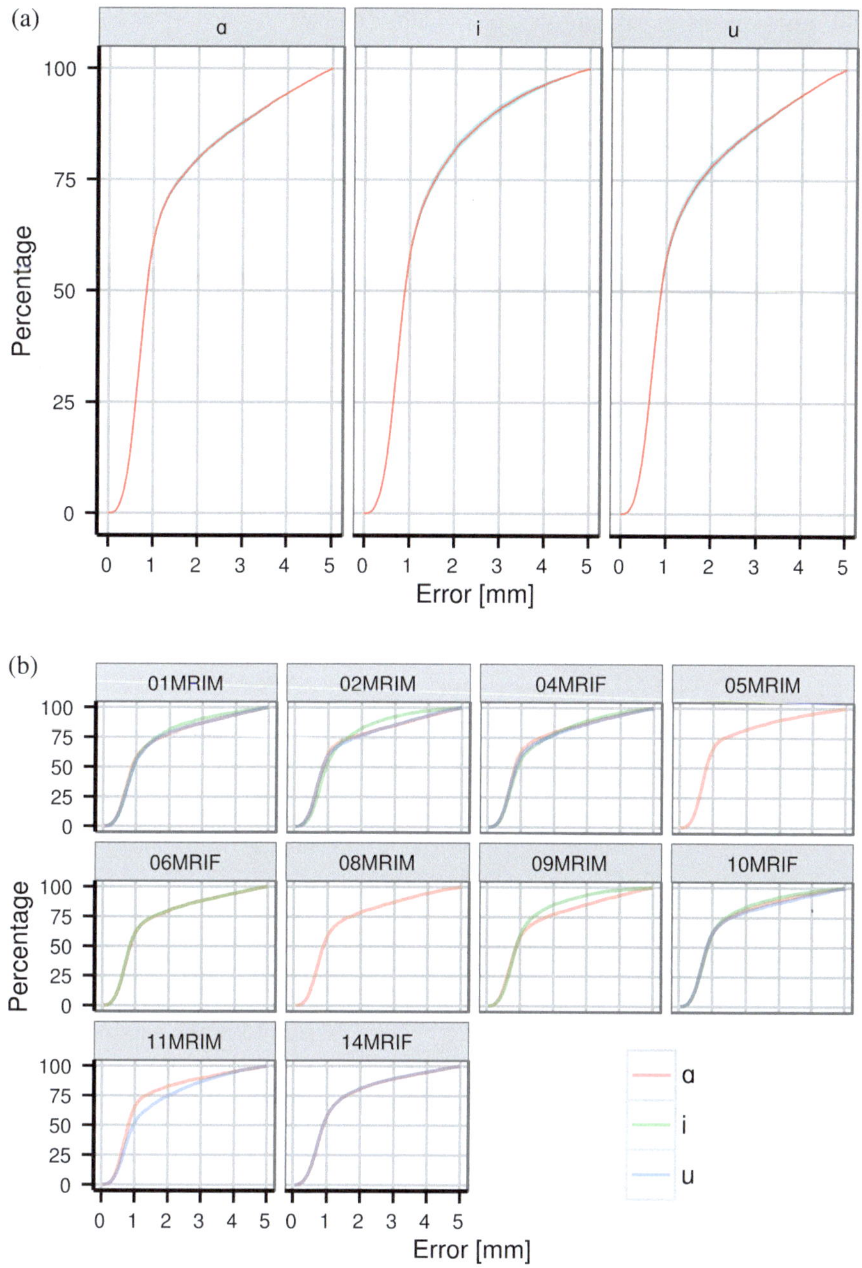

Fig. 16.7 Visualizations of the cumulative error for the 22 scans where our approach succeeded. (**a**) Cumulative error of the results grouped by phone. The plot shows the mean error (*line*) and the standard deviation (*ribbon*) of all results belonging to the corresponding phone. (**b**) Cumulative error grouped by speaker. Missing lines indicate that no result was obtained for the specific phone

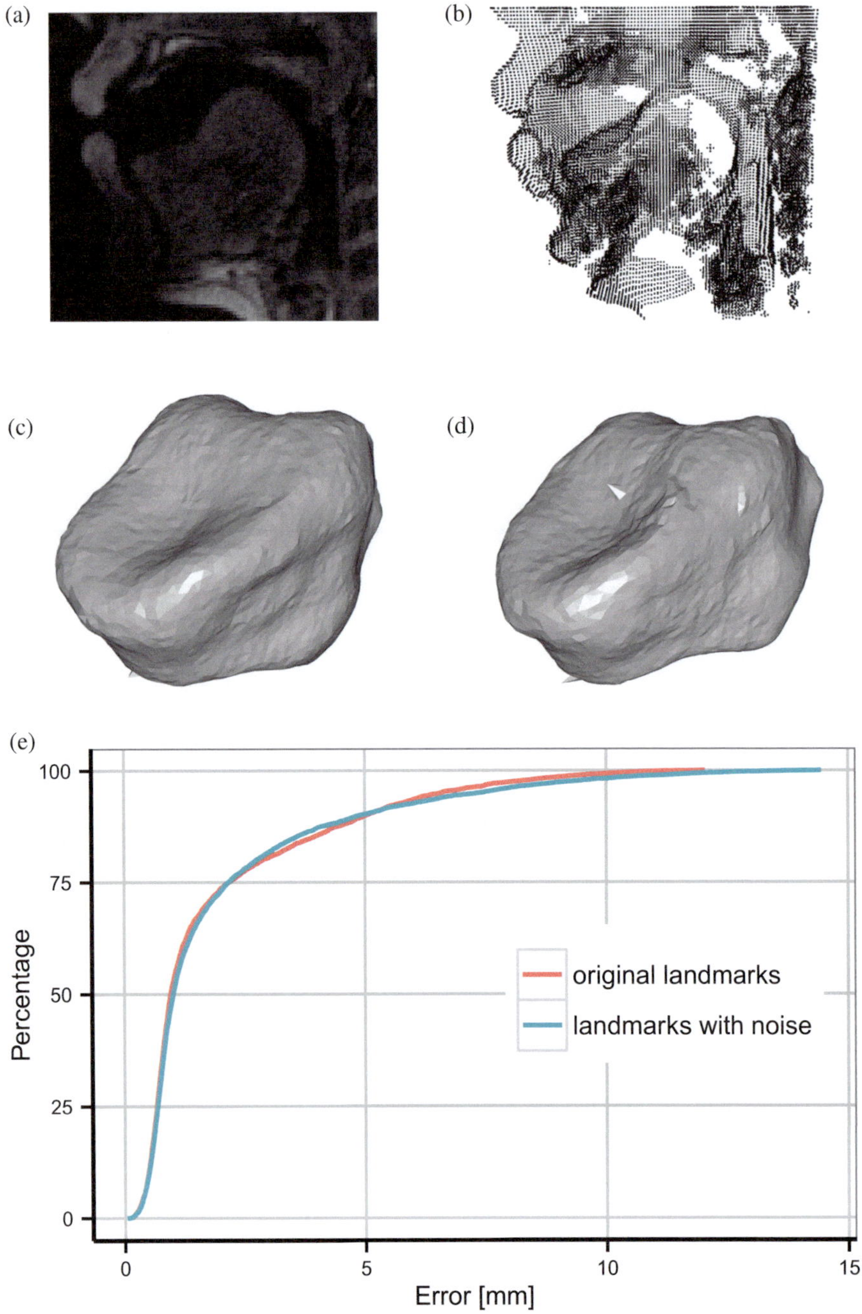

Fig. 16.8 Example showing the effect of noise in the landmarks. (**a**) Sagittal slice of the used MRI scan. (**b**) Generated point cloud. (**c**) Result for the original landmarks. (**d**) Result for landmarks with added noise. (**e**) Cumulative error functions for the two results

16.5.4 Discussion

Our approach succeeded in 75 % of the selected scans for a fixed template and fixed parameter settings. Furthermore, the proposed framework did not show any significant bias towards a specific phone or speaker, which indicates that it is phone- and speaker-independent. Here, we want to note that in the study of Harandi et al. [12] only the speaker-independence of their approach was analyzed. In particular, they only considered the tongue in the resting position and evaluated their method across 18 speakers. Moreover, our approach is robust against errors in the landmarks provided by the user. Thus, even an inexperienced user may obtain acceptable results using our method.

The observed failure cases stem from three main causes. Issues with the segmentation approach forced us to discard data from certain speakers completely, or prevented our framework from producing acceptable results. Using more than one segmentation technique may help to overcome these problems. Multiple segmentation results could be generated, and the user could then select the best one to use in the subsequent processing steps.

Furthermore, for scans where a contact between tongue and palate occurred, finding surface information of the tongue in the contact area is difficult because it may not be visible, which leads to a hole in the point cloud. Note that if we reconstruct the hard palate surface in this region, it may be used to represent the portion of the tongue surface in contact with the palate. For a point cloud P where such a hole is present due to a contact in the region of the hard palate, we explored the following approach to reconstruct the palate. First, a scan is selected where the hard palate is clearly visible, and the subset of points H representing the palate surface is extracted. Second, the hard palate is manually aligned to match the vocal tract configuration in P, which results in the set of transformed points H^*. Note that this alignment is easy to perform manually because the hard palate can only undergo rigid body motions. Third, P and H^* are merged into a single point cloud, which is used in the template fitting. This palate reconstruction can improve the results in cases where palate contact results in incomplete point clouds.

Finally, for the scans where the template fitting failed, we suspect that using more landmarks could help to align the template correctly to the point cloud.

16.6 Conclusion

In this chapter, we presented a minimally supervised approach to extract mesh representations of the human tongue from MRI data of the vocal tract. The experiments performed revealed promising results, as the presented approach leads to results of high quality in 75 % of our tests. An important feature of the approach is its independence with respect to changes in tongue shape due to different speakers and phones. Furthermore, the approach is robust to noise in the manually placed landmarks.

We leave the following open problems for future work. A palate reconstruction could help to significantly increase the number of scans that can be processed successfully by our approach. Hence, it is important to facilitate the process of palate reconstruction. We plan to replace the process of manually aligning the palate surface to the MRI data with a rigid alignment approach based on landmarks that are not necessarily located on the tongue.

Our template fitting could be improved by including more information, such as the sublingual part of the tongue, more annotated landmarks, or typical MR-values at the vertices. Such modifications may improve the performance of the template fitting.

Moreover, the evaluation of our approach could be made more thorough by using more datasets and comparing the results to other methods in literature. However, datasets in literature are in general not easy to access due to privacy concerns for the recorded subjects.

For the future, we also think that it would be worthwhile to explore the performance of robust unsupervised methods, like for example [24], in the segmentation part of the framework. Detecting the position of the landmarks automatically would be another interesting modification. Both improvements could make the framework more accurate and further reduce the input required by the user or even make it fully automatic.

Acknowledgements This study uses data from work supported by EPSRC Healthcare Partnerships Grant number EP/I027696/1 ("Ultrax").

References

1. Ultrax: Real-time tongue tracking for speech therapy using ultrasound (2014). http://www.ultrax-speech.org/. Accessed 5 May 2015
2. Allen, B., Curless, B., Popović, Z.: The space of human body shapes: reconstruction and parameterization from range scans. ACM Trans. Graph. **22**(3), 587–594 (2003). doi:10.1145/1201775.882311
3. Baker, A.: A biomechanical tongue model for speech production based on MRI live speaker data (2011). http://www.adambaker.org/qmu.php. Accessed 5 May 2015
4. Blandin, R., Arnela, M., Laboissière, R., Pelorson, X., Guasch, O., Hirtum, A.V., Laval, X.: Effects of higher order propagation modes in vocal tract like geometries. J. Acoust. Soc. Am. **137**(2), 832–843 (2015). doi:10.1121/1.4906166
5. Botsch, M., Kobbelt, L., Pauly, M., Alliez, P., Levy, B.: Polygon Mesh Processing. A K Peters/CRC Press, Natick (2010)
6. Boykov, Y., Funka-Lea, G.: Graph cuts and efficient ND image segmentation. Int. J. Comput. Vis. **70**(2), 109–131 (2006). doi:10.1007/s11263-006-7934-5
7. Brunton, A., Salazar, A., Bolkart, T., Wuhrer, S.: Review of statistical shape spaces for 3D data with comparative analysis for human faces. Comput. Vis. Image Underst. **128**, 1–17 (2014). doi:10.1016/j.cviu.2014.05.005

8. Chan, T.F., Vese, L.A.: Active contours without edges. IEEE Trans. Image Process. **10**(2), 266–277 (2001). doi:10.1109/83.902291
9. Engwall, O.: Can audio-visual instructions help learners improve their articulation? – an ultrasound study of short term changes. In: 9th Annual Conference of the International Speech Communication Association (Interspeech), Brisbane, pp. 2631–2634 (2008)
10. Eryildirim, A., Berger, M.O.: A guided approach for automatic segmentation and modeling of the vocal tract in MRI images. In: European Signal Processing Conference (EUSIPCO), Barcelona, pp. 61–65 (2011)
11. Grady, L.: Random walks for image segmentation. IEEE Trans. Pattern Anal. Mach. Intell. **28**(11), 1768–1783 (2006). doi:10.1109/TPAMI.2006.233
12. Harandi, N.M., Abugharbieh, R., Fels, S.: 3D segmentation of the tongue in MRI: a minimally interactive model-based approach. Comput. Methods Biomech. Biomed. Eng. Imaging Vis. **3**(4), 178–188 (2015). doi:10.1080/21681163.2013.864958
13. Hewer, A., Steiner, I., Wuhrer, S.: A hybrid approach to 3D tongue modeling from vocal tract MRI using unsupervised image segmentation and mesh deformation. In: 15th Annual Conference of the International Speech Communication Association (Interspeech), Singapore, pp. 418–421 (2014)
14. Kazhdan, M., Bolitho, M., Hoppe, H.: Poisson surface reconstruction. In: Eurographics Symposium on Geometry Processing (SGP), Cagliari, pp. 61–70 (2006). doi:10.2312/SGP/SGP06/061-070
15. Ladefoged, P.: A Course in Phonetics, 2nd edn. Harcourt Brace Jovanovich, New York (1982)
16. Lee, J., Woo, J., Xing, F., Murano, E.Z., Stone, M., Prince, J.L.: Semi-automatic segmentation of the tongue for 3D motion analysis with dynamic MRI. In: IEEE 10th International Symposium on Biomedical Imaging (ISBI), San Francisco, pp. 1465–1468 (2013). doi:10.1109/ISBI.2013.6556811
17. Li, C., Kao, C.Y., Gore, J.C., Ding, Z.: Implicit active contours driven by local binary fitting energy. In: IEEE Conference on Computer Vision and Pattern Recognition (CVPR), Minneapolis, pp. 1–7 (2007). doi:10.1109/CVPR.2007.383014
18. Li, H., Adams, B., Guibas, L.J., Pauly, M.: Robust single-view geometry and motion reconstruction. ACM Trans. Graph. **28**(5), 175:1–175:10 (2009). doi:10.1145/1618452.1618521
19. Liu, D.C., Nocedal, J.: On the limited memory BFGS method for large scale optimization. Math. Program. **45**(1–3), 503–528 (1989). doi:10.1007/BF01589116
20. Liu, J., Udupa, J.K.: Oriented active shape models. IEEE Trans. Med. Imaging **28**(4), 571–584 (2009). doi:10.1109/TMI.2008.2007820
21. Osher, S., Sethian, J.A.: Fronts propagating with curvature-dependent speed: algorithms based on Hamilton-Jacobi formulations. J. Comput. Phys. **79**(1), 12–49 (1988). doi:10.1016/0021-9991(88)90002-2
22. Peng, T., Kerrien, E., Berger, M.O.: A shape-based framework to segmentation of tongue contours from MRI data. In: IEEE International Conference on Acoustics, Speech and Signal Processing (ICASSP), Dallas, pp. 662–665 (2010). doi:10.1109/ICASSP.2010.5495123
23. Raeesy, Z., Rueda, S., Udupa, J.K., Coleman, J.: Automatic segmentation of vocal tract MR images. In: IEEE 10th International Symposium on Biomedical Imaging (ISBI), San Francisco, pp. 1328–1331 (2013). doi:10.1109/ISBI.2013.6556777
24. Witten, D.M.: Penalized unsupervised learning with outliers. Stat. Interface **6**(2), 211–221 (2013). doi:10.4310/SII.2013.v6.n2.a5
25. Wuhrer, S., Lang, J., Tekieh, M., Shu, C.: Finite element based tracking of deforming surfaces. Graph. Models **77**, 1–17 (2015). doi:10.1016/j.gmod.2014.10.002

Index

M. Breuß et al. (eds.), *Perspectives in Shape Analysis*, Mathematics and Visualization, DOI 10.1007/978-3-319-24726-7

MIX
Papier aus verantwortungsvollen Quellen
Paper from responsible sources
FSC® C105338
FSC
www.fsc.org

If you have any concerns about our products,
you can contact us on
ProductSafety@springernature.com

In case Publisher is established outside the EU,
the EU authorized representative is:
Springer Nature Customer Service Center GmbH
Europaplatz 3, 69115 Heidelberg, Germany

Printed by Libri Plureos GmbH
in Hamburg, Germany